POLYESTER: PROPERTIES, PREPARATION AND APPLICATIONS

POLYESTER: PROPERTIES, PREPARATION AND APPLICATIONS

HINA YAMASHITA AND YUI NAKANO
EDITORS

Nova Science Publishers, Inc.
New York

LIBRARY OF CONGRESS CATALOGING-IN-PUBLICATION DATA

Yamashita, Hina.
Polyester : properties, preparation & applications / Hina Yamashita and Yui Nakano.
p. cm.
ISBN 978-1-60456-753-3 (hardcover)
1. Polyesters. I. Nakano, Yui. II. Title.
TP1180.P6Y36 2008
668.4'225--dc22

2008026982

Published by Nova Science Publishers, Inc. ; New York

CONTENTS

PREFACE

Polyester (aka Terylene) is a category of polymers which contain the ester functional group in their main chain. Although there are many forms of polyesters, the term "polyester" is most commonly used to refer to polyethylene terephthalate (PET). Other forms of polyester include the naturally-occurring cutin of plant cuticles as well as synthetic polyesters such as polycarbonate and polybutyrate.

Polyester may be produced in numerous forms. For example, polyester as a thermoplastic may be heated and processed into different forms and shapes, e.g., fibers, sheets and three-dimensional shapes. While combustible at high temperatures, polyester tends to shrink away from flames and self-extinguishes.

This book provides leading edge research on this field from around the globe.

Chapter 1 - Processes were developed in recent years in order to recover raw materials from poly(ethylene terephtalate) (PET) and polycarbonate (PC) by hydrolysis. The main focus was the recovery of terephthalic acid and ethylene glycol besides other products such as benzene, salts of terephthalic acid and oxalic acid. Processes developed for PET are also valid for polyesters such as poly(butylene terephthalate) (PBT) and poly(ethylene 2,6-napthalene dicarboxylate) (PEN). The recovery of bisphenol-A (BPA) from PC requires more sophisticated methods due to the low stability of BPA at high temperatures. Often phenol and isopropenyl phenol are obtained as degradation products of BPA.

Chapter 2 - This chapter presents the preparation of polymer nanocomposites and the effects of multiwall carbon nanotube (MWCNT) on the structure and properties of poly(ethylene 2,6-naphthalate) (PEN) nanocomposites. The combination of a very small quantity of expensive MWCNT with conventional cheap thermoplastic polymers provides attractive possibilities for improving the physical properties of polymer composites using a cost-effective method, from a commercial perspective. MWCNT-reinforced PEN nanocomposites were prepared by a melt blending process in a twin screw extruder to create advanced materials for possible practical applications in numerous industrial fields. There are significant dependence of the crystallization behaviors and their kinetics of the PEN/MWCNT nanocomposites on the MWCNT content, the cooling rate, and the crystallization temperature. The MWCNT in the PEN nanocomposites exhibited much higher nucleation activity than any nanoreinforcing filler. In the PEN/MWCNT nanocomposites, the incorporation of the MWCNT promoted the nucleation and the growth with higher crystallization rate of the PEN/MWCNT nanocomposites, and simultaneously reduced the fold surface free energy and the works required in folding macromolecular chains in the

PEN/MWCNT nanocomposites. The non-terminal behavior observed in the PEN/MWCNT nanocomposites was related to the dominant nanotube-nanotube interactions at higher MWCNT content, leading to the formation of the interconnected or network-like structures of the MWCNT in the PEN nanocomposites. The incorporation of very small quantity of the MWCNT significantly improved the mechanical properties of the PEN/MWCNT nanocomposites. There is a significant dependence of the thermal stability and degradation behavior of the PEN/MWCNT nanocomposites on the MWCNT content and heating rate. The interconnected network-like structures of the MWCNT resulted in the physical barrier effect against thermal degradation both by retarding the rate of thermal degradation and by hindering the transport of volatile decomposed products in the PEN nanocomposites, leading to the improvement in the thermal stability of the PEN/MWCNT nanocomposites. This Chapter attempts for the first time to summarize the preparation, the non-isothermal crystallization kinetics, the crystallization and melting behavior, the rheological and mechanical properties, the thermal stability, and the thermal degradation kinetics of MWCNT-reinforced PEN nanocomposites.

Chapter 3 - Cyanate Ester Resins (CER) offer a variety of excellent thermal and good mechanical properties, which commend them for use in high performance technology (e.g. as matrices for composites for high-speed electronic circuitry and transportation). The product of CER curing process polycyanurates (PCN) are synthesized by a catalytic high temperature polycyclotrimerization reaction of cyanate esters of bisphenols For the electronics market, attractive features of PCN are their low dielectric loss characteristics, dimensional stability at molten solder temperatures (220-270°C), high purity, inherent flame-retardancy (giving the potential to eliminate brominated flame retardants) and excellent adhesion to conductor metals at temperatures up to 250°C . Since the late 1970s, cyanate ester resins have been used with glass or aramid fibers in high-speed multilayer circuit boards and this remains their primary application. Several reviews collecting the numerous publications (papers and patents) in the field of PCN synthesis, processing, characterization, modification and application have appeared since 1990s. In addition, like conventional FR-4 diepoxides, cyanate ester laminates retain the desirable (ketone) solution processing characteristics and the ability to be drilled, making possible to employ them in printed circuit board manufacture. In the last three decades, aerospace composites have evolved into damage-tolerant primary and secondary structures utilizing both thermoset and thermoplastic resins. PCN homopolymers develop approximately twice the fracture toughness of multifunctional epoxies while qualifying for service temperatures of at least 150°C, intermediate between epoxy and bismaleimides capabilities. PCN have already flown in prototype radomes and high gain antennae, with possible applications in primary and secondary structures of the High Speed Civil Transport (HSCT) and European Fighter Aircraft. PCN are also being qualified for satellite truss and tube structures and cryogenic, radiation-resistant components in the Superconducting Supercollider. This is indeed the problem, to convince a traditionally conservative industry that the superior performance of PCN (which surpass the glass transition temperature and hydrophobicity of epoxies while matching their processability and are easily toughened) makes them worthy of further investigation in spite of their price, which is currently higher than the price of the epoxies. PCN must be traditionally cured at high temperatures in order to achieve complete conversion, which increases manufacturing cost, but reactive modification of PCN allows decreasing the high temperature of PCN post-curing.

The primary drawback of PCN, which hinders more extensive application of the cured materials, however, is low room temperature toughness..

Chapter 4 - Among biodegradable polymers, polyesters derived from aliphatic dicarboxylic acids and diols are of special importance. Polyesters of 1,3-propanediol were overlooked till recently, since the specific monomer was not available in the quantities and price that might enable production of polymers. However, in recent years more attractive processes have been developed for the production of 1,3-propanediol from renewable resources. Nowadays, research on biodegradable poly(1,3-propylene alkanedioate)s, such as poly(propylene succinate) (PPSu), poly(propylene adipate) (PPAd) and poly(propylene sebacate) (PPSe), has gained increasing interest, due to their fast biodegradation rates and their potential uses in biomedical or pharmaceutical applications, such as drug delivery systems. The odd number of methylene units in the diol segment is responsible for the lower melting points, lower degree of crystallinity and higher biodegradation rates of the specific polymers compared with their homologues based on ethylene-glycol or 1,4-butanediol. In this chapter synthesis and properties of the 1,3-propanediol based aliphatic polyesters and especially their biodegradation characteristics are reviewed. Specific attention has been paid to preparation of related copolymers and blends with other important polymers, since these techniques may offer routes for optimizing properties and producing tailor-made materials. Copolymerization of 1,3-propanediol with mixtures of aliphatic or even aromatic acids, leads to linear polyesters with improved or balanced biodegradation and mechanical properties. Blends with other biodegradable polymers have been studied recently. Finally, potential pharmaceutical applications of poly(1,3-propylene alkanedioate)s as solubilizing and stabilizing carriers for drugs are exemplified.

Chapter 5 - Large institutions, such as hotels and hospitals, often use specialized industrial laundries for laundering sheets, towels or uniforms. The main purpose of this study was to determine the effect of industrial laundering procedures on the durability of cotton warp knitted towels with a synthetic ground structure of either nylon or polyester. The durability of cotton/nylon and cotton/polyester terry towelling fabric samples that were subjected to repeated industrial laundering procedures, was compared by measuring the tensile strength of fabric samples after 50 washing cycles and 50 washing/tumble-drying cycles. The difference between the tensile strengths of the cotton/polyester and cotton/nylon terry towelling samples after washing alone was not significant. The tensile strength of the cotton/nylon samples, however, was significantly less than that of the cotton/polyester samples after tumble-drying. It was concluded that industrial laundering procedures, especially tumble-drying, have a more detrimental effect on the durability of the nylon ground structure than on the polyester ground structure of warp-knitted terry towelling fabrics.

Chapter 6 - Polyester type shape memory polymers were synthesized to improve their mechanical and shape memory properties and used for the preparation of sandwich type composite materials. Especially, poly(ethylene terephthalate) (PET) and poly(ethylene glycol) (PEG) copolymers with shape memory ability were prepared. After selecting the best composition of PET-PEG copolymer in mechanical properties, cross-linking agent such as glycerine, sorbitol, or maleic anhydride (MAH) was included for cross-linked copolymer, followed by analysis of its effect on mechanical, shape memory, and damping properties. The highest shape recovery was observed for copolymer with 2.5 mol% of glycerine, and the best damping effect indicating vibration control ability was from copolymer with 2.5 mol% of sorbitol. With the optimum copolymers in hand, sandwich-structured epoxy beam composites fabricated from epoxy beam laminate and cross-linked PET-PEG copolymer showed that impact strength increased from 1.9 to 3.7 times depending on the type of copolymer, and damping effect also increased as much as 23 times for the best case as compared to epoxy laminate beam alone. PET-PEG copolymers cross-linked with glycerol and sulfoisophthalate (SP) were also prepared to investigate the feasibility of vibration-control of composite laminate by additional ionic interaction. Composition of glycerol and SP was varied in order to get a copolymer with the best mechanical and shape memory properties. The highest shape recovery was observed for the copolymer with 2.5 mol% of glycerol and 2.5 mol% of SP. The sandwich-type copolymer composite showed improved impact strength (3.5 times) and damping effect (2.6 times) as compared to epoxy laminate beam alone. The resultant sandwich-structured epoxy beam composite can be utilized as structural composite material with vibration control ability and its glass transition temperature can be controlled by adjustment of hard segment content and cross linking agent composition.

Chapter 7 - Special attention to these polymers is defined by their specific feature, which is orientation in the melt, mostly associated with the intense development in computer technologies. Owing to this property such polymers are devoted to the "family" of liquid-crystal polymers. The liquid-crystal properties are also observed for PAI with an uneven number of CH_2-groups.It should be noted that polyalkanimide (PA-12), discussed in, also displays liquid-crystal properties under definite processing modes.

Liquid-crystal aromatic copolyesters (LCP) were studied. They were derived from dioxydiphenyl diacetate, acetoxybenzoic, iso- and terephthalic acids (IPA and TPA, respectively): 100/0, 75/25, 50/50, 25/75, 0/100

Chapter 8 - The thermal decomposition of halogenated and non-halogenated unsaturated polyester resins (UPR`s), fire retarded by zinc hydroxystannate (ZHS) and cross-linked with styrene, has been investigated by thermogravimetry (TG) and TG coupled on-line with Fourier transform infra red spectroscopy (TG-FTIR) or mass spectroscopy (TG-MS).

In this Chapter, thermal analysis of the decomposition process has been performed – hence, the flame retardancy and thermal stabilization of halogenated and non- halogenated polyester resins by ZHS may be explained by the formation of surface-localized spherical barriers which are growing according to the nucleation growth mechanism and which attenuate the transfer of heat from the decomposition zone to the substrate. This effect was found as dominating in the flame-retardancy mode of action.

In: Polyester: Properties, Preparation and Applications ISBN: 978-1-60456-753-3
Editors: Hina Yamashita, Yui Nakano

Chapter 1

HYDROLYSIS OF POLYESTERS AND POLYCARBONATES

Toshiaki Yoshioka* and Guido Grause
Graduate School for Environmental Studies, Tohoku University,
Aramaki Aza Aoba 6-6-07 Aoba-ku, Sendai 980-8579, Japan

ABSTRACT

Processes were developed in recent years in order to recover raw materials from poly(ethylene terephtalate) (PET) and polycarbonate (PC) by hydrolysis. The main focus was the recovery of terephthalic acid and ethylene glycol besides other products such as benzene, salts of terephthalic acid and oxalic acid. Processes developed for PET are also valid for polyesters such as poly(butylene terephthalate) (PBT) and poly(ethylene 2,6-napthalene dicarboxylate) (PEN). The recovery of bisphenol-A (BPA) from PC requires more sophisticated methods due to the low stability of BPA at high temperatures. Often phenol and isopropenyl phenol are obtained as degradation products of BPA.

ABBREVIATIONS

BPA	Bisphenol-A
c	Concentration (*c*(*COOH*): acid groups, *c*(*EL*): ester links, *c*(*water*): water, c_A: acid
CHA	Cyclohexylamine
DMSO	Dimethylsulfoxide
E_A	Activation energy
EG	Ethylene glycol
k	reaction rate constant
K	Equilibrium constant

* E-mail: yoshioka@env.che.tohoku.ac.jp

NPA	2,6-Naphthalene dicarboxylic acid
PBT	Poly(butylene terephthalate)
PC	Polycarbonate
PEN	Poly(ethylene napthalenate)
PET	Poly(ethylene terephthalate)
r	Particle radius
t	Time
TGA	Thermogravimetric analysis
THF	Tetrahydrofurane
TPA	Terephthalic acid
X	Conversion
ρ	density

1. Introduction

Polyesters and polycarbonates are plastics produced by the polycondensation of dialcohols with diprotonic acids, either a dicarboxylic acid or carbonic acid in the form of its acid dichloride phosgene. The resulting ester links are sensitive to hydrolysis. Especially during processing, it is necessary to work with dry polymers. Even the presence of small quantities of water causes the shortage of the chain length at the processing temperature of more than 250°C, resulting in an unwanted change of properties.

However, this effect can be employed after use to degrade polyesters and polycarbonates in order to obtain monomers or other valuable products. Today, the focus is on the reuse of waste materials. Land filling is limited by space and ecological concerns. In the European Union, land filling is even limited by law to materials with a loss by combustion of less than 5wt% [1]. Other ways of waste treatments are necessary. Combustion and gasification are used for the energy recovery of mixed wastes from municipal households or agriculture. Materials can be separately collected or separated from other wastes by sorting. In Japan, several recycling laws are enforced for the construction of a "Sound Material-Cycle Society", such as:

- the "Law for Promotion of Sorted Collection and Recycling of Containers and Packaging"
- the "Home Appliance Recycling Law"
- the "Food Recycling Law"
- the "Construction Material Recycling Law" and
- the "Automobile Recycling Law".

Both physical and chemical recycling are possibilities for these materials. However, the physical recycling of polymers is always connected with the loss of properties (downcycling), due to the use of materials with different chain lengths, different additives and a different history. Chemical recycling (feedstock recycling) has to be developed for every single material according to its structure and reactivity. Processes suitable for polymers as polyolefins, polystyrene or poly(methyl methacrylate) are usually not appropriate for polycondensates. Polymers can be easily degraded by pyrolysis resulting in monomers such as styrene, methyl methacrylate, ethylene or propylene. Waxes and hydrocarbon oils can also be obtained depending on the reaction temperature. Random scission of the polymer backbone releases mainly ethylene derivatives from polymerizates. Mixed plastics deliver oils with characteristics similar to crude oil. However, polycondensates tend to carbonize during the degradation. The chemical structure of the monomer is destroyed by the degradation of the macromolecule releasing the functional group, e.g. CO_2 and CO in the case of polyesters [2] and polycarbonates [3] or HCN in the case of polyamides [4]. The high ratio of aromatic structures included in the structure of many polycondensates promotes the formation of carboneous residues. Solvolysis is a more efficient way for the degradation of polycondensates. At temperatures below pyrolysis temperature, the material is ceased by a nucleophilic agent. The original monomer or a monomer derivative is recovered. The absence of radical reactions below the pyrolysis temperatures prevents the formation of carboneous residue.

The focus of this review is the degradation of polyesters and polycarbonates in the presence of water or hydroxide ions as a method of recycling. Different methods of hydrolysis using different reaction media and the products obtained from these shall be introduced including the modification of the primary hydrolysis products in further reaction steps. Moreover, it should be provided an overlook about the mechanisms and kinetics as far as available.

2. POLYESTERS

The polyesters considered in this review are polyesters synthesized from aromatic dicarbonic acids, as terephthalic acid (TPA) and 2,6-naphthalene dicarboxylic acid (NPA). Only these plastics are produced in large quantities and have to be treated after use. Aliphatic polyesters feature a low hydrolytic stability and a high biodegradability. The way of after use treatment differs fundamentally from those mentioned before [5].

The hydrolysis of polyesters in their simplest form is the inversion of their synthesis reaction:

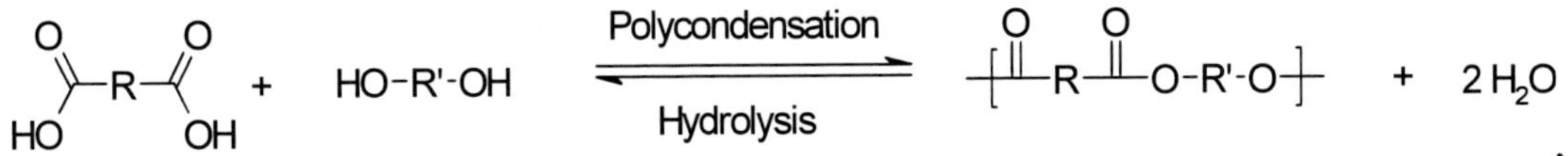

where R is aromatic and R' is usually an aliphatic rest. Both polycondensation and hydrolysis occurs mainly in the melt of the polyester. At temperatures below the melting point of the plastic, the bulk material is only attacked at the surface if no solvents are involved.

The most important polyester is poly(ethylene terephthalate) (PET). It can be used in one, two and three dimensional applications as fibres, films and injection molded bodies (e.g. bottles). Especially PET-bottles are of interest in view of recycling due to the size of the pieces, the purity of the material and the homogeneity of the properties. There are already collection systems introduced in different parts of the world and mechanical recycling is established. However, mechanical recycling causes an erosion of mechanical and optical properties due to thermal and mechanical stress during lifetime and processing. The use for food applications is also restricted due to hygienic concerns. Though, the demand of recycled PET is limited.

The recycling of PET-fibres is more difficult. They are often blended with other synthetic and natural fibres, containing additives as dyes and fire retardants. They are used for various different applications as clothes, seats or tire cord with the needs of different chain lengths and properties. Recycling of such materials has to be sophisticated and expensive.

Recycling of poly(butylene terephthalate) (PBT) faces similar problems. PBT is preferable used for injection molding applications due to its fast crystallization and the low uptake of water. It is often blended with PET, polycarbonate (PC) or polystyrene, containing various kinds of additives as fire retardants, dyes and fillers, as well. Any kind of recycling concept has to deal with this after-use disadvantages.

Poly(ethylene 2,6-naphthalene dicarboxylate) (PEN) is considered as an alternative to PET for packaging applications due to its better barrier properties against gases as CO_2 and oxygen. However, the higher price for 2,6-naphthalene dicarboxylic acid prevents the wide range use of this material.

2.1. Hydrolysis of PET in Water

The hydrolysis of PET resulting in terephthalic acid (TPA) and ethylene glycol (EG) in hot water shows some interesting properties. Firstly, the protolytic constant increases with the temperature and more protons are available even in the absence of acids and bases catalysing the reaction. Secondly, the dielectric constant of water is decreasing with increasing temperature making it easier to dissolve organic molecules in hot water. This is especially important for the dissolution of TPA which is almost insoluble at room temperature, however, showing a good solubility in water at elevated temperatures. TPA is gained by precipitation during cooling the reaction solution. EG is separated from water by distillation in industrial processes.

The hydrolysis of PET in laboratory scale experiments is usually accomplished in a stirred batch reactor [6-9]. The reaction velocity is slow below the melting point of PET and increases rapidly above this temperature (Table 1) [10]. PET melts at temperatures higher than 235°C. However, evidence was found for a lower melting point under pressure [7]. The agitation maintains a good contact between the water phase and the PET above the melting point, which is necessary to avoid a diffusion controlled kinetic due to the migration of water through a TPA layer to the PET surface. However, TPA is dissolved at the water-melt interface and transported into the water phase by a proper agitation.

The pressure is strictly dependent on the temperature under autogenous conditions. An increasing temperature accelerates the reaction velocity, but also the pressure in the reactor increases. A complete hydrolysis can be achieved at 250°C after 30min and at 265°C after 10min reaction time [7]. The reduction of the temperature from 250°C to 200°C reduces the yield by 30% at a reaction time of two hours [11]. A complete hydrolysis can be achieved after six hours reaction time [12].

Table 1. Rate constants of the hydrolysis of PET at different temperatures

Temperature [°C]	Rate constant k [$mol^{-1}min^{-1}$]
100	$3.0*10^{-7}$
186	$2.0*10^{-4}$
250	0.244
265	0.352
280	0.487

Due to costs and safety requirements, it is a concern of research to reduce the temperature necessary for the complete reaction. High temperature is also responsible for the loss of product. While TPA is not affected over a period of one hour at temperatures up to 350°C, a significant reduction of the yield of TPA can be observed at 400°C. The yield of EG is reduced even more due to its lower stability. The degradation of EG begins at 300°C. At 400°C, EG is completely degraded after one hour. It is indicated that the dehydration of EG is catalysed by protons [6].

Table 2. Equilibrium constant $K = \frac{c_{COOH}^2}{c_{EL}c_{WATER}}$ with the concentration of the hydrolysed ester links c_{COOH}, the concentration of the remaining ester links c_{EL} and the water concentration c_{WATER} under equilibrium conditions

Temperature [°C]	*K* [7]	*K* [8]	*K* [9]
225	-	-	20.86
235	7.09	-	-
240	-	-	12.37
250	3.99	1.43	-
255	-	-	8.74
265	1.53	0.664	-
270	-	-	6.92
280	-	0.384	-

Next to the temperature, the water/PET ratio has the strongest effect on hydrolysis reactions. An initial water/PET ratio (w/w) of 5.1 [8] and 6 [7,9] was found to be necessary for a complete hydrolysis between 265°C and 270°C. Below this water concentration, an equilibrium is established (Table 2). The data for the equilibrium constant *K* differ strongly between the experimental set-ups. However, a reduction of *K* with increasing temperature was found in all experiments indicating a shift of the equilibrium to the free acid at lower temperatures.

The variation of the pressure at a constant temperature shows some effects on the reaction behavior. While the yield slightly increases with the pressure, the reaction velocity shows a more sophisticated behavior. At 150°C, the reaction rate is accelerated by increasing pressure, while at 200°C the influence is negligible. At 250°C, the reaction rate is even decreasing [11].

Divalent metal salts known as effective transesterfication catalysts have little effect on the hydrolysis of PET. The addition of zinc acetate increases the hydrolysis rate to a maximum of about 20% [13].

Other methods were employed for the hydrolysis of PET, as well. PET is hydrolysed completely at a pressure of 2MPa indicating a temperature of 210°C in two hours by using microwave (MW) radiation. The MW radiation heated the water without directly affecting the PET. The employed power of the MW radiation shows no effect on the hydrolysis rate. The shortage of the heat-up time is negligible compared to the reaction time. In contrast to reactions in a stirred batch reactor, a maximum yield for TPA was found with an initial water/PET charge ratio of 10. Higher charge ratios cause a reduction of the yield [14].

The reaction mechanism is the same as for monomeric esters [15]:

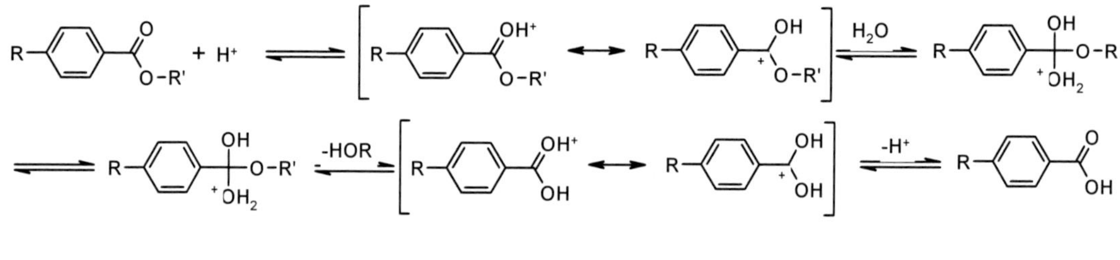

It can be seen that the pathway consists only of balance reactions and the mechanism proceeds in random scission mode due to the late appearance of free EG in the solution [8,14]. This is in agreement with the idea of the inversion of the PET synthesis where PET is obtained only at a high conversion. Water and EG are removed constantly to force the equilibrium to high molecular weights. Equilibria are formed during the hydrolysis between the different PET oligomeric species dependent on the water concentration. Transesterfication reactions between the oligomers most likely happen in the melt. Detailed studies to the distribution statistic are not available up today. Investigations on the hydrolysis of solid PET at 205°C show the reduction of the chain length of residual PET with the reaction time [12]. The finding of unchanged chain lengths of residual PET contradicts the random scission mechanism and the results mentioned earlier [9].

According to the mechanism, the hydrolysis of PET is described best by a second-order model. Besides the forward reaction and the backward reaction, the autocatalytic effect of the produced TPA must be considered, as well [7].

The Resulting Reaction Velocity Law

$$\frac{dc(COOH)}{dt} = kc(EL)c(water)c^{\frac{1}{2}}(COOH) - \frac{c^{\frac{5}{2}}(COOH)}{K} \qquad \text{Eq. 1}$$

with the concentrations of the acid groups $c(COOH)$, the ester links $c(EL)$, water $c(water)$ and the equilibrium constant K leads after integration to

$$kt = \frac{S - S_0}{2B(1-K)} \qquad \text{Eq. 2}$$

with

$$S = (B-A)^{-\frac{1}{2}} \ln \frac{c^{\frac{1}{2}}(COOH) - (A+B)^{\frac{1}{2}}}{c^{\frac{1}{2}}(COOH) + (B-A)^{\frac{1}{2}}} - \ln \frac{c^{\frac{1}{2}}(COOH) - (-A-B)^{\frac{1}{2}}}{c^{\frac{1}{2}}(COOH) + (-A-B)^{\frac{1}{2}}},\ K > 1$$

Eq. 3a

$$S = (B-A)^{-\frac{1}{2}} \ln \frac{c^{\frac{1}{2}}(COOH) - (A+B)^{\frac{1}{2}}}{c^{\frac{1}{2}}(COOH) + (B-A)^{\frac{1}{2}}} - \tan^{-1} \frac{c^{\frac{1}{2}}(COOH)}{(A+B)^{\frac{1}{2}}},\ \mathrm{K}<1 \qquad \text{Eq. 3b}$$

$$A = -\frac{c(EL)_i + c(H_2O)_i}{2(1-K^{-1})} \qquad \text{Eq. 4}$$

$$B = \sqrt{\frac{A^2 - c(EL)_i c(H_2O)_i}{1-K^{-1}}} \qquad \text{Eq. 5}$$

S_0 is S at the time $t = 0$.

When k is calculated using concentrations instead of the mass of PET, k shows an independency from the initial water/PET charge ratio (Table 3). The results show clearly the dependence of the rate constant from the volume due to the fact that both the ester link concentration and the water concentration have to be recognized. In contrast, rate constants related to the mass are not universal valid. They can just be used for the special conditions they were obtained. Nevertheless, values for k can vary from sample to sample due to its properties and from experimental set-up to set-up, as well.

The activation energy ranges from 56kJ/mol [8] to 123kJ/mol [7] for the hydrolysis of molten PET. Both values were obtained in a similar temperature range. However, the initial water/PET charge ratio was 10 and 6, respectively. Migration effects are strongly affected by the viscosity of the mixture and the viscosity is strongly affected by the temperature, as well, indicating that a reduced viscosity at high water/PET rates reduces also the activation energy. Higher resistance for the material transport is also expected for solid PET and high activation energies are found. At 200°C, the activation energy shows a distinct dependence on the pressure. It drops from 9.1MJ/mol at 1.4MPa to about 6MJ/mol at 3.4MPa and remains nearly constant up to 5.5MPa [11]. It can be seen that pressure is able to overcome partly the resistance. Probably, a better contact between water and PET is achieved by pressing water between the PET chains. The formation of hydrogen bonds between the water and the ester groups can be seen as the first step of hydrolysis.

Table 3. Rate constant for the hydrolysis of PET with initial water/PET charge ratios between 2 and 10

Temperature [°C]	235	250	265
Rate constant $[10^{-5}(l/mol)^{3/2}s^{-1}]$	3	7	18

2.2. Acidic Hydrolysis of PET

It was shown that the hydrolysis of molten PET takes place in a satisfactory time scale, however, the hydrolysis velocity is reduced drastically below the melting point of PET. The hydrolysis is, however, accelerated in the presence of concentrated strong mineral acids even at moderate temperatures and atmospheric pressure.

The hydrolysis of solid PET in the presence of acids is a heterogeneous reaction catalyzed by protons. Sulfuric acid, nitric acid and hydrochloric acid were employed to obtain TPA, EG and other products. The reaction is affected by the diffusion of reagents and products, the heat transfer inside the PET related to the low heat conductivity and the crystallinity of the PET, as well.

Early works on the hydrolysis of PET in hydrochloric acid show the influence of different crystallinities and HCl concentrations on the reaction [16]. The hydrolysis rate increases with the HCl concentration due to the catalytic effect of the protons. However, oriented crystalline fibres show a higher resistance against hydrolysis than unoriented crystalline fibres and amorphous fibres, as well. It is assumed that HCl is mainly absorbed in the amorphous areas and hydrolysis is restricted to the amorphous PET and the surface of crystallites. The particle size shows no effect on the hydrolysis.

The results show the accelerating effect of hydrochloric acid on the hydrolysis. However, it was not thought as a recycling process. Therefore, just small parts of the PET were hydrolysed and transport processes were negligible. These processes have to be considered when nitric acid is used and a complete recovery of TPA and EG shall be established [17,18]. The hydrolysis in nitric acid is limited to the surface of the PET particle due to the larger molecule size and the smaller diffusion coefficient of HNO_3. The solubility of TPA in nitric acid is low and TPA is deposited on the surface of the PET particle making a further attack of the surface difficult (Figure 1). The hydrolysis rate decreases with the time and a complete hydrolysis is not achieved. The reaction is accelerated by an increase in temperature and acid concentration. However, a maximum yield of 91% for TPA at 100°C in 13M nitric acid is just obtained after 24h reaction time.

EG is mainly degraded to oxalic acid and CO_2:

$$\mathrm{HOCH_2CH_2OH} + \frac{8}{3}\mathrm{HNO_3} \longrightarrow \mathrm{HOOC-COOH} + \frac{8}{3}\mathrm{NO} + \frac{10}{3}\mathrm{H_2O}$$

$$3\mathrm{HOOC-COOH} + \frac{2}{3}\mathrm{HNO_3} \longrightarrow 6\mathrm{CO_2} + 2\mathrm{NO} + 4\mathrm{H_2O}$$

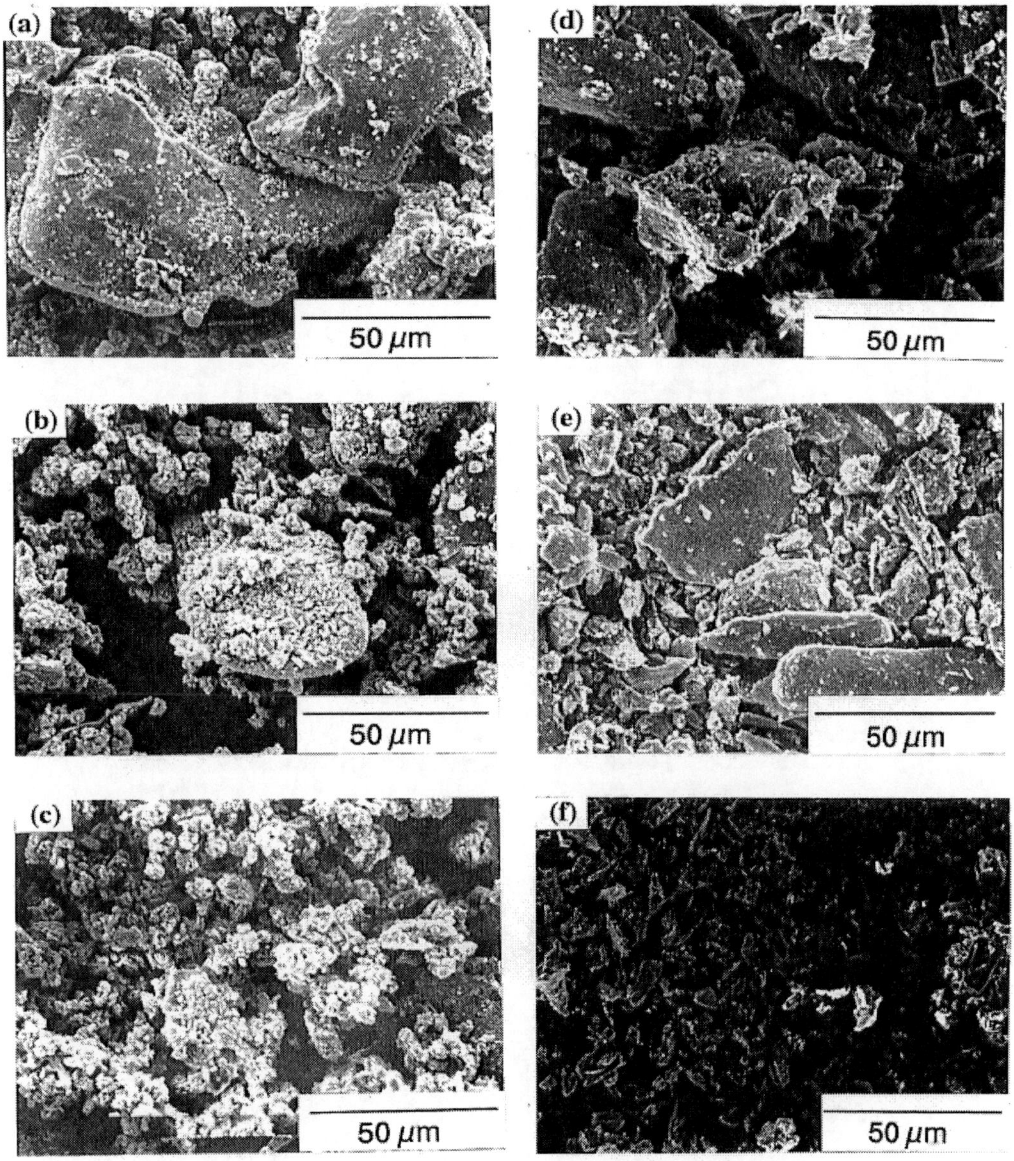

Figure 1. SEM photographs of the PET surface covered with TPA after a) 2h, b) 8h and c) 12h hydrolysis with nitric acid and after the removal of TPA after d) 2h, e) 8h and f) 12h reaction time.

Yields of 30% to 40% of oxalic acid are obtained in 13M nitric acid at 100°C. The conversion of EG to oxalic acid might be acceptable due to the higher price of oxalic acid even if yield is lost by the oxidation to CO_2.

The hydrolysis of PET in nitric acid is improved by the addition of dimethylsulfoxide (DMSO) to the reaction slurry [19]. DMSO is one of few solvents with good properties for the dissolution of TPA. DMSO improves the contact between the nitric acid and the PET surface. The reaction time is reduced to acceptable 160min. The conversion of EG to oxalic acid can be avoided by the addition of $NaSO_4$ as a salting-out agent. EG is separated from nitric acid by the reduced stabilization of the EG in the aqueous solution caused by the hydration of the $NaSO_4$ during its dissolution.

The incomplete hydrolysis of PET with nitric acid leaves a material with cation exchange properties [20]. After hydrolyzing PET for 30min in 10mol/l HNO_3 under reflux, a material is obtained containing about 0.5mmol/g of carboxylic groups able to adsorb Cd^{2+}-ions and cationic organic molecules, as well. Figure 2 shows the surface of PET after the hydrolysis with NaOH and HNO_3. The surface is covered with cracks and pores providing a large surface where the acidic sites after the hydrolysis are located. On the other hand, NaOH leaves a much smoother surface with just a small number of acidic sites.

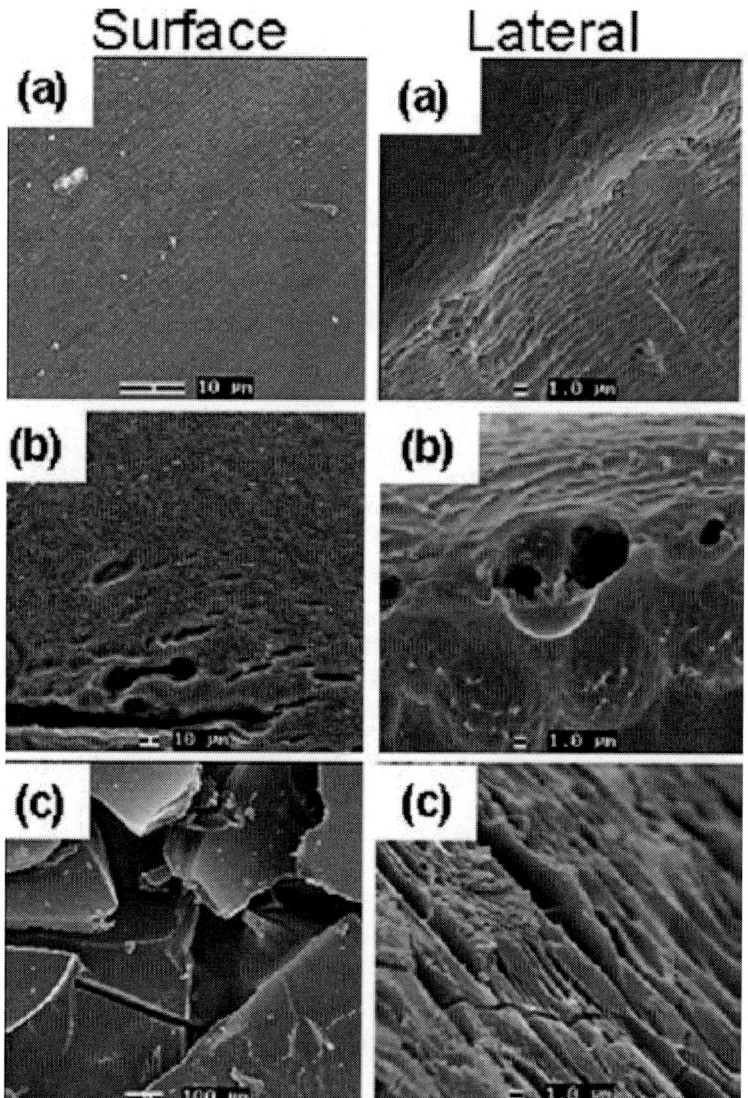

Figure 2. SEM photographs of PET a) without hydrolysis, after hydrolysis under reflux for 30min in b) 10mol/l NaOH and c) 10mol/l HNO_3 (reprinted from [20], with permission from Wiley).

It is assumed that the reaction of PET in nitric acid is a reaction occurring just at the surface of the PET particles [18]. Nitric acid does not swell the particles and caves and pores are not formed, though the attack of deeper particle layers is excluded. Further, the solution of PET is negligible due to the long time particles resist in the solution. In order to deal with the deposition of TPA on the particle surface a modified shrinking core model for PET powder was developed (Figure 3).

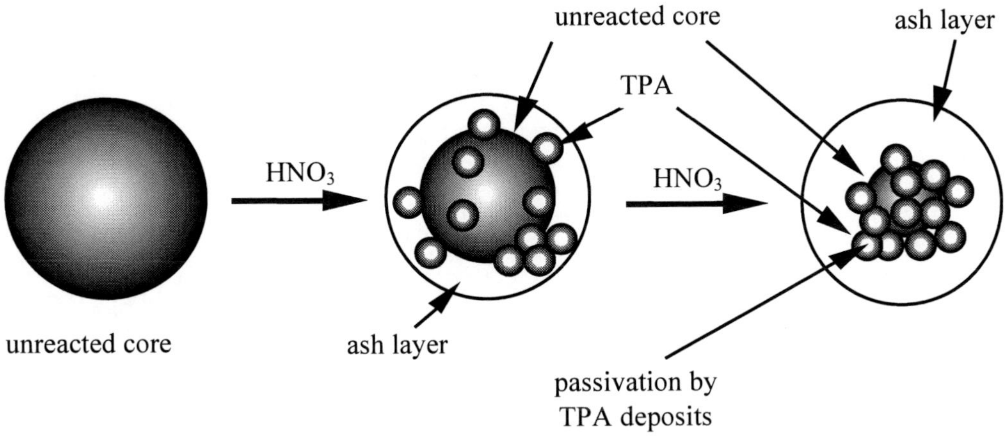

Figure 3. Modified shrinking core model of the hydrolysis of PET in nitric acid. The PET particle is covered by TPA during the process. The complete hydrolysis of the particle is inhibited.

The reaction is no longer dependent on the PET concentration, but on the surface area of the PET particles, which is decreasing with the time due to the hydrolysis reaction. An additional reduction of the surface area accessible to the acid is observed by the TPA deposition. The resulting reaction velocity law is expressed by the change of the conversion *X*:

$$-\frac{d(1-X)}{dt} = \frac{3}{\rho r_0}(1-X)^{\frac{5}{3}} k c_{A_0} \qquad \text{Eq. 6}$$

with the density of PET ρ, the initial PET particle radius r_0, the rate constant k and the initial acid concentration c_{A_0}. After integration

$$(1-X)^{-\frac{2}{3}} - 1 = k't \qquad \text{Eq. 7}$$

is obtained with k' as the new reaction rate constant including all relevant constants.

However, the kinetics is dependent on the particle shape and size and for other PET particles other solutions have to be found, as for larger particles where a proportional relationship between the effective surface area S_{eff} and the radius of the unreacted core r_c leads to a closer approximation [21]:

$$-\frac{dr}{dt} = \frac{3b}{\rho}\frac{r}{r_0}kc_{A_0} \qquad \text{Eq. 8}$$

with b as the stoichiometric factor of PET and r as the particle radius at the time t.

After integration

$$\ln(1-X)^{-\frac{1}{3}} = k''t \qquad \text{Eq. 9}$$

is achieved (Figure 4).

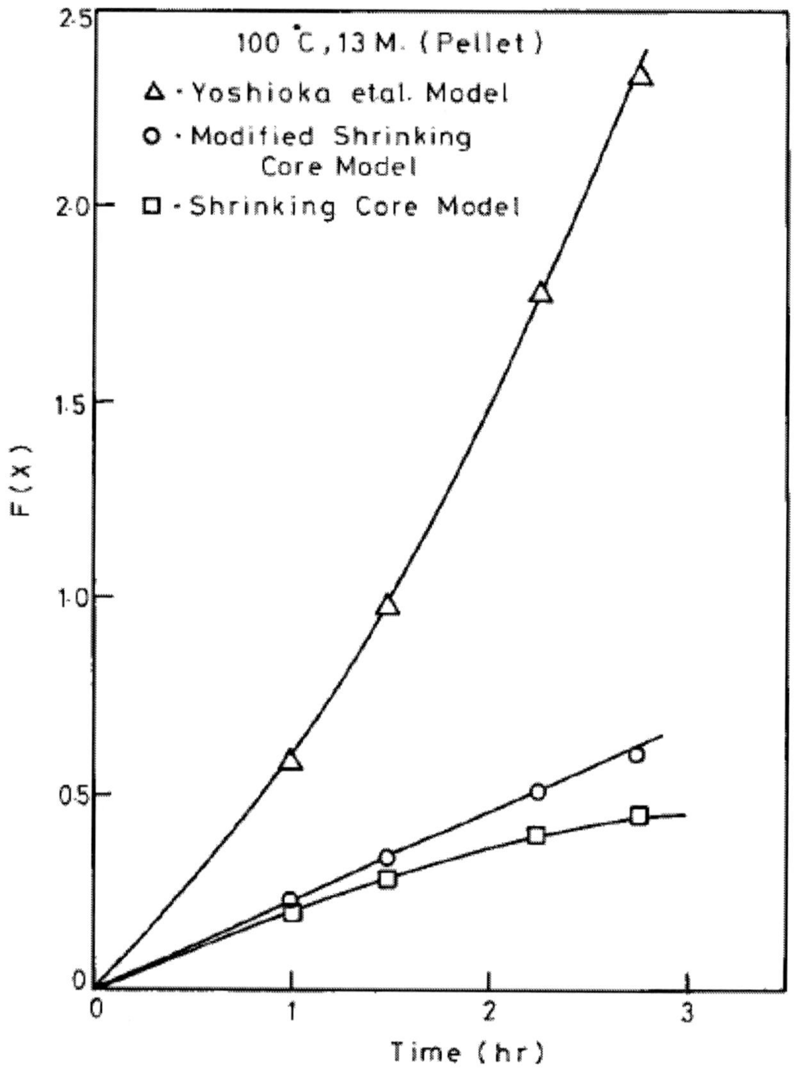

Figure 4. Comparison of different models for the hydrolysis of PET in nitric acid (reprinted from [21], with permission from Wiley).

Both methods described above are independent from the particle size in the end of the calculation. The first one describes the hydrolysis of small PET particles, while the second one describes the hydrolysis of pellets. The second one is stronger affected by the passivation from the deposition of TPA on the surface of the larger particles. Therefore the reaction rate is reduced.

According to the stronger passivation of larger particles, a rise of the activation energy can be expected due to the limited mass transfer. In fact, the activation energy of the hydrolysis of small particles of about 100μm is with 101kJ/mol [18] smaller than for pellets with a diameter of 3.5mm (135kJ/mol) [21]. Vice versa, the removal of the TPA deposit from the surface of the PET particles reduces the activation energy to 88kJ/mol [19].

Compared with other acids is the hydrolysis of PET in sulfuric acid a fast reaction. PET is already dissolved in concentrated sulfuric acid at room temperature in short times. TPA is precipitated by diluting the acid with water.

The reaction time is reduced when the sulfuric acid concentration rises from 3M and 9M [22,23]. It is necessary to use a temperature of 150°C with 9M and 190°C with 3M sulfuric acid to achieve a complete hydrolysis in one hour.

The reaction products are soluble in sulfuric acid and immediately transported into the solution. In the difference to nitric acid, there is no disposal of TPA on the particle surface inhibiting the material transport. Instead of that, sulfuric acid penetrates the amorphous areas of the PET leading to the fission of the chains [24]. The chain fragments are able to rearrange forming new crystalline areas. In concentrated sulfuric acid at 30°C, less ordered areas are formed in the first 60min of the reaction. After 60min, higher orientated areas are observed due to the dissolution and precipitation of material. At the end of the process, the crystalline areas left from the hydrolysis are also dissolved.

The attack of the sulfuric acid on the amorphous areas of the PET leaves cracks and pores reaching deep into the material [23] (Figure 5). The hydrolysis rate is slow for the first five hours at 150°C with 3M sulfuric acid and accelerates after that due to the increasing surface area by the formation of pores. The long inhibition time might be related to the hindered dissolution of the chain fragments inside the pores. The kinetics can be explained by a modified shrinking core model which acknowledges the growing surface area during the hydrolysis (Figure 6). The reduction of the surface area with increasing conversion X is regarded by $(1-X)$ and the growth of the surface area related to the formation of pores is regarded by $(1-aX)$ with a constant factor a representing the linear growth of the surface area of the residual PET (figure 7). It follows the rate law:

$$-\frac{d(1-X)}{dt} = \frac{3}{\rho r_0}(1-X)(1+aX)kc_A \qquad \text{Eq. 10}$$

After integration

$$\frac{1}{c+1}\ln\frac{1+cX}{1-X} = k't \qquad \text{Eq. 11}$$

is obtained.

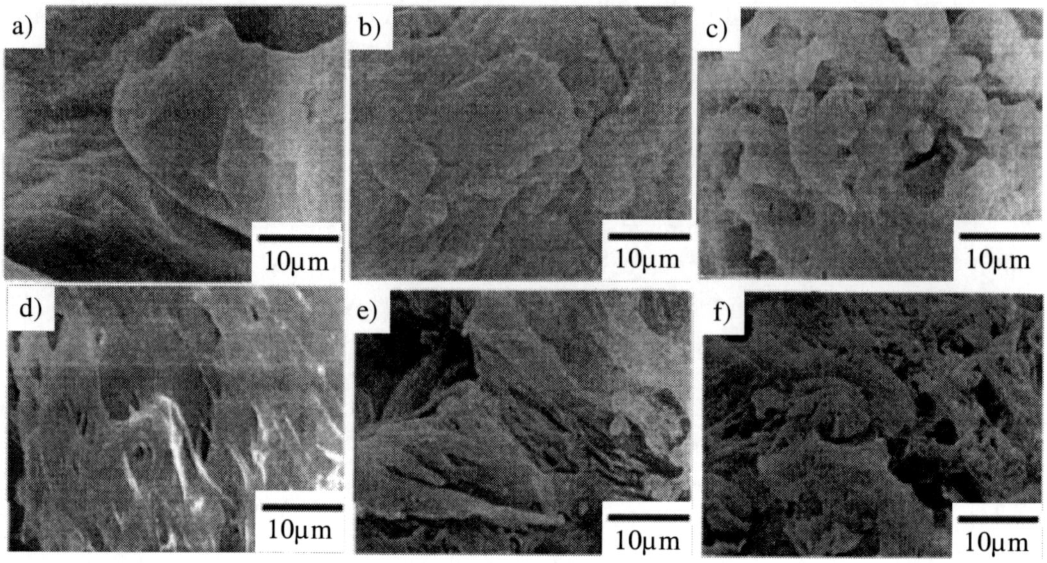

Figure 5. SEM photographs of the hydrolysis of PET with 3M sulfuric acid at 150°C. a) before the reaction, b) after 5h (conversion X=6.6%), c) after 6h (X=12%), d) after 7h (X=27%), e) after 8h (X=46%), f) after 9h (X=67%).

The velocity law and the agreements made in the modified shrinking core model show good results for the determination of the hydrolysis rate constant. The activation energy was found to be 89kJ/mol.

EG is dehydrated under the conditions described above. The slurry changes its color from colorless to brown indicating the carbonization of EG.

The acidic hydrolysis of PET is able to reduce the temperature and the pressure during the reaction. Accurate chosen process conditions may be able to avoid pressurized systems at all. Sulfuric acid and nitric acid are able to satisfy the claim of a complete hydrolysis of PET. Nitric acid requires long reaction times due to the deposit of TPA on the PET surface. The recovery of EG requires special arrangements in order to avoid losses of product by oxidation or dehydration. The precipitation of TPA from sulfuric acid requires the dilution with water and an additional recycling step for the acid.

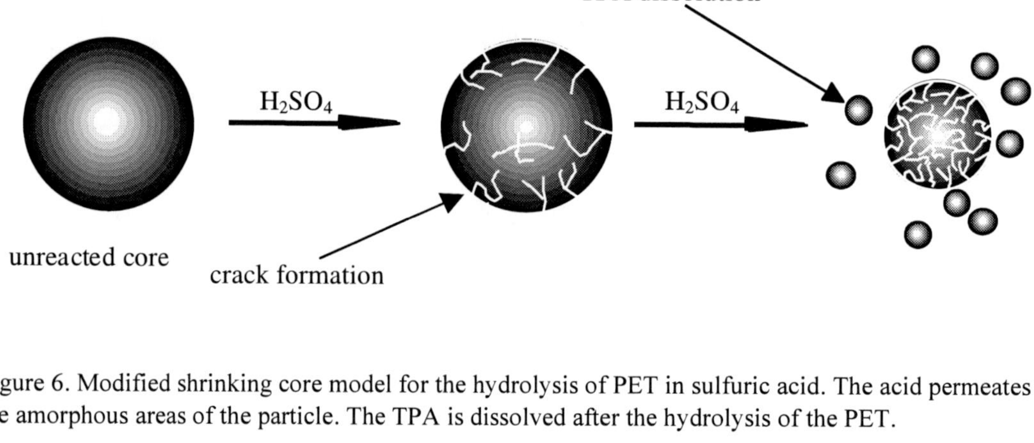

Figure 6. Modified shrinking core model for the hydrolysis of PET in sulfuric acid. The acid permeates the amorphous areas of the particle. The TPA is dissolved after the hydrolysis of the PET.

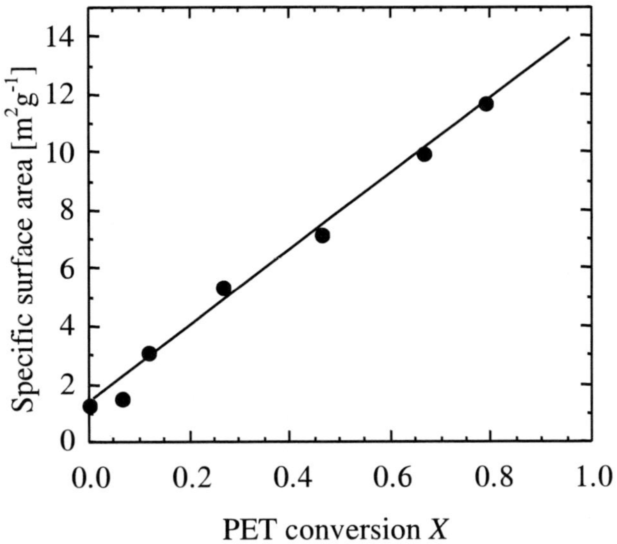

Figure 7. Dependence of the specific surface area on the conversion of PET in H_2SO_4 measured by BET.

2.3. Alkaline Hydrolysis of PET

Various methods for the alkaline hydrolysis (saponification) of PET were investigated in recent years. Both sodium hydroxide and potassium hydroxide were used in different solvent systems, as water, EG or other kinds of alcohols. The mechanism of the saponification is slightly different from the acid and neutral hydrolysis and the sodium or potassium salt of TPA is always obtained as the first product:

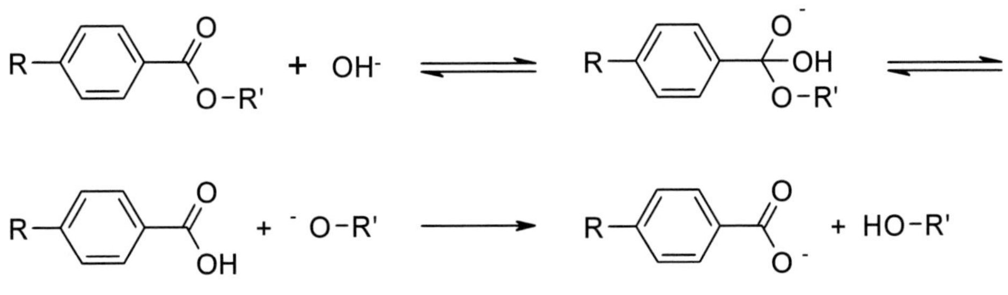

TPA is recovered from the solution by precipitation after the acidification with hydrochloric acid or sulfuric acid. An economic reasonable process is only possible if there is also a market for the residual salts (NaCl, Na_2SO_4, etc.) which can be gained in reagent quality.

Most experiments are carried out to obtain TPA from waste bottle PET or virgin material. However, the separation of different fibers is also possible by the alkaline hydrolysis of PET fibers. Rayon and cotton fibers remain unaffected and TPA is recovered as well [25].

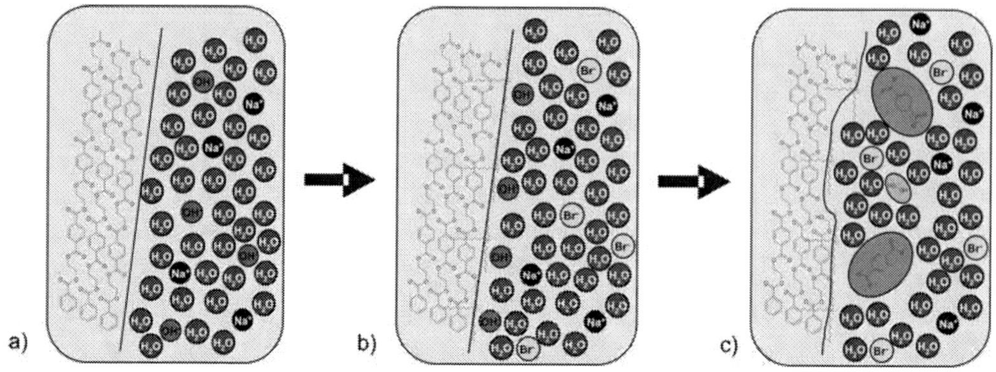

Figure 8. Alkaline hydrolysis in the presence of methyltrioctylammonium bromide as a phase transfer catalyst: a) without catalyst, b) after the addition of the catalyst, c) hydrolysis reaction.

The use of temperatures higher than the boiling point of the alkaline solution (about 100°C) requires the employment of autoclaves. However, the temperature can be reduced below the melting point of PET and less pressure is required. An almost complete hydrolysis of PET flakes can be achieved in 1h at 200°C [26] compared to 6h in pure water [12]. The reaction time increases clearly when the temperature is reduced.

In order to avoid autoclaves, the reaction can be carried out at temperatures below the boiling point of water [27]. The reaction time is extended to several hours in order to obtain high yields of TPA. NaOH (70g/l) is able to hydrolyse PET in 2.5h at 99°C with a conversion of 85%. Another limitation of the reaction is given by the NaOH concentration which shows a maximum for the PET conversion. It is assumed that the increasing viscosity of the NaOH solution limits the removal of the products from the surface of the PET particle.

The reaction time is also reduced by catalysts [28]. Phase transfer catalysts are common agents used for the hydrolysis of esters. Catalysts such as methyltrioctylammonium bromide act as a transportation medium for the hydroxide anion. The hydrophobic character of the aliphatic groups of the phase transfer catalyst reduces the solvation of the catalyst in the water phase. However, the hydrophobic surface of the PET particle is easily penetrated by the catalyst increasing the hydroxide concentration at the PET surface (Figure 8). The reaction temperature is reduced from 200°C [26] to 95°C [28] for a TPA yield of 98% and 84%, respectively, after 1h. However, be aware that the particle size without catalyst was 20 times smaller than with catalyst.

Similar results are achieved when aqueous KOH solutions are used [29]. However, a direct comparison of the results is difficult due to the different reaction conditions used in the experiments.

The reaction rate is strongly influenced by the choice of the solvent. The hydroxide ion is solvated and stabilized in aqueous media. Hence, the charge is more dislocated during the transition state of the saponification the reaction rate is increased by solvents with a lower polarity which lowers the solvation of the hydroxide ion. Solvents with a low polarity solvate also the hydrophobic PET chains causing swelling of the material and improve the access of the hydroxide ion. Methanol, ethanol and EG fulfill the requirements of a lower polarity than water and the ability to dissolve NaOH and KOH. Furthermore, the boiling point of EG (198°C) is high enough to avoid any kind of pressurized techniques. Aprotic solvents should also have a strong effect on the reaction rate due to their ability to solvate the transition state stronger than the charged hydroxide ion [30].

The solvent effects described above are confirmed by the replacement of water by EG [31-33]. A complete hydrolysis of PET can be achieved in 1min at 185°C with NaOH. Even if pressurized reactors are avoided, the higher viscosity of EG solutions requires a strong agitation in order to overcome migration effects and the deposition of TPA on the PET surface. It is observed that an unsatisfactory agitation causes a drop in the reaction rate, however, the activation energy remains constant, assuming the formation of an equilibrium between the deposition of new formed Na-TPA and its dissolution which does not interfere with the hydrolysis reaction. It is also observed that higher alkaline concentrations reduce the reaction rate due to the increasing viscosity. This effect is overcome by using a solvent with a lower viscosity. The hydrolysis with 9M KOH in 2-methoxy ethanol (methyl cellosolve) after 2.5h results in a TPA yield of 82% [26].

The reaction is strongly accelerated by using the primary alcohols methanol and ethanol while the reaction rate is slower in the secondary alcohol isopropanol [34,35]. Methanol and ethanol deliver already at moderate temperatures between 50°C and 60°C high yields of Na-TPA and K-TPA. The addition of an aprotic solvent such as dioxane, THF or dimethoxy ethane accelerates the reaction strongly. A complete hydrolysis is achieved in less than 1h.

The expected effect of aprotic solvents on the reaction rate for the hydrolysis of PET is also observed for the addition of DMSO to EG (20/80). The reaction is 12 times faster than in pure EG. The same effect is visible with THF and dimethoxyethane [32,33]. The alkaline hydrolysis in the aprotic solvent 1,3-dimethyl-2-imidazolidinone carried out at 150°C leads to the a complete reaction in 1h [34]. However, the high boiling point makes the separation of EG from the solvent difficult.

The history of the material has also a high impact on the reaction rate. A high water content of the material reduces the reaction time. Thermopressing at 260°C of water saturated PET leads to a primarily hydrolysis and shortening of the chain length. The subsequent hydrolysis in nonaqueous solution of NaOH in glycol at 170°C reduced the reaction time to one quarter [31].

The quality of the TPA recovered after acidification of the basic solution is reported to be very similar to that of reagent TPA [27]. However, the appearance of about 2% isophthalic acid in the product was reported. Nevertheless, it is possible to obtain new PET from the polycondensation with EG [26,28].

The kinetic models for alkaline hydrolysis are very similar to those for the acidic hydrolysis. The shrinking core model is used as base due to the physical insolubility of PET in alkaline solutions. According to the findings, the model is modified dependent on the shape of the particles, the particle size and the experimental setup.

Flakes are only hydrolysed at the surface while deeper areas remain intact. DSC and viscosity investigations show no significant difference between the original and the residual flakes. The molar mass of the residual PET drops slightly due to the formation of oligomers on the PET surface [29]. However, it was found that in the beginning of the hydrolysis of small particles of 300µm or less the acid value of the obtained product (mainly TPA) is far below the theoretically expected value, assuming that small particles decay during the alkaline hydrolysis releasing fragments and oligomers into the solution before finally TPA is obtained [27] (Figure 9).

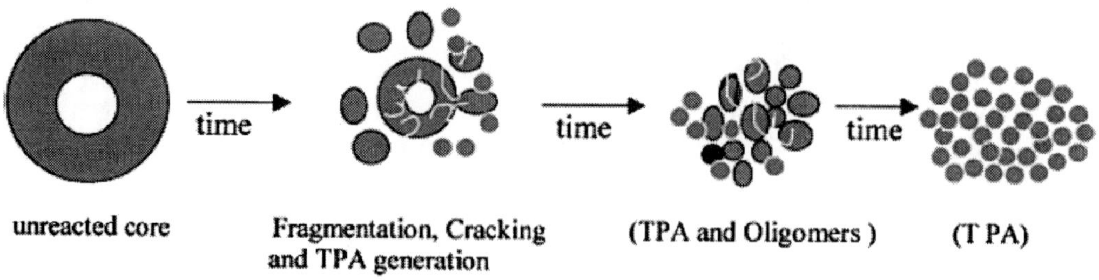

Figure 9. Fragmentation of PET during the alkaline hydrolysis of small particles (reprinted from [27], with permission from Wiley).

EG supports the fragmentation of the PET particles. Pores and cavities are formed during short reaction times accelerating the reaction and causing the break down of the structure [31] (Figure 10). However, high agitation speed is required to overcome the transport limitations caused by the viscosity of the solution.

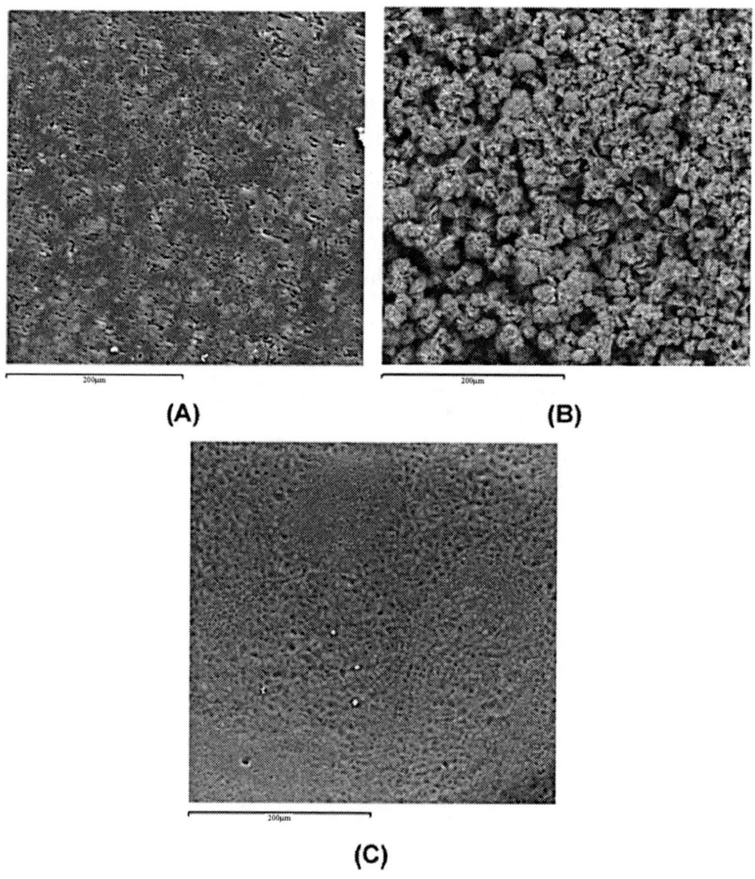

Figure 10. SEM of PET sheets after the hydrolysis with NaOH in EG at 170°C after (A) 1min (30% conversion), (B) 2min (50%) and (C) 4min (70%). The magnification for (A) and (C) is 500 and for (B) 100 (reprinted from [31], with permission from ACS).

Table 4. Comparison of activation energies E_A under different conditions during the alkaline hydrolysis of PET

Reactant	Solvent	E_A [kJ/mol]
NaOH [25]	Water	99
KOH [35]	Water	29
KOH [28]	Water	69
NaOH [27]	Water/Methyltrioctylammoniumbromide	83
NaOH [30]	Ethylene glycol	173
KOH [32]	Ethylene glycol, CHA, THF, DMSO	24
KOH [34]	Methanol	24

Researchers used also the diffusion coefficients of the products [35] and the molar distribution coefficient of the hydroxide [36] for the modification of the shrinking core model.

Some researchers [29,36] assume a chain-end scission mechanism for this reaction due to the first order kinetic regarding the hydroxide concentration. It is assumed that random-scission only occurs when both ester linkages of one TPA unit are undergo hydrolysis at the same time. However, chain-end ester linkages are a minority of less than 2% in a PET material with a molar mass of M_n=20,000g/mol and the chemical difference between chain-end and inner-chain ester-linkages is negligible. If the chain-end scission is the main path way of the reaction, the alkaline hydrolysis would be much slower than the hydrolysis in neutral water where the random scission mechanism is suggested. Furthermore, the GPC and HPLC analysis of oligomers after the hydrolysis of PET fabrics in methanolic NaOH shows the formation especially of oligomers with a molar mass of about 2400g/mol, while larger fragments of the hydrolysis are not observed [37]. The length of these oligomers is comparable with the length of crystallized units, while the surrounding amorphous area is hydrolysed. This is more likely expected for a random-scission mechanism than for a chain-end scission mechanism.

The activation energy depends also in the case of the alkaline hydrolysis strongly on the reaction conditions. It ranges from 24kJ/mol for the hydrolysis of PET powder in aqueous KOH to 173kJ/mol for the hydrolysis with NaOH in EG (Table 4). Values below 30kJ/mol seem to be very low indicating that the reaction rate is controlled by physical processes. The values indicate lower activation energies for the hydrolysis with KOH compared with NaOH as it is also visible in the more rapid hydrolysis of PET is KOH solutions. Also the use of catalysts shows a reduction on the activation energy. However, the wide range of activation energies indicates the strong influence of effects which are not related to the fundamental hydrolysis reaction. It is essential to identify and eliminate these effects in order to obtain reliable values.

2.4. Other Processes

Besides solvolysis, processes are employed working at high temperatures and using steam as water source in order to reduce the pressure necessary for the reaction. Such processes are performed in fluidized beds, rotary kiln reactors or by reactive extrusion.

TGA shows in the absence of steam the occurrence of pyrolysis and the formation of 16wt% of carboneous residue [38]. In the presence of steam the degradation starts at a lower temperature and the formed residue is reduced to 1wt%. Hydrolysis and pyrolysis are competitive reactions with activation energies of 160kJ/mol and 200kJ/mol, as well. The lower activation energy enables the hydrolysis to repress the pyrolysis. In this way, TPA is recovered from PET quantitatively at 450°C in steam atmosphere by using a fixed bed reactor. The IR-spectra show no significant differences between the recovered TPA and reagent TPA.

The hydrolysis of PET in a fluidized bed at 450°C with steam as fluidizing gas and quartz sand as bed material shows similar results [39]. In fluidized bed reactors, a material of fine particles is fluidized by a gas stream from the bottom of the reactor. The behavior of a fluidized bed is comparable with a liquid and shows excellent properties for heat and mass transport. Products are removed immediately from the hot reactor zone by the fluidizing gas. Typical residence times ranges from seconds to 1min.

In this way, TPA is produced in high yields (72%) and some low mass oligomers are obtained as by products (yield: 24%) indicating an incomplete hydrolysis due to the short residence time. However, EG is mainly degraded in the presence of water and CO_2 is formed. The applicable temperature ranges from 410°C to 470°C. At lower temperatures, TPA deposits are formed in the reactor and product gas lines due to the sublimation point of TPA at 402°C. At higher temperatures, the product color changes from white to yellow indicating the start of pyrolysis reactions. TPA is easily separated from water and by products during condensation by its high sublimation point. However, an appropriate filter system has to be established for the removal of TPA in order to avoid blocking problems.

Blocked pipes and corrosion are problems waste treatment facilities have to deal with during processing of PET containing waste due to the formation of TPA and benzoic acid. The addition of FeOOH as a catalyst to PET during the hydrolysis with steam at 500°C reduces the quantities of TPA and benzoic acid significantly [40]. As a result mainly acetophenone, benzene and CO_2 are produced. During the treatment of FeOOH with steam at the processed temperature a highly porous Fe_2O_3 is obtained while commercial Fe_2O_3 shows no activity.

A concept was found to degrade corrosive and sublime products from the pyrolysis of PET and PBT containing wastes and convert them into fuels [41]. In a first step, the waste (PE/PET 15/2) is pyrolysed and hydrolysed in a circulated-spheres reactor in the presence of steam at 500°C (Figure 11). The reactor is filled with glass beads acting as heat transfer medium and increasing the surface area of the molten plastic. The beads are lifted by a stirring system in order to distribute the plastic and to increase to hold-up time. The gaseous products enter a tank reactor where the degradation is completed. Sublime products from polyesters are degraded over FeOOH in a fixed bed reactor at the same temperature. Mainly waxes are obtained after the last degradation step. The product oil is converted into low boiling hydrocarbons (C_6-C_{12}) over Rey-Ni-catalyst at 400°C.

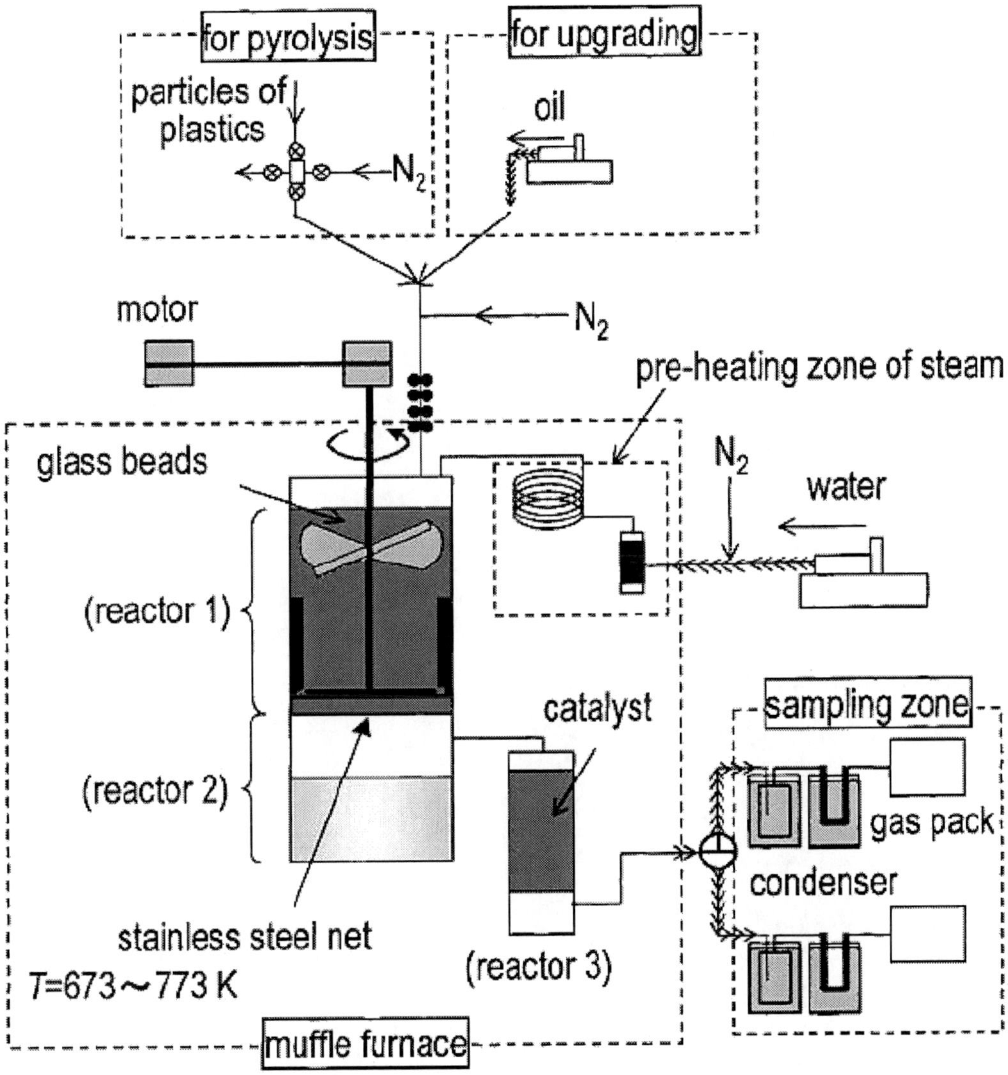

Figure 11. Treatment of PET and PBT containing waste in a circulated-spheres reactor (reprinted from [41], with permission from Elsevier).

The hydrolysis in the presence of $Ca(OH)_2$ leads to the decarboxylation of TPA and the formation of benzene [42]. A benzene yield of 88% is obtained at 700°C with a benzene purity of 79% in the oil (Figure 12). By products are mainly biphenyl and other aromatics. EG is completely degraded to carbon oxides. Similar results are obtained from the hydrolysis of PET in the presence of CaO and steam assuming that CaO and $Ca(OH)_2$ undergo in the first step the same reaction forming Ca-TPA which is decarboxylized leaving benzene and CO_2 and restoring CaO. It is necessary to mention that $Ca(OH)_2$ is dehydrated at 550°C. However, if the reaction is performed at lower temperatures, similar results with longer reaction times are obtained.

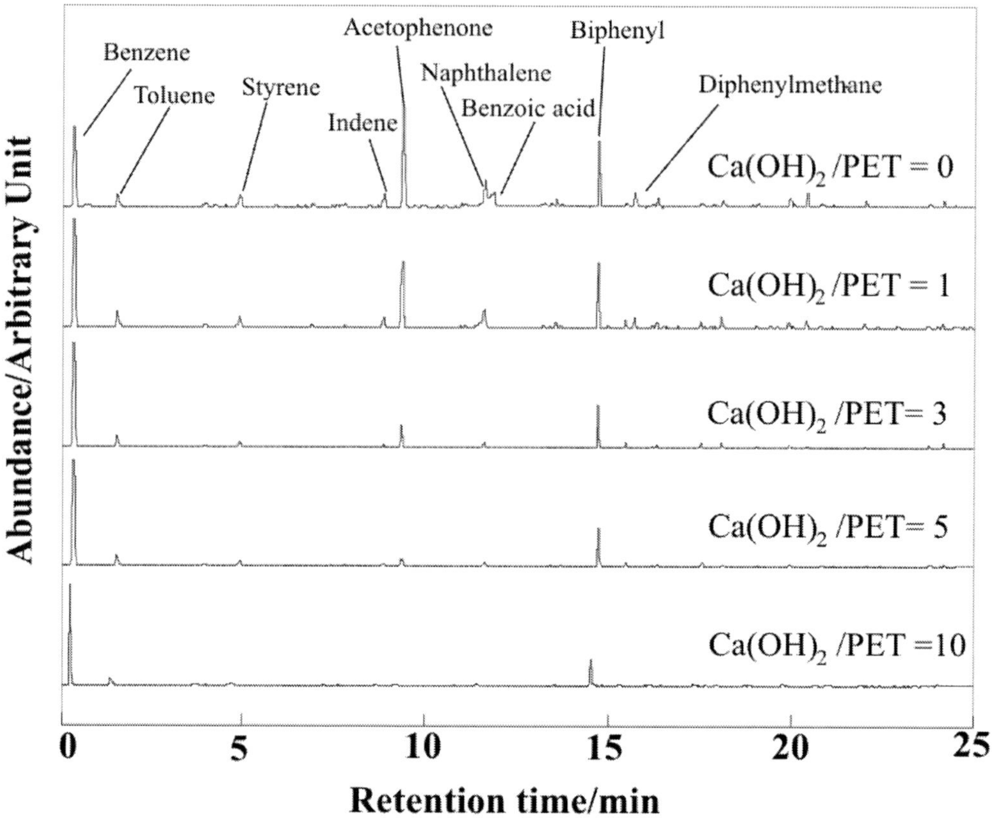

Figure 12. GC of the product oil from the hydrolysis of PET in the presence of $Ca(OH)_2$.

The local separation of the PET hydrolysis from the decarboxylation by a two chamber reactor with two different heating zones at temperatures below 500°C helps to avoid the formation of by products. In the first chamber, the PET is hydrolysed in the absence of any catalyst using a slow heating rate from 300°C to 500°C. The resulting TPA is decarboxylized in the presence of CaO at 500°C. Benzene is obtained with a yield of 74% and a purity of 97% (unpublished results). The realization of a rotary kiln pilot plant using these results is in progress.

The addition of Fe_2O_3 and TiO_2 to the $Ca(OH)_2$/PET system has little effect. However, NiO is an effective catalyst for the production of synthesis gas from PET [43]. Especially bottle PET is an excellent source for the gasification due to its purity, the lack of catalyst poisoning substances and the amount of waste available.

The hydrolytic degradation of PET in a twin extruder using steam under high pressure enables the reduction of the viscosity of the material due to the reduction of the chain length [44]. However, a complete hydrolysis is not obtained and oligomers with an average molar mass of M_n=325g/mol are found. Best results are obtained at 300°C, high pressure (4800kN/m^2) and a low water/PET ratio (wt/wt: 1/2) resulting in a low residence time of the melt. Decomposition of TPA occurs at higher temperatures. The activation energy is given with 75kJ/mol.

2.5. Hydrolysis of PEN and PBT

The hydrolysis properties of PEN and PBT are very similar to those of PET. Both PBT and PEN undergo the same reactions as PET. However, PBT and PEN show a higher hydrolysis resistance than PET below their melting areas due to the lower water uptake caused by the stronger hydrophobic character of these plastics.

Experiments on fibres have shown that PBT is less affected by aqueous and alcoholic NaOH solutions than PET [45]. While PET loses about one third of its weight during 5min of hydrolysis with 40wt% aqueous NaOH solution at 97°C, PBT loses just about 3wt%. Both PET and PBT are stronger affected by methanolic and ethanolic NaOH solutions, however, the resistance of PBT against the alkaline attack is still stronger than that of PET.

Similar results are observed during the hydrolysis of PEN and PET films in saturated steam between 120°C and 160°C [46]. PEN shows a stronger resistance against hydrolysis than PET. The reaction rate is about 2-5 times smaller for PEN than for PET. The effect of the lower diffusion rate of water in PEN is negligible since the diffusion rate is 4-6 orders of magnitude larger than the reaction rate.

All three polyesters, PET, PBT and PEN are quantitatively hydrolysed in neutral water above their melting point. All effects described before for PET are found for PBT [47] and PEN [6], as well. Differences in the hydrolysis behavior are related to the difference in the melting point. PBT has the lowest melting point (about 220°C) of the common polyesters and it shows the fastest hydrolysis. Under the same conditions, a complete hydrolysis of PBT is achieved at 245°C after 10min, while for PET (mp~245°C) a temperature of 270°C is necessary [9] even if the equilibrium is more advantageous for the hydrolysis of PET.

Still higher temperatures are observed in the case of PEN due to the melting point at about 270°C. PEN conversions of more than 90% are achieved at temperatures of more than 300°C. However, degradation of EG occurs at these temperatures as mentioned earlier.

The decarboxylation of PBT and PEN occurs in the same way as for PET in the presence of $Ca(OH)_2$ in a fixed bed reactor [48]. Best results are obtained at 700°C with a $Ca(OH)_2$/polyester repeating unit ratio of 5:1 with a benzene yield of 67% from PBT and a naphthalene yield of 80% from PEN, as well. 1,4-Butanediol undergoes secondary reactions and the subsequent dehydration leads to the formation of tetrahydrofuran and butadiene (Figure 13).

2.6. Industrial Applications

Most processes used in laboratory scale experiments are batch processes. Industrial plants, however, have to be developed as continuous working facilities in order to reduce costs and time for processing. Several types of reactors are used as hydrolysators and also combined methods were developed for the hydrolysis of polyesters. All described processes are valid for PET, PBT and PEN, as well.

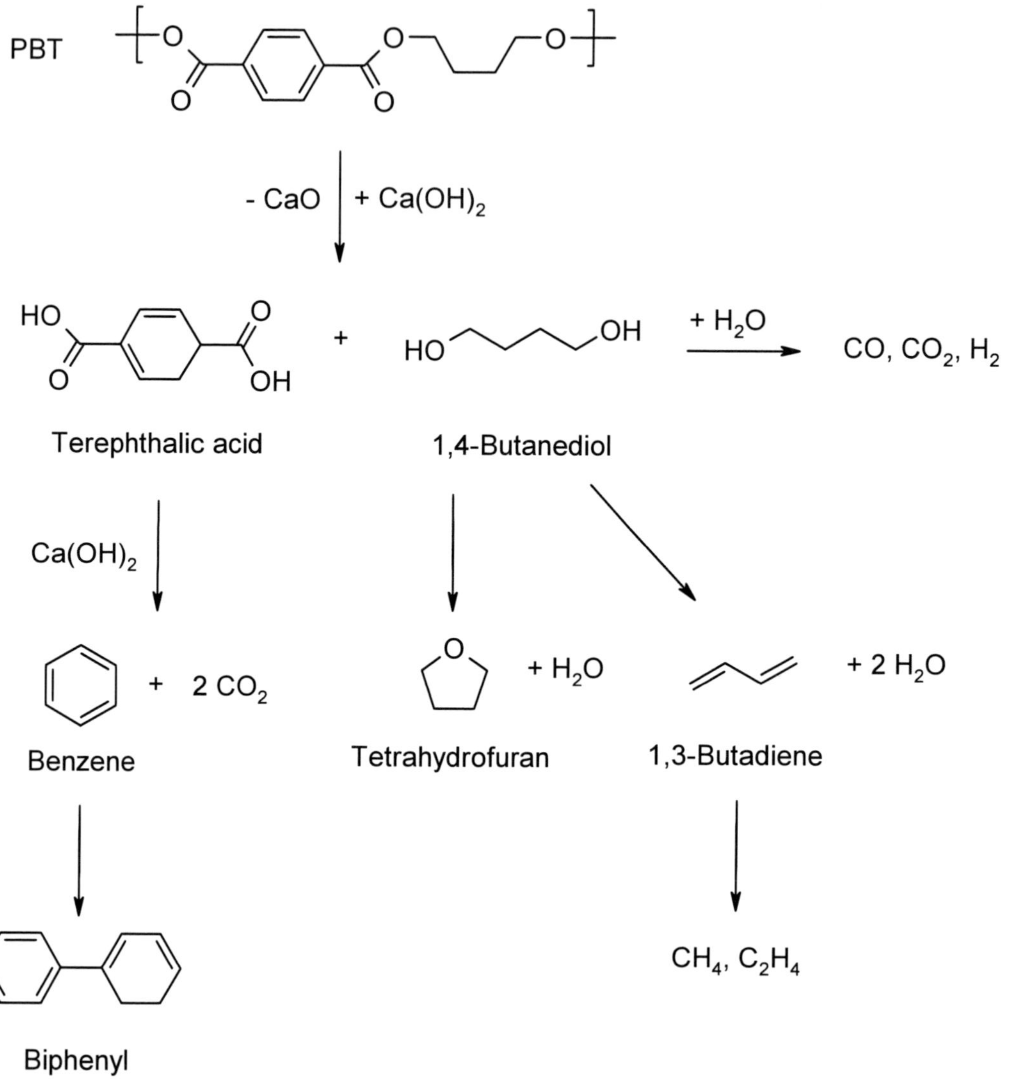

Figure 13. Scheme of the hydrolysis of PBT in the presence of $Ca(OH)_2$.

Water and PET are heated separately and then mixed and introduced into a tube reactor in a process developed by Asahi Chem. Ind. Turbulent conditions provide a good heat and material transfer necessary for the heterogeneous slurry. At 300°C, the reaction takes less than 4min and is fast enough to obtain high yields of TPA and EG, as well, since the heat transfer inside the low heat conductive PET melt is minimized [49].

Celanese chose another way to perform a continuous process. PET is fed by an extruder from the top into the hydrolyzer. The PET melt is heated by steam introduced from below. The steam provides also agitation for the slurry. The condensed steam forms a water phase at the top of the hydrolyzer containing the products TPA and EG. The products are removed by withdrawing the water phase from the reactor. At 248°C and 4.2MPa, a residence time of 2h is achieved for a complete hydrolysis [50].

The reaction is accelerated by manganese, zinc and calcium acetate [51] or tin and lead compounds [52]. However, catalysts have to be removed carefully from the obtained products, since they are able to interfere also during the PET polycondensation. Even if the catalysts remain quantitatively in the mother liquid, especially heavy metal salts cause concerns in the relation with daily life applications.

Another possibility is the use of a two step process. In the first step, PET is hydrolysed for 30min at 220°C to 280°C in a twin extruder. The partly hydrolysed slurry is cooled and the precipitated solid is separated from the mother liquid. The solid is dissolved in an ammonia solution after that, and the hydrolysis is finished at 180°C to 230°C after 30min reaction. TPA is precipitated by adding sulfuric acid [53].

The most processes described above are also under consideration as industrial relevant processes. Both methods, using acidic [54,55] and basic hydrolysis with the subsequent neutralization with an adequate mineral acid were developed, as well. The basic hydrolysis is performed with alcoholic KOH [56] and NaOH [57,58] solutions in the presence of aprotic solvents or in aqueous NaOH including the oxidation of contaminants with air obtaining an insoluble inorganic residue [59]. The hydrolysis of PET in 7wt% ammonia solution results in a diammonium terephthalate yield of 80% after 30min reaction at 190°C [60].

The technical realization of fluidised bed plants [61] and the hydrolysis using extruders [62] are also considered.

3. POLYCARBONATE

The term polycarbonate (PC) is mainly used for bisphenol-A-PC (Figure 14) as the diester of carbonic acid and bisphenol-A (BPA) as the diol component. It is the most important PC used for applications which require high transparency (CDs, DVDs, headlight covers, windows for planes and vehicles) and impact strength (casings for computers and other electric devices).

The ester structure makes PC sensitive to hydrolysis (Figure 15) and processes as described above for polyesters should be practicable for PC, too. However, some disadvantages have to be considered. Firstly, the low thermal stability of BPA, which degrades under formation of phenol and isopropenyl phenol in hot water at temperatures higher than 200°C [63-65], makes it necessary to remove the product quickly from the hot reactor. This might be the reason that there is no work available dealing with the hydrolysis of PC in autoclaves at high temperatures due to the fast degradation of BPA under reaction conditions. The second reason might be that real waste PC contains usually high amounts of additives (dyes, fire retardants, etc.) and fillers and it is also frequently blended with other plastics such as PBT or polystyrene.

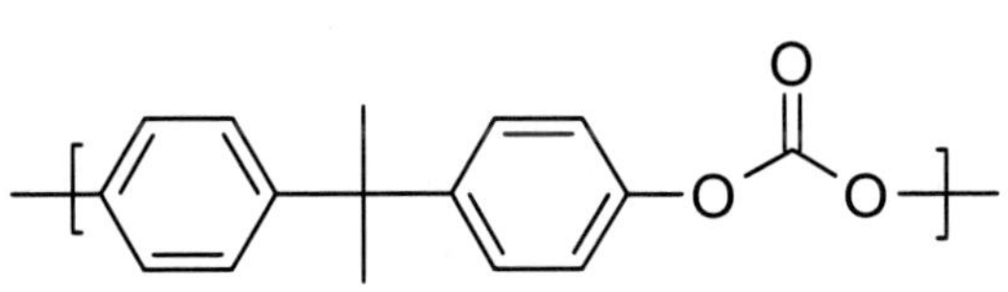

Figure 14. Bisphenol-A-Polycarbonate.

Table 5. Product yields of the hydrolysis of PC in the presence of MgO and CaO at different temperatures

The yields are related to the reactions:
PC → BPA
PC → Phenol + Isopropenyl phenol

	MgO		CaO	
Temperature [°C]	300	500	300	500
Bisphenol-A [%]	78	16	49	21
Phenol [%]	0	57	9	41
Isopropenyl phenol [%]	2	49	7	27

TGA shows the start of the hydrolysis in steam atmosphere at about 300°C and the end at 500°C leaving just low amounts of residue [41]. The reaction is accelerated by MgO and CaO and BPA is obtained in high yields at 300°C when the products are removed quickly from the hot reactor zone (unpublished results). At 500°C, BPA is mainly destroyed and phenol and isopropenyl phenol are produced as degradation products [66]. The reaction mechanism given in Figure 16 offers different path ways for the formation of phenol resulting in high phenol yields at 500°C. 2-(4-Hydroxy phenol)-2-propanol which is observed during the fission of BPA in hot water [63] is not found as product due to the equilibrium located on the left side of the formula. However, traces of acetone are found in the GC-MS spectra at least indicating the possibility of the formation of (HPP). Table 5 shows the product yields from the hydrolysis of PC in the presence of CaO and MgO achieved in a fixed bed reactor as described elsewhere [42].

Similar to PET, the hydrolysis of PC occurs also after a random scission mechanism. The choice of the end-group has an important impact on the degradation behavior. Phenol capped PC shows an extraordinary stability of the terminal phenol groups at the beginning of the degradation in saturated steam at 125°C [67]. After a delay time the hydrolysis of the end groups accelerates and overtakes even the hydrolysis rate of the BPA-ester links indicating a more sophisticated kinetics.

4. Conclusion

In recent years, the interest in the recycling of polyesters and polycarbonates has increased noticeably. New regulations let to the necessity to reduce the amount of waste and to recover raw materials from plastics. In this view, the hydrolysis of polyesters and PC shows promising results. PET is hydrolysed under different conditions in the range from several minutes to several hours. TPA and EG are recovered in high yields close to 100%. Neutral hydrolysis requires high temperatures and pressures for the recovery of TPA.

Extreme conditions are avoided using acids or alkaline solutions. However, the recovery of TPA requires additional steps. All processes developed for PET are also valid for PBT and PEN, as well.

Figure 15. Hydrolysis of PC.

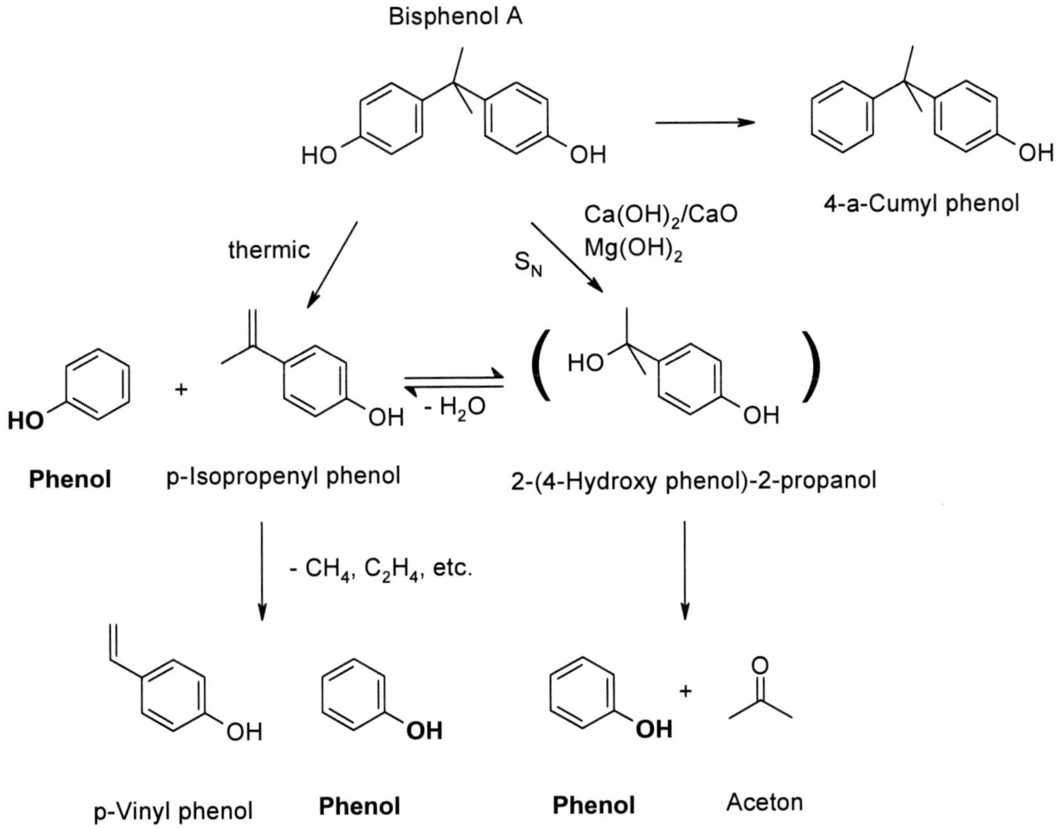

Figure 16. Reaction mechanism of the degradation of BPA after the hydrolysis of PC in the presence of MgO and CaO.

On the other hand, the determination of kinetic and thermodynamic data faces serious problems. Data for the activation energy for PET vary in a wide range depending on the experimental conditions. The causes for this behavior are not clear in detail. More investigations are required to come to more universal data.

Principally, the hydrolysis of PC proceeds after the same rules than PET. However, the low stability of BPA at high temperatures and the use of additives in PC require more sophisticated processes than for polyesters. Just a few methods have been developed for the recovery of monomers from PC until now.

References

[1] 2003/33/EC: Council Decision of 19 December 2002 establishing criteria and procedures for the acceptance of waste at landfills pursuant to Article 16 of and Annex II to Directive 1999/31/EC.

[2] Yoshioka, T.; Grause, G.; Eger, C.; Kaminsky, W.; Okuwaki, A.: Pyrolysis of Poly(ethylene terephthalate) in a Fluidised Bed Plant. *Polym. Degrad. Stab.* 2004, *86*, 499-504.

[3] Puglisi, C.; Sturiale, L.; Montaudo, G.: Thermal Decomposition Processes in Aromatic Polycarbonates Investigated by Mass Spectrometry. *Macromolecules*. 1999, *32*, 2194-2203.

[4] Jellinek, H.H.G.; Dunkle, S.R.: Kinetics and Mechanism of Hydrogen Cyanide Evolution from Nylon 66 and Design of Apparatus. *J. Polm. Sci., Polm. Chem. Ed.* 1982, *20*, 85-101.

[5] Hassan, M.K.; Abou-Hussein, R.;Zhang, X.; Mark, J.E.; Noda, I.: Biodegradable Copolymers of 3-Hydroxybutyrate-co-3-hydroxyhexanoate (Nodax^{TM}), Including Recent Improvements in their Mechanical Properties. *Mol. Cryst. Liq. Cryst.* 2006, *447*, 23/[341]-44/[362].

[6] Sato, O.; Arai, K.; Shirai, M.: Hydrolysis of poly(ethylene terephthalate) and poly(ethylene 2,6-naphthalene dicarboxylate) using water at high temperature: Effect of porton on low ethylene glycol yield. *Catalysis Today*. 2006, *111*, 297-301.

[7] Kao, C.-Y.; Wan, B.-Z.; Cheng, W.-H.: Kinetics of Hydrolytic Depolymerization of Melt Poly(ethylene terephthalate). *Ind. Eng. Chem. Res.* 1998, *37*, 1228-1234.

[8] Campanelli, J.R.; Kamal, M.R.; Cooper, D.G.: A Kinetic Study of the Hydrolytic Degradation of Polyethylene Terephthalate at High Temperatures. *J. Appl. Polym. Sci.* 1993, *48*, 443-451.

[9] Goje, A.S.; Thakur, S.A.; Diware, V.R.; Patil, S.A.; Dalwale, P.S.; Mishra, S.: Hydrolytic Depolymerization of Poly(Ethylene Terephthalate) Waste at High Temperature Under Autogenous Pressure. *Polymer-Plastics Technology and Engineering*. 2004, *43*, 1093-1113.

[10] Yoshioka, T.; Okuwaki, A.: *Recycling of PET bottles*. Ryusan to Kogyo. 1994, Feb., 21-29.

[11] Mishra, S.; Zope, V.S.; Goje, A.S.: Kinetics and Thermodynamics of Hydrolytic Depolymerization of Poly(ethylene terephthalate) at High Pressure and Temperature. *J. Appl. Polym. Sci.* 2003, *90*, 3305-3309.

[12] Mancini, S.D.; Zanin, M.: Optimization of Neutral Hydrolysis Reaction of Postconsumer PET for Chemical Recycling. *Progress in Rubber, Plastics and Recycling Technology*. 2004, *20*, 117-132.

[13] Campanelli, J.R.; Cooper, D.G.; Kamal, M.R.: Catalyzed Hydrolysis of Polyethylene Terephthalate Melts. *J. Appl. Polym. Sci.* 1994, *53*, 985-991.

[14] Liu, L.; Zhang, D.; An, L.; Zhang, H.; Tian, Y.: Hydrolytic Depolymerization of Poly(ethylene terephthalate) Under Microwave Irradiation. *J. Appl. Polym. Sci.* 2005, *95*, 719-723.

[15] Schnabel, W.: *Polymer Degradation*. Hanser Int. Munich. 1981, 179-181.

[16] Ravens, D.A.S.: The Chemical Reactivity of Poly(ethylene terephthalate): Heterogeneous Hydrolysis by Hydrochloric Acid. *Polymer*. 1960, *1*, 375-383.

[17] Yoshioka, T.; Okayama, N.; Okuwaki, A.: Chemical Recycling of Post Consumer PET Powder by Hydrolysis and Simultaneous Oxidation in Nitric Acid. *The Third International Symposium on East Asian Resources Recycling Technology, Nov. 21-24, 1995, Taipei.* 519-525.

[18] Yoshioka, T.; Okayama, N.; Okuwaki, A.: Kinetics of Hydrolysis of PET Powder in Nitric Acid by a Modified Shrinking-Core Model. *Ind. Eng. Chem. Res.* 1998, *37*, 336-340.

[19] Mishra, S.; Goje, A.S.; Zope, V.S.: Chemical Recycling, Kinetics, and Thermodynamics of Poly(Ethylene Terephthalate) (PET) Waste Powder by Nitric Acid Hydrolysis. *Polymer Reaction Engineering*. 2003, *11*, 79-99.

[20] Rosmaninho, M.G.; Jardim, E.; Moura, F.C.C.; Ferreira, G.L.; Thom, V.; Yoshida, M.I.; Araujo, M.H.; Lago, R.M.: Surface Hydrolysis of Postconsumer Polyethylene Terephthalate to Produce Adsorbents for Cationic Contaminants. *J. Appl. Polym. Sci.* 2006, *102*, 5284-5291.

[21] Kumar, A.; Rao, T.R.: Kinetics of Hydrolysis of Polyethylene Terephthalate Pellets in Nitric Acid. *J. Appl. Polym. Sci.* 2003, *87*, 1781-1783.

[22] Yoshioka, T.; Sato, T.; Okuwaki, A.: Hydrolysis of Waste PET by Sulfuric Acid at 150°C for a Chemical Recycling. *J. Appl. Polm. Sci.* 1994, *52*, 1353-1355.

[23] Yoshioka, T.; Motoki, T.; Okuwaki, A.: Kinetics of Hydrolysis of Poly(ethylene terephthalate) Powder in Sulfuric Acid by a Modified Shrinking-Core Model. *Ind. Eng. Chem. Res.* 2001, *40*,75-79.

[24] de Carvalho, G.M.; Muniz, E.C.; Rubira, A.F.: Hydrolysis of post-consume poly(ethylene terephthalate) with sulfuric acid and product characterization by WAXD, ^{13}C NMR and DSC. *Polym. Degrad. Stab.* 2006, *91*, 1326-1332.

[25] Kwon, H.; Negulescu, I.I.; Collier, B.J.: Complete Recycling of Textile Components from Cotton/Polyester Fabrics. *Book of Papers: 1996 International Conference & Exhibition,* September 15-18, Opryland Hotel, Nashville, Tenn. 480-488.

[26] Karayannidis, G.P.; Chatziavgoustis, A.P., Achilias, D.S.: Poly(ethylene terephthalate) Recycling and Recovery of Pure Terephthalic Acid by Alkaline Hydrolysis. *Advances in Polymer Technology*. 2002, *21*, 250-259.

[27] Mishra, S.; Zope, V.S.; Goje, A.S.: Kinetic and thermodynamic studies of depolmerisation of poly(ethylene terephthalate) by saponification reaction. *Polym. Int.* 2002, *51*, 1310-1315.

[28] Kosmidis, V.A.; Achilias, D.S.; Karayannidis, G.P.: Poly(ethylene terephthalate) Recycling and Recovery of Pure Terephthalic Acid. Kinetics of a Phase Transfer Catalyzed Alkaline Hydrolysis. *Macromol. Mater. Eng.* 2001, *286*, 640-647.

[29] Wan, B.-Z.; Kao, C.-Y.; Cheng, W.-H.: Kinetics of Depolymerization of Poly(ethylene terephthaltate) in a Potassium Hydroxide Solution. *Ind. Eng. Chem. Res.* 2001, *40*, 509-514.

[30] Smith, M.B.; March, J.: *March's Advanced Organic Chemistry*; 5th Edition; Wiley: New York, 2001, p. 450.

[31] Ruvolo-Rilho, A.; Curti, P.S.: Chemical Kinetic Model and Thermodynamic Compensation Effect of Alkaline Hydrolysis of Waste Poly(ethylene terephthalate) in Nonaqueous Ethylene Glycol Solution. *Ind. Eng. Chem. Res.* 2006, *45*, 7985-7996.

[32] Oku, A.; Hu, L.-C.; Yamada, E.: Alkali Decomposition of Poly(ethylene terephthalate) with Sodium Hydroxide in Nonaqueous Ethylene Glycol: A Study on Recycling of Terephthalic Acid and Ethylene Glycol. *J. Appl. Polym. Sci.* 1997, *63*, 595-601.

[33] Goje, A.S.; Thakur, S.A.; Diware, V.R.; Chauhan, Y.P.; Mishra, S.: Chemical Recycling, Kinetics, and Thermodynamics of Hydrolysis of Poly(Ethylene Terephthalate) Waste with Nonaqueous Potassium Hydroxide Solution. *Polymer-Plastics Technology and Engineering.* 2004, *43*, 369-388.

[34] Hu, L.-C.; Oku, A.; Yamada, E.; Tomari, K.: Alkali-Decomposition of Poly(ethylene terphthalate) in Mixed Media of Nonaqueous Alcohol and Ether. Study on Recycling of Poly(ethylene terephthalate). *Polym. J.* 1997, *29*, 708-712.

[35] Goje, A.S.; Diware, V.R.; Patil, T.M.; Mishra, S.: Microkinetics and Mass Transfer Study of Treatment of Aqueous and Alcoholic Potassium Hydroxide on Poly(Ethylene Terphthalate) Waste. *Polymer-Plastics Technology and Engineering.* 2004, *43*, 889-912.

[36] Kumar, S.; Guria, C.: Alkaline Hydrolysis of Waste Poly(Ethylene Terephthalate): A Modified Shrinking Core Model. *Journal of Macromolecular Science, Part A: Pure and Applied Chemistry.* 2005, *42*, 237-251.

[37] Collins, M.J.; Zeronian, S.H.: The Molecular Weight Distribution and Oligomers of Sodium Hydroxide Hydrolysed Poly(ethylene terephthalate). *J. Appl. Polym. Sci.* 1992, *45*, 797-804.

[38] Masuda, T.; Miwa, Y.; Tamagawa, A.; Mukai, S.R.; Hashimoto, K.; Ikeda, Y.: Degradation of waste poly(ethylene terephthalate) in a steam atmosphere to recover terephthalic acid and to minimize carbonaceous residue. *Polym. Degrad. Stab.* 1997, *58*, 315-320.

[39] Grause, G.; Kaminsky, W.; Fahrbach, G.: Hydrolysis of poly(ethylene terephthalate) in a fluidised bed reactor. *Polym. Degrad. Stab.* 2004, *85*, 571-575.

[40] Masuda, T.; Miwa, Y.; Hashimoto, K.; Ikeda, Y.: Recovery of oil from waste poly(ethylene terephthalate) without producing any sublimate materials. *Polym. Degrad. Stab.* 1998, *61*, 217-224.

[41] Masuda, T.; Kushino, T.; Matsuda, T.; Mukai, S.R.; Hashimoto, K.; Yoshida, S.: Chemical recycling of mixture of waste plastics using a new reactor system with stirred heat medium particles in steam atmosphere. *Chemical Engineering Journal.* 2001, *82*, 173-181.

[42] Yoshioka, T.; Kitagawa, E.; Mizoguchi, T.; Okuwaki, A.: High Selective Conversion of Poly(ethylene terephthalate) into Oil Using $Ca(OH)_2$. *Chem. Lett.* 2004, *33*, 282-283.

[43] Yoshioka, T.; Handa, T.; Grause, G.; Lei, Z.; Inomata, H.; Mizoguchi, T.: Effects of metal oxides on the pyrolysis of poly(ethylene terephthalate). *J. Anal. Appl. Pyrolysis.* 2005, *73*, 139-144.

[44] Yalçinyuva, T.; Kamal, M.R.; Lai-Fook, R.A.; Özgümüs, S.: Hydrolytic Depolymerization of Polyethylene Terephthalate by Reactive Extrusion. *Intern. Polymer Processing.* 2000, *15*, 137-146.

[45] Shukla, S.R.; Mathur, M.R.: Action of Alkali on Polybutylene Terephthalate and Polyethylene Terephthalate Polyesters. *J. Appl. Polym. Sci.* 2000, *75*, 1097-1102.

[46] Zhang, H.; Ward, I.M.: Kinetics of Hydrolytic Degradation of Poly(ethylene naphthalene-2,6-dicarboxylate. *Macromolecules.* 1995, *28*, 7622-7629.

[47] Goje, A.S.: Auto-Catalyzed Hydrolytic Depolymerization of Poly(Butylene Terephthalate) Waste at High Temperature. *Polymer-Plastics Technology and Engineering*. 2006, *45*, 171-181.
[48] Yoshioka, T.; Grause, G.; Otani, S.; Okuwaki, A.: Selective production of benzene and naphthalene from poly(butylene terephthalate) and poly(ethylene naphthalene-2,6-dicarboxylate) by pyrolysis in the presence of calcium hydroxide. *Polym. Degrad. Stab.* 2006, *91*, 1002-1009.
[49] Matsubara, K.; Kuroda, Y.: Apparatus and method for catalyst-free degradation of thermoplastic polyesters using high-pressure hot water with high recovery of monomers. Japanese Patent JP2000309663.
[50] Mandoki, J.W.: Depolymerization of condensation polymers. US Patent US 4,605,762.
[51] Michalshi, A.; Urbanowsi, A.; Czarnecki, J.; Dajewska, M.; Pozniak, A.; Felbur, M.; Boninski, W.: Recovery of terephthalic acid and ethylene glycol from waste PET polyester. Polish Patent PL 185,810.
[52] Gau, J.-Y.; Jeng, W.-S.; Wan, B.-R.: Catalytic depolymerization of polyethylene terephthalate with Sn or Pb compounds. Taiwanese Patent TW 448,204.
[53] Sirek, M.; Vavra, M.; Vesely, V.; Drahos, J.: The method of chemical recycling of polyethylene terephthalate waste. World Patent WO 2006-039,872.
[54] Cho, D.H.; Jang, T.S., Jung, T.O.; Kim, T.S.; Lee, D.G.: Method for preparing high purity terephthalic acid from polyethylene terephthalate. Korean Patent KR 2002078366.
[55] Okuwaki, A.; Sato, T.; Yoshioka, T.: Recovery of dibasic acid and/or glycol components from polyesters. Japanese Patent JP 7,082,409.
[56] Oku, A.; Ebisu, R.; Yamada, E.: Recovery of terephthalic acid and ethylene glycol by hydrolysis of poly(ethylene terehthalate) in alcohol-ether mixed solvent system. Japanese Patent JP 10,087,529.
[57] Tindall, G.W.; Pery, R.L.; Spaugh, A.T.: Rapid depolymerisation of substantially amorphous polyesters by alkaline hydrolysis in alcohol-aprotic solvent mixtures. World Patent WO 9,410,121.
[58] Oku, A.; Ko, R.; Yamada, E.: Recovery of terephthalic acid and ethylene glycol from waste PET containers. Japanese Patent JP 9,286,744.
[59] Rollick, K.L.: Process for recovering dicarboxylic acid with reduced impurities from polyester polymer. World Patent WO 9,510,499.
[60] Wulf, G.: Decomposition of poly(ethylene terephthalate) wastes. GDR Patent DD 96,966.
[61] Grause, G.; Kaminsky, W.; Fahrbach, G.: Production of chemical raw materials from polycondensates and polyaddition compounds. German Patent DE 10,206,321.
[62] Yamamoto, S.: Method and apparatus for recovering of monomers from waste poly(ethylene terphthalate). Japanese Patent JP 9,077,905.
[63] Adschiri, T.; Shibata, R.; Arai, K.: Phenol recovey by Bisphenol-A (BPA) tar hydrolysis in supercritical water. *Sekiyu Gakkaishi*. 1997, *40*.291-297.
[64] Hunter, S.E., Felczak, C.A.; Savage, P.E.: Synthesis of p-isopropenyphenol in high-temperature water. *Green Chem.* 2004, *6*, 222-226.
[65] Hunter, S.E.; Savage, P.E.: Kinetics and Mechanism of p-Isopropenylphenol Synthesis via Hydrothermal Cleavage of Bisphenol A. *J. Org. Chem.* 2004, *69*,4724-4731.

[66] Yoshioka, T.; Sugawara, K.; Mizoguchi, T.; Okuwaki, A.: Chemical Recycling of Polycarbonate to Raw Materials by Thermal Decomposition with Calcium Hydroxide/Steam. *Chem. Lett.* 2005, *34*, 282-283.
[67] Schilling, F.C.; Ringo, W.M.; Sloane, N.J.A.; Bovey, F.A.: Carbon-13 Nuclear Magnetic Resonance Study of the Hydrolysis of Bisphenol A Polycarbonate. *Macromolecules.* 1981, *14*, 532-537.

In: Polyester: Properties, Preparation and Applications ISBN: 978-1-60456-753-3
Editors: Hina Yamashita, Yui Nakano

Chapter 2

MULTIWALL CARBON NANOTUBE REINFORCED POLYESTER NANOCOMPOSITES

Jun Young Kim[1] and Seong Hun Kim

Department of Fiber and Polymer Engineering,
Hanyang University, Korea

ABSTRACT

This chapter presents the preparation of polymer nanocomposites and the effects of multiwall carbon nanotube (MWCNT) on the structure and properties of poly(ethylene 2,6-naphthalate) (PEN) nanocomposites. The combination of a very small quantity of expensive MWCNT with conventional cheap thermoplastic polymers provides attractive possibilities for improving the physical properties of polymer composites using a cost-effective method, from a commercial perspective. MWCNT-reinforced PEN nanocomposites were prepared by a melt blending process in a twin screw extruder to create advanced materials for possible practical applications in numerous industrial fields. There are significant dependence of the crystallization behaviors and their kinetics of the PEN/MWCNT nanocomposites on the MWCNT content, the cooling rate, and the crystallization temperature. The MWCNT in the PEN nanocomposites exhibited much higher nucleation activity than any nanoreinforcing filler. In the PEN/MWCNT nanocomposites, the incorporation of the MWCNT promoted the nucleation and the growth with higher crystallization rate of the PEN/MWCNT nanocomposites, and simultaneously reduced the fold surface free energy and the works required in folding macromolecular chains in the PEN/MWCNT nanocomposites. The non-terminal behavior observed in the PEN/MWCNT nanocomposites was related to the dominant nanotube-nanotube interactions at higher MWCNT content, leading to the formation of the interconnected or network-like structures of the MWCNT in the PEN nanocomposites. The incorporation of very small quantity of the MWCNT significantly improved the mechanical properties of the PEN/MWCNT nanocomposites. There is a significant

[1] Present address: Corporate R&D Center, Samsung SDI Co., Korea E-mail address: junykim74@hanmail.net (J. Y. Kim).

dependence of the thermal stability and degradation behavior of the PEN/MWCNT nanocomposites on the MWCNT content and heating rate. The interconnected network-like structures of the MWCNT resulted in the physical barrier effect against thermal degradation both by retarding the rate of thermal degradation and by hindering the transport of volatile decomposed products in the PEN nanocomposites, leading to the improvement in the thermal stability of the PEN/MWCNT nanocomposites. This Chapter attempts for the first time to summarize the preparation, the non-isothermal crystallization kinetics, the crystallization and melting behavior, the rheological and mechanical properties, the thermal stability, and the thermal degradation kinetics of MWCNT-reinforced PEN nanocomposites.

1. INTRODUCTION

1.1. Multiwall Carbon Nanotube (MWCNT)

Carbon nanotubes (CNTs) have attracted a great deal of interest, both as advanced materials for next generation and in a broad range of potential scientific and industrial applications since they were discover by Iijima [1]. Moreover, this discovery has created a high level of activity in materials research for a practical realization of the extraordinary properties of the CNT, with their almost infinite number of possibilities for new materials. The CNT was first synthesized as a by-product in arc-discharge method in the synthesis of fullerenes and are currently being prepared by various methods, including arc-discharge [2-4], laser ablation [5-7], high-pressure CO conversion (HiPCO) [8, 9], chemical vapor deposition (CVD) [10-14], electrolysis [15], solar energy [16] methods.

The CNT that consists of concentric cylinders of graphite layers are a new form of carbon, and they can be classified into three types: single-walled carbon nanotubes (SWNT), double-walled carbon nanotubes (DWNT), and multi-walled carbon nanotubes (MWNT). The SWNT consist of a single layer of carbon atoms through the thickness of the cylindrical wall with diameters of 1.0-1.4 nm, two such concentric cylinders forms the DWNT, and the MWNT consist of several layers of coaxial carbon tubes, the diameters of which range from 10 to 50 nm with the length of more than 10 μm [17-19]. The comparison of the diameters of various carbon materials are shown in Figure 1. The graphite nature of the nanotube lattice results in a fiber with high strength, stiffness, and conductivity, and higher aspect ratio represented by very small diameter and long length makes it possible for the CNT to be promising nanoreinforcing fillers in advanced polymer nanocomposite [20]. Both theoretical and experimental approaches investigating the mechanical properties of the CNT suggest that the elastic modulus of the CNT may exceed 1.0 TPa with a tensile strength in the range of 10-50 GPa [21-24].

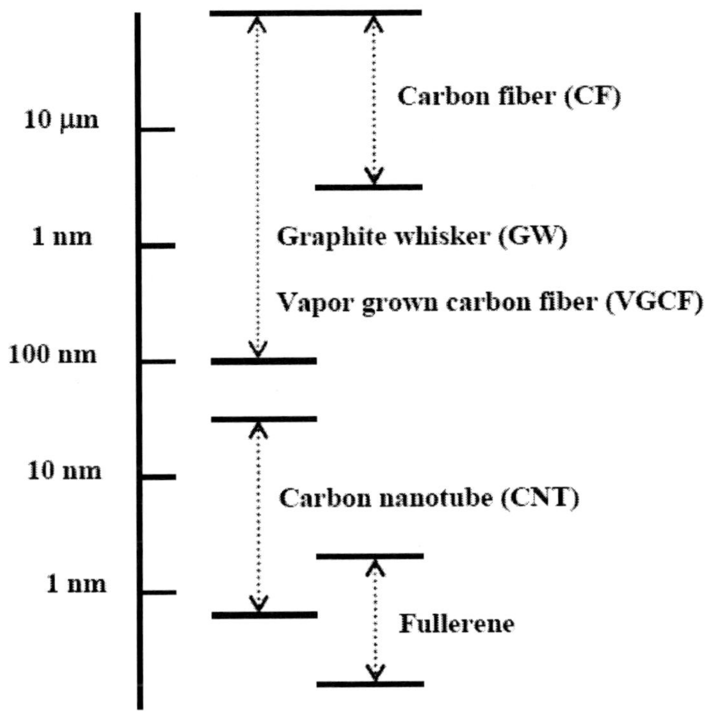

Figure 1. Comparison of the diameters of various carbon materials.

In general, the MWNT have inferior mechanical performance compared with the SWNT. However, the MWNT have a cost advantage, in that they can be produced in much larger quantities at lower cost compared with the SWNT. In addition, the MWNT are usually individual, longer than the SWNT, and more rigid because of their larger cross section. Due to their high aspect ratio and excellent axial strength, the MWNT are regarded as prospective reinforcements in high performance polymer nanocomposites. For these reasons, extensive research and development have been directed towards the potential applications of the CNT as novel materials for use in a broad range of industry. The fundamental research progressed to date on applications of the CNT suggests that the CNT can be utilized as a promising nanoreinforcing filler in new kinds of polymer nanocomposites. They are on the scale of nanometers and exhibit remarkable physical characteristics, such as high aspect ratio and high mechanical properties [25]. However, because of their high cost and limited availability, only a few applications in industrial fields have been realized to date.

1.2. CNT/Polymer Nanocomposites

One of major changes for high performance polymer nanocomposites is to optimize the processing of CNT-reinforced polymer nanocomposites with low costs. Four processing techniques are in common use to fabricate the CNT/polymer nanocomposites: direct mixing, solution method, *in situ* polymerization, and melt compounding [26-35]. Among these processing techniques, melt compounding has been accepted as the simplest and the most effective method from an industrial perspective, because this process makes it possible to fabricate high-performance polymer nanocomposites at low process cost, and facilitates commercial scale-up.

Up to now, much research on the CNT/polymer nanocomposites have been carried out based on the possible applications of the CNT as advanced polymer nanocomposites, and most of them have focused on the preparation, the morphology, and the thermal, rheological, and mechanical properties of their nanocomposites [26-45]. The major research field of the CNT-based polymer composites performed by various research groups and the experimental results of various CNT/polymer nanocomposites are presented in Tables 1 and 2, respectively.

As the mechanical properties of polymer nanocomposites are influenced by their morphology and crystallization behavior [46, 47] and the incorporation of the CNT plays a significant role in that, it is very instructive to characterize the crystallization and melting behavior of the CNT/polymer nanocomposites, which should be required both for optimizing the processing conditions. The crystallization behavior of polymer nanocomposites and their crystallization kinetics as a function of processing conditions are of great importance in polymer processing, particularly for the analysis and design of processing operations. Therefore, the crystallization behavior and structural development of CNT-reinforced polymer nanocomposites should be analyzed to realize the full potential of the CNT for application in thermoplastic matrix-based polymer nanocomposites.

From a commercial perspective, it is very important to understand the non-isothermal crystallization behavior of polymer nanocomposites, particularly if processing techniques for the fabrication of engineering plastics under non-isothermal conditions are being considered. The processing of polymer nanocomposites involves complex deformation behaviors, which may affect the nucleation and crystallization behavior of polymer nanocomposites. Thus, it is also important to characterize the nucleation and crystallization behavior of polymer nanocomposites to optimize the process conditions. Despite extensive studies on the crystallization and melting behavior of the CNT/polymer nanocomposites [48-51], a generalized understanding of their behavior induced by the CNT has not been fully achieved yet. Furthermore, few reports regarding the effect of the CNT on the nucleation, the non-isothermal crystallization kinetics, and the crystallization and melting behavior of conventional polyester resins can be rarely found in the literature to date.

The rheological behavior of polymer nanocomposites as a function of processing conditions is of great importance in polymer processing, particularly for the analysis and design of processing operations, as well as understanding structure-property relationships of polymer nanocomposites. In this regard, the rheological properties of CNT-reinforced polymer nanocomposites should be characterized both to realize the full potential of the CNT for application in the thermoplastic matrix-based polymer nanocomposites and to optimize the processing conditions for achieving high performance polymer nanocomposites. However, the rheological behavior of CNT-reinforced conventional polyester nanocomposites has rarely been investigated to date, and few reports can be found in the literature regarding the effect of the CNT on the rheological properties of polyester-based nanocomposites.

Table 1. Major study on CNT-based polymer composites performed by various research groups [36]

Groups	Research fields
S. Kumar (GIT)	- CNT/polymer fiber spinning - Holds the champion mechanical property data
E. Barrea (Rice Univ.)	- CNT/polymer extrusion - Mechanical properties, high CNT concentration
J. Fisher and K. Winey (Univ. Penn)	- SWNT/polymer - Goof physical background
O. Zhou (UNC)	- Physical biased work for filed emission display
M.Shaffe and A. Windle (Univ. Cambridge)	- CNT/polymer - Incomplete dispersion of CNT
Rud Ruoff (Northwestern Univ.)	- CNT, CNT/polymer - Micro and nano scale mechanical testing
P. Ajayan (Rensselaer Polytech)	- Long SWNT bundles - Microscopic study
C. Park (ICASE/NASA)	- SWNT/polymer dispersion and fiber spinning - Electrospinning, alignment, multifunctional applications - Synthesis, microscopy, modeling, processing
Frank Ko (Drexel Univ.)	- CNT/polymer, electrospinning - graphite nanoplate/polymer (spider silk)
T. Chou (Univ. Delaware)	- Aligned MWNT/polymer composites - Engineering based works for structural materials
D. Wagner (Weizmann Institute)	- CNT/polymer interfacial characterization - Raman spectroscopy

As thermal stability of polymer nanocomposites is one of the most important factors for their processing and applications, it is very instructive to characterize the thermal degradation behavior that is affected their final properties such as the upper-limit use temperature and dimensional stability, for developing polymer nanocomposites with better balance in processing and performance. Thermogravimetric analysis (TGA) is in common use both to characterize thermal degradation behavior and to estimate the kinetic parameters of thermal degradation processes, due to its simplicity and useful information about thermal degradation [52, 53].

Table 2. The research results of various CNT/polymer nanocomposites

Materials	Results	Ref.
SWNT/PEO	- 40% increase in storage modulus (4wt% SWNT) - Decrease in the T_m of PEO	78
MWNT/PS	- Relationships between processing conditions and mechanical and electrical properties - Lower strains at yield stress than neat PS films - Double tensile modulus at 2.5 vol.% CNT	79
SWNT/PP	- No noticeable mechanical reinforcement - Dependence of crystallization on SWNT	80
CNT/PP	- 50% and 100% increase in the modulus and strength at 5 wt% CNT, respectively	81
MWNT/PS	- 49% increase in the storage modulus at 5 wt% MWNT	82
CNF/PP	- Working temp. increased by 100 °C - 350% increase in the tensile modulus - Increase in the degree of crystallinity	83
SWNT/epoxy	- Loading behavior in tension and compression - CNT slipping within the bundles	84
CNT/polymer	- Role of an ionic surfactant - Increase in the T_g and modulus	85
MWNT/PS	- 42% and 25% increase in the elastic modulus and break stress at 1 wt% CNT, respectively.	86
SWNT/PMMA	- Increase in the modulus and strength with draw ratio - Higher electrical conductivity	68
SWNT/epoxy	- Higher compressive strength than any other fiber (about 0.5 GPa)	87

Thus, the understanding of the thermal stability and thermal degradation kinetics of polymer nanocomposites makes it possible to develop and extend their applications in a broad range of industry. However, fundamental studies on the thermal stability and thermal degradation kinetics of polymer nanocomposites have been rarely investigated to date, and only a few reports may be found in the literature regarding the roughly thermal degradation behavior of CNT/polymer nanocomposites [54-56]. Furthermore, the thermal stability, thermal degradation, and thermal degradation kinetics of CNT-reinforced aromatic polyester nanocomposites have been rarely investigated to date.

In this present paper, we prepared aromatic polyester nanocomposites based on a very small quantity of the CNT and the poly(ethylene 2,6-naphthalate) (PEN) by direct melt compounding in a twin-screw extruder to create high performance composite materials with low process cost for possible practical applications in various industrial fields. Furthermore, the combination of a very small quantity of relatively expensive CNT with conventional cheap polyester resins provides attractive possibilities for improving the physical properties of polymer nanocomposites from a commercial perspective. To the best of our knowledge, attempts to disperse the CNT in the PEN matrix and to fabricate the PEN/MWCNT nanocomposites by simple melt blending have not been previously investigated, and the study on the characterization of the PEN/CNT nanocomposites have not yet been reported in the literature. Thus, we expect that this study will help in preliminary evaluation and understanding of the characterization of the PEN/CNT nanocomposites. It is the objective of this study both to determine the possibilities for the fabrication of high performance polyester nanocomposites and to attempt to overcome the limitation of conventional polyester composites.

2. Processing of CNT/Polymer Nanocomposites

2.1. General Features

We prepared the CNT-reinfroced PEN nanocomposites using a direct melt blending process in a twin-screw extruder to create advanced materials for possible practical applications in various industrial fields. There are significant dependence of the crystallization behaviors and their kinetics of the PEN/CNT nanocomposites on the CNT content, the cooling rate, and the crystallization temperatures. The incorporation of a very small quantity of the CNT into the PEN matrix accelerated the mechanism of nucleation and crystal growth of the PEN/CNT nanocomposites. Combined Avrami and Ozawa analysis was found to be effective in describing the non-isothermal crystallization of the PEN/CNT nanocomposites. The introduced CNT in the PEN/CNT nanocomposites exhibited much higher nucleation activity than any nanoreinforcing filler. The storage modulus (G') and loss modulus (G'') of the PEN/CNT nanocomposites increased with increasing frequency and this enhancing was more significant at low frequency-region. The terminal zone slopes of G' and G'' for the PEN/CNT nanocomposites decreased increasing CNT content, and the non-terminal behavior of those was related to the dominant nanotube-nanotube interactions at higher CNT content, leading to the formation of the interconnected or network-like structures of the CNT in the PEN/CNT nanocomposites. In the PEN/CNT nanocomposite system, the incorporation of the CNT promoted the nucleation and the growth with higher crystallization rate of the polymer nanocomposites, and simultaneously reduced the fold surface free energy and the work required in folding polymer chains in the polymer nanocomposites, which was confirmed by the isothermal crystallization analysis. There is a significant dependence of the thermal stability and degradation behavior of the PEN/CNT nanocomposites on the CNT content. The variations of the activation energy for thermal degradation revealed that a small quantity of the MWCNT significantly enhanced the thermal stability of the PEN/CNT nanocomposites in terms of physical barrier effect induced by the CNT. The interconnected network-like structures of the CNT resulted in physical barrier effect against thermal degradation by hindering the transport of volatile decomposed products in the PEN/CNT nanocomposites as well as by retarding the rate of thermal degradation, leading to the enhancement of the thermal stability of the PEN/CNT nanocomposites. The incorporation of very small quantity of the CNT significantly improved the mechanical properties of the PEN/CNT nanocomposites due to the nanoreinforcing effect of the CNT.

2.2. CNT-Reinforced PEN Nanocomposites

The thermoplastic polymer used was PEN with an intrinsic viscosity of 0.97 dL/g, supplied by Hyosung Corp., Korea. The nanotubes used were multi-walled CNT (degree of purity > 95%) synthesized by a thermal chemical vapor deposition process, purchased from Iljin Nanotech Co., Korea. According to the supplier, their length and diameter were in the range of 10-50 μm and 10-30 nm, respectively, indicating that their aspect ratio reached 1000.

All materials were dried at 120 °C *in vacuo* for at least 24 h before use, to minimize the effects of moisture. The PEN/CNT nanocomposites were prepared by a melt blending process in a Haake rheometer (Haake Technik GmbH, Germany) equipped with a twin-screw (non-intermeshing co-rotating type). The temperature of the heating zone, from the hopper to the die was set to 280, 290, 295, and 285 °C, and the screw speed was fixed at 20 rpm. For the fabrication of PEN/CNT nanocomposites, PEN was melt blended with the addition of various CNT content, specified as 0.1, 0.5, 1.0, and 2.0 wt% in the polymer matrix, respectively.

The thermal behavior of PEN/CNT nanocomposites was measured with a TA Instrument 2010 DSC over a temperature range of 30 to 295 °C at a scan rate of 10 $^{o}C{\cdot}min^{-1}$. The samples were heated to 295 °C at a heating rate of 10 $^{o}C{\cdot}min^{-1}$, held at 295 °C for 8 min to eliminate any previous thermal history, and then cooled to room temperature at a cooling rate of 10 $^{o}C{\cdot}min^{-1}$. Non-isothermal crystallization kinetics was investigated by cooling samples from 295 to 30 °C at constant cooling rates of 2.5, 5, 10, 15, and 20 $^{o}C{\cdot}min^{-1}$. The isothermal crystallization measurements for PEN/CNT nanocomposites were performed with a Perkin-Elmer DSC7 over the temperature range of 30~295 °C at a scan rate of 10 $^{o}C{\cdot}min^{-1}$. The samples of approximately 8.5 mg placed inside an aluminum sample holder were melted to 295 °C for 8 min to eliminate any previous thermal history in the materials. Then, they rapidly cooled to the desired crystallization temperatures, and maintained at that temperatures for the period to complete crystallization process of the materials. After the isothermal crystallization finished, the samples were heated to 295 °C at heating rate of 10 $^{o}C{\cdot}min^{-1}$. Dynamic mechanical analysis of the PEN/CNT nanocomposites was performed with a TA Instrument Q-800 dynamic mechanical thermal analyzer (DMTA) using a tensile mode at a fixed frequency of 1 Hz, over a temperature range of 30 to 250 °C at a heating rate of 5 $^{o}C{\cdot}min^{-1}$. The morphology of pristine CNT and the PEN/CNT nanocomposites was observed using a JEOL JSM-6340F scanning electron microscope (SEM), and detailed morphological observations were performed using a JEOL 2000FX transmission electron microscope (TEM). The rheological properties of PEN/CNT nanocomposites were measured at 285, 295, and 305 °C, covering the temperature processing windows of the polymer nanocomposites, with an ARES (Advanced Rheometric Expansion System) rheometer (Rheometric Scientific, Inc.). The dynamic shear measurements were performed in oscillation mode and parallel-plate geometry using the plate diameter of 25 mm and the plate gap setting of approximately 1 mm, by applying a time-dependent strain, $\gamma(t) = \gamma_0 \sin(\omega t)$ and measuring resultant shear stress, $\gamma(t) = \gamma_0 [G'\sin(\omega t) + G''\cos(\omega t)]$, where G' and G'' are storage and loss moduli, respectively. The frequency ranges were varied between 0.05 and 450 rad/s, and the strain amplitude was applied to be within the linear viscoelastic ranges. The mechanical properties of PEN/CNT nanocomposites were measured at room temperature using an Instron 4465 testing machine, according to the procedures in the ASTM D 638 standard. The gauge length and the crosshead speed were set to 20 mm and 10 mm·min^{-1}, respectively. The films of testing samples were prepared in a hydrolytic press at 295 °C. Thermogravimetric analysis of PEN/CNT nanocomposites was performed with a TA Instrument SDF-2960 TGA over a temperature range of 30 to 800 °C at a heating rate of 10 °C·min^{-1}. Dynamic TGA measurements were conducted on the PEN/CNT nanocomposites at different heating rates of 5, 10, and 20 °C·min^{-1} under nitrogen and air. WAXD measurements for the crystallized PEN/CNT nanocomposites were performed using a Rigaku Denki X-ray diffractometer with Ni-fillltered CuKα X-rays ($\lambda = 0.1542$ nm). The diffracting intensities were recorded at steps of $2\theta = 0.05°$ over the range of $5° < 2\theta < 35°$.

3. Unique Nucleation of CNT and PEN Nanocomposites during Non-Isothermal Crystallization

3.1. Morphology

Representative SEM and TEM images of pristine CNT are shown in Figure 2. The CNT exhibits highly curved and random coiled features in the PEN/CNT nanocomposites, which may be attributed to hydrogen bonding and van der Waals attractions between the individual nanotubes [57, 58]. The diameters of the CNT were approximately 10-30 nm, with a length of several micrometers, implying a high aspect ratio for the CNT. The morphologies of the PEN/CNT nanocomposites are shown in Figure 3. From the SEM image of the fractured PEN/CNT nanocomposite, it can be seen that the CNT form entangled structures in the PEN matrix. In addition, the TEM image shows that the CNT were randomly dispersed in the PEN matrix, with some entanglements or bundles of the CNT, indicating highly aggregated CNT and weak interactions with the PEN matrix. On a larger scale, however, the CNT were uniformly dispersed in the PEN matrix, despite some aggregated CNT structures.

3.2. Thermal Behavior

The thermal stability of polymer nanocomposites plays a critical role in determining their processing and industrial application, because it affects the final properties of polymer nanocomposites such as the upper-limit working temperature and dimensional stability for targeted applications. The TGA thermograms of the PEN/CNT nanocomposites with the CNT content are shown in Figure 4. The incorporation of the CNT into the PEN matrix increased the thermal decomposition temperatures and residual yields of the PEN/CNT nanocomposites and this enhancing effect was more significant at higher CNT content, implying that the thermal decomposition was retarded by incorporating CNT into the PEN matrix with higher residual yield.

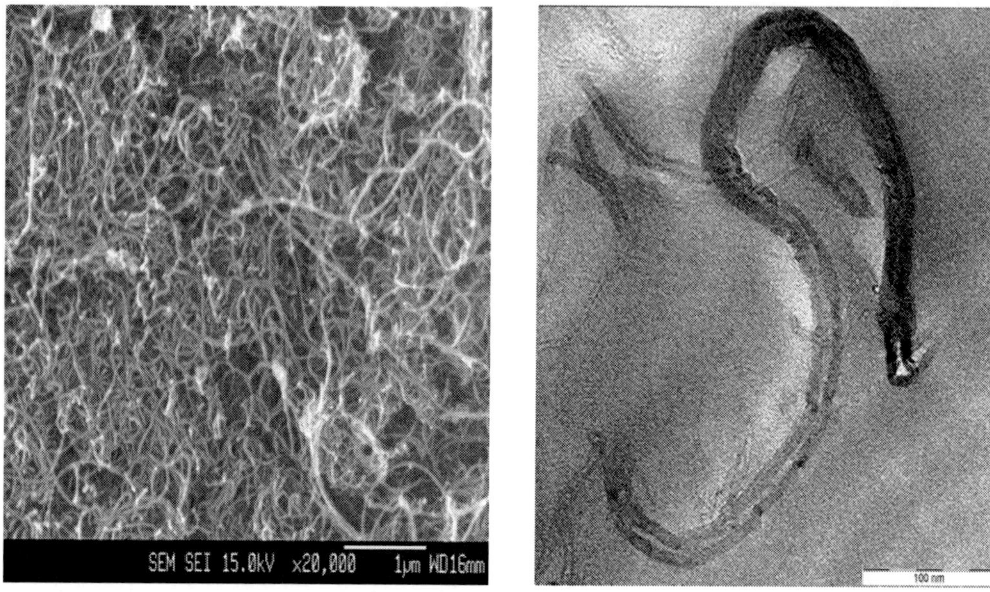

(a) **(b)**

Figure 2. (a) SEM and (b) TEM images of pristine CNT

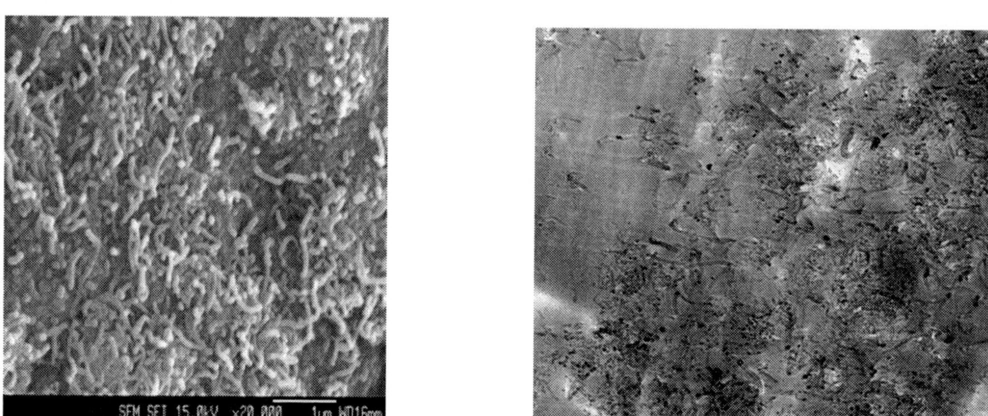

(a) **(b)**

Figure 3. (a) SEM and (b) TEM images of the PEN/CNT 2.0 nanocomposites.

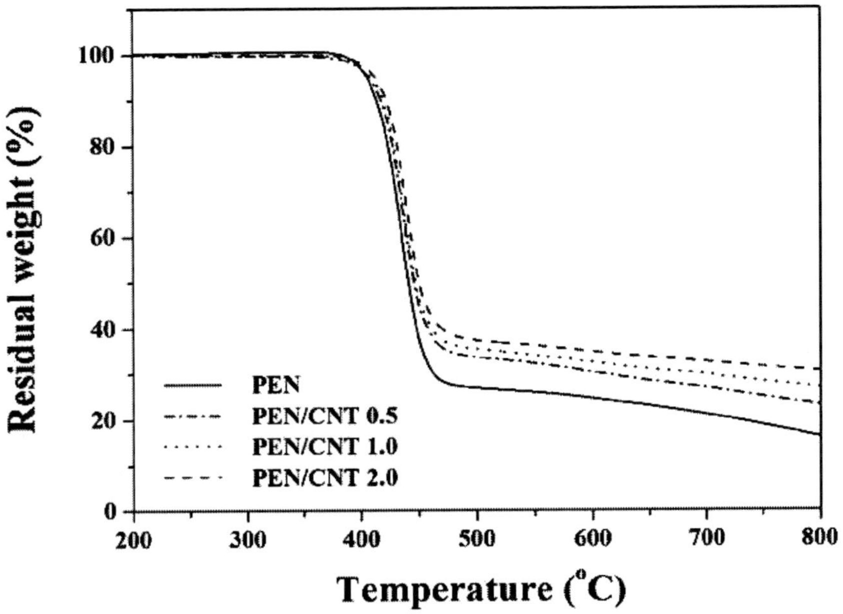

Figure 4. TGA thermograms of the PEN/CNT nanocomposites with the CNT content.

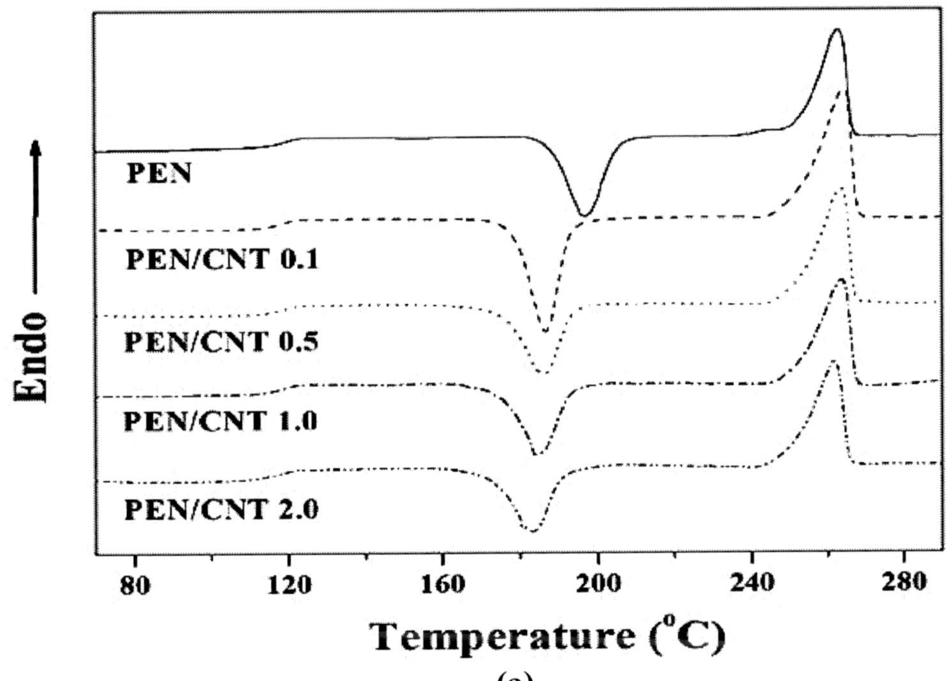

Figure 5. DSC (a) heating and (b) cooling curves of the PEN/CNT nanocomposites as a function of CNT content.

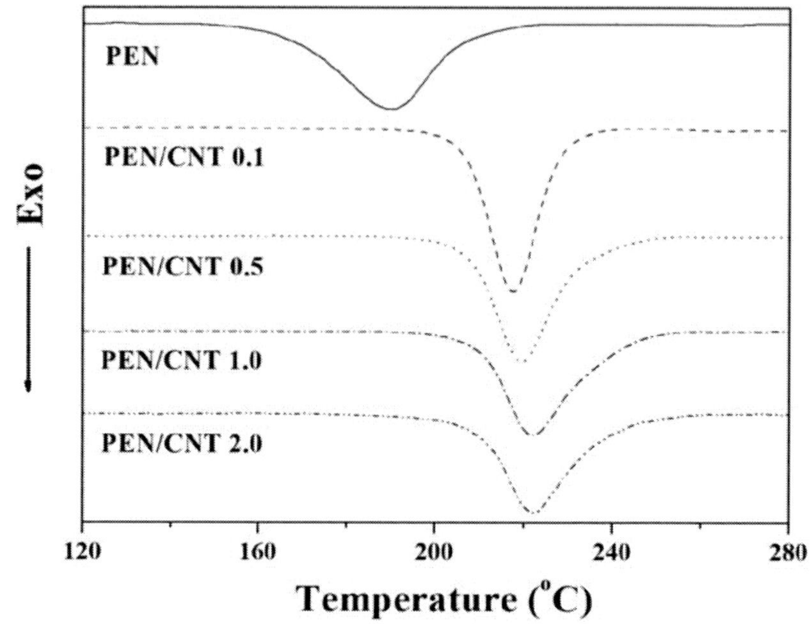

(b)

Figure 5. (Continued)

The presence of the CNT could stabilize the PEN matrix, resulting in the enhancement of the thermal stability of the PEN/CNT nanocomposites. In the PEN/CNT nanocomposites, the introduced CNT can effectively act as physical barriers to prevent the transport of volatile decomposed products in the PEN/CNT nanocomposites [59, 60]. Therefore, the incorporation of a very small quantity of CNT significantly improved the thermal stability of the PEN/CNT nanocomposites.

DSC heating and cooling traces for the PEN/CNT nanocomposites as a function of CNT content at a scan rate of 10 $^{o}C \cdot min^{-1}$ are shown in Figure 5. The incorporation of the CNT had little effect on the glass transition and melting temperatures of the PEN/CNT nanocomposites. The incorporation of CNT into the PEN matrix increased the crystallization temperature of the PEN/CNT nanocomposites, with this increment being greatest with lower CNT content. This result confirms that the incorporation of a very small quantity of the CNT enhances the nucleation of PEN crystallization. The CNT promote the formation of heterogeneous nuclei, with lower energy consumption required to reach critical stability for crystal growth [61], resulting in them functioning as effective nucleating agents in the PEN matrix. The crystallization peak temperatures of the PEN/CNT nanocomposites at various cooling rates are shown in Figure 6.

As the cooling rate increases, the crystallization peak temperature range becomes broader and shifts to lower temperatures for pure PEN and the PEN/CNT 0.1 nanocomposites. When the specimens were cooled quickly, more supercooling was required to initiate crystallization, because the motion of the PEN molecules could not follow the cooling rate [62, 63].

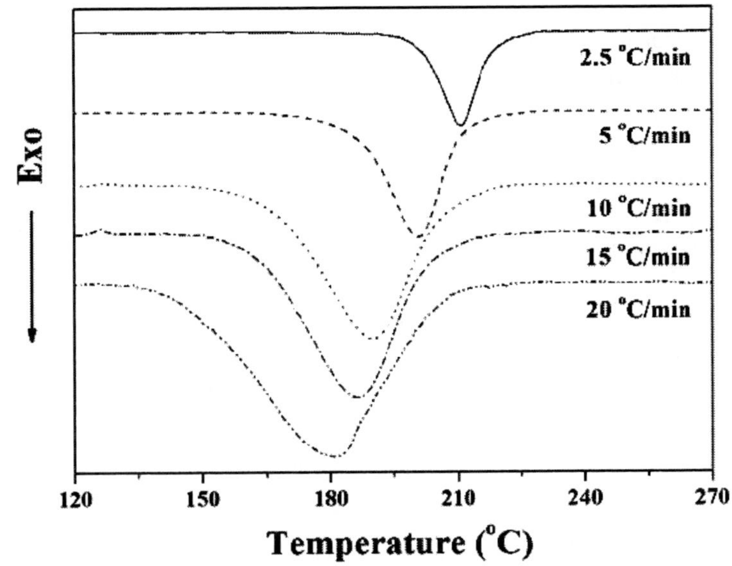

(a) PEN

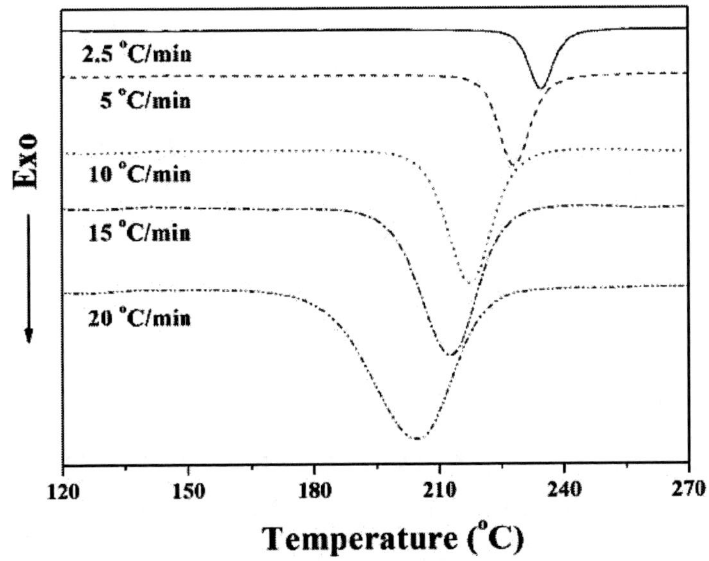

(b) PEN/CNT 0.1

Figure 6. Non-isothermal crystallization curves of (a) PEN and (b) the PEN/CNT 0.1 nanocomposites at various cooling rates.

With increasing the cooling rate, the crystallization peak temperature corresponding to the maximum crystallization rate shifted to lower temperature, indicating that the lower the cooling rate, the earlier crystallization occurs. In addition, at a given cooling rate, the peak temperature of the PEN/CNT nanocomposites were higher than that of pure PEN, and the overall crystallization time decreased with the introduction of the CNT. In general, homogeneous nucleation started spontaneously below the melting temperature and required longer times, whereas heterogeneous nuclei formed as soon as the specimens reached the crystallization temperature [64-66]. Thus, the crystallization of the PEN/CNT nanocomposites proceeds through heterogeneous nucleation by the CNT. DSC results for the PEN/CNT nanocomposites as a function of CNT content are shown in Table 3. The increase in the crystallization temperature of the PEN/CNT nanocomposites with increasing CNT content, together with the fact that the PEN/CNT nanocomposites have a lower degree of supercooling ($\Delta T = T_m - T_c$) for crystallization with increasing CNT content, suggests that the CNT can effectively act as strong nucleating agents in the PEN/CNT nanocomposites. As shown in Table 3, the degree of crystallinity of the PEN/CNT nanocomposites increased with the introduction of the CNT, which may be explained by the supercooling temperature. The CNT acts as a strong nucleating agent in the PEN matrix under the non-isothermal crystallization process, and the crystallization temperature shifts to higher temperature, implying that the supercooling of the PEN/CNT nanocomposites at a given cooling rate decreased by the presence of the CNT. When a polymer crystallized with less supercooling, it crystallized more perfectly than with more supercooling [64, 66], and thus, the degree of crystallinity of the PEN/CNT nanocomposites increased with the CNT content at a given cooling rate.

Table 3. DSC results for PEN/CNT nanocomposites with the CNT content

Materials	T_g (°C)	T_c [a]		T_m		T_c [b] (°C)	ΔT [c] (°C)	X_c [d] (%)
		Peak (°C)	ΔH_c (J/g)	Peak (°C)	ΔH_m (J/g)			
PEN	119.9	199.9	17.4	266.6	38.9	203.8	62.8	20.8
PEN/CNT 0.1	118.1	188.8	16.5	266.9	39.6	228.0	38.9	22.3
PEN/CNT 0.5	118.8	187.7	14.1	266.8	40.9	228.2	38.1	25.9
PEN/CNT 1.0	118.4	186.8	11.8	266.3	41.5	229.2	37.6	28.7
PEN/CNT 2.0	118.0	186.1	9.9	265.8	42.1	229.7	36.1	31.1

[a] Crystallization temperature measured on the second heating at 5 °C/min.
[b] Crystallization temperature measured on the second cooling at 5 °C/min.
[c] Degree of supercooling, $\Delta T = T_m - T_c$.
[d] Degree of crystallinity, X_c (%) = $[(\Delta H_f - \Delta H_c)/\Delta H_f^0 \times 100]$ (ΔH_f0 = 103.4 J/g) is the heat of fusion of and infinitely thick crystal) [66].

3.3. Non-Isothermal Crystallization Behavior

From a practical industrial point of view, it is very important to characterize the non-isothermal crystallization behavior of polymeric materials, because the processing of polymers or polymer composites is performed under non-isothermal conditions during fabrication of engineering plastics. The relative degree of crystallinity, *X(T)* as a function of temperature, can be defined as follows:

$$X(T) = \int_{T_0}^{T} \left(\frac{dH_c}{dT}\right) dT \Big/ \int_{T_0}^{T_\infty} \left(\frac{dH_c}{dT}\right) dT \qquad (1)$$

where T_0 and T_∞ are the initial and final crystallization temperature, respectively. The relative degree of crystallinity as a function of temperature for pure PEN and the PEN/CNT 0.1 nanocomposites at various cooling rates is shown in Figure 7.

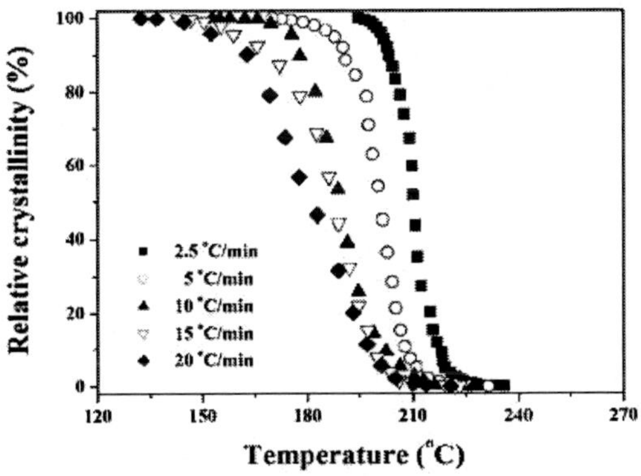

(a) PEN

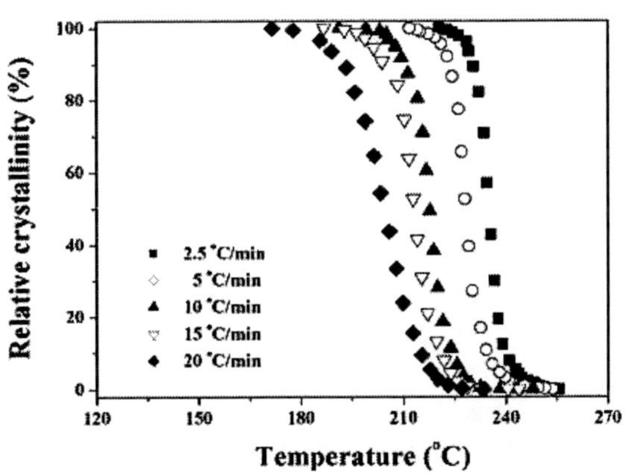

(b) PEN/CNT 0.1

Figure 7. Relative degree of crystallinity of (a) PEN and (b) the PEN/CNT 0.1 nanocomposites as a function of temperature at various cooling rates.

All the curves have similar sigmoidal shapes and the crystallization occurred at lower temperature with increasing cooling rate, indicating that at slower cooling rates there is sufficient time to activate nuclei at higher temperatures, and thus, crystallization nucleates at higher temperatures with slower cooling rates [67]. During non-isothermal crystallization, the relationship between crystallization temperature (T) and crystallization time (t) can be represented as follows:

$$t = \frac{T_0 - T}{a} \tag{2}$$

where T_0 is the initial temperature at which crystallization begins ($t = 0$) and a is the cooling rate. Thus, the abscissa of the temperature in Figure 7 can be transformed into the time scale as shown in Figure 8, according to the equation (2).

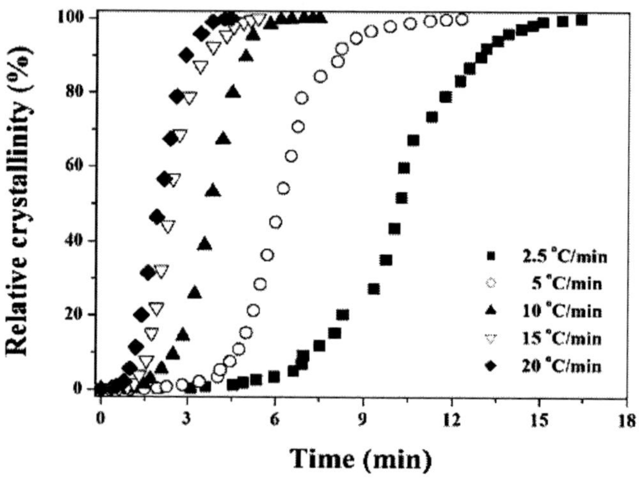

(a) PEN

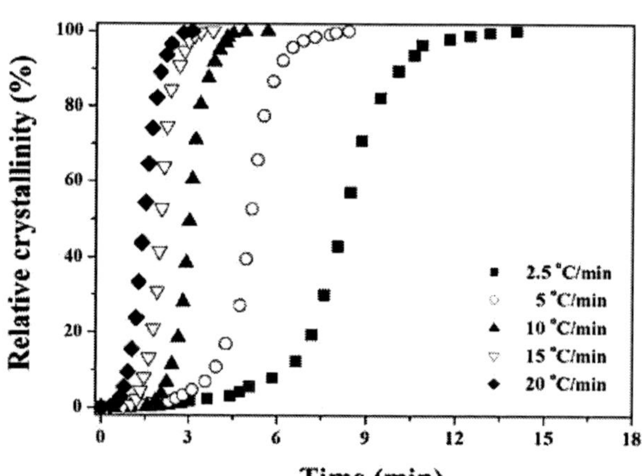

(b) PEN/CNT 0.1

Figure 8. Relative degree of crystallinity of (a) PEN and (b) the PEN/CNT 0.1 nanocomposites as a function of time at various cooling rates.

Crystallization of the PEN/CNT nanocomposites occurred at higher temperature and over a longer time with decreasing cooling rate, suggesting that the crystallization may be controlled by a nucleation. Crystallization occurred at higher temperature with decreasing cooling rate, indicating that crystallization nucleated at higher temperature with slower cooling rates. The time taken to complete crystallization reduced with increasing cooling rate, and the relative degree of crystallinity of the PEN/CNT nanocomposites was higher than that of pure PEN at the same time for complete crystallization. The values for the peak temperature (T_p) and the crystallization half-time ($t_{1/2}$) obtained from the non-isothermal crystallization thermograms of PEN/CNT nanocomposites at various cooling rates are shown in Table 4. The crystallization half-time can be defined as the time taken to complete half of the non-isothermal crystallization process, i.e., the time required to attain a relative degree of crystallinity of 50%. The values of T_p and $t_{1/2}$ for the PEN/CNT nanocomposites decreased with increasing cooling rate, indicating that the higher the cooling rate, the shorter the time for complete crystallization.

Table 4. Kinetic parameters of PEN/CNT nanocomposites during non-isothermal crystallization

Materials	Cooling rate (°C/min)	T_p (°C)	Z_c	n	$t_{1/2}$ (min)
PEN	2.5	210.8	9.59×10^{-7}	5.58	10.25
	5	201.0	4.31×10^{-6}	5.81	6.12
	10	190.0	1.11×10^{-3}	4.82	3.98
	15	187.0	1.13×10^{-2}	4.70	2.36
	20	181.3	5.89×10^{-2}	3.89	2.02
PEN/CNT 0.1	2.5	234.6	2.64×10^{-3}	5.81	8.25
	5	228.1	9.65×10^{-2}	5.96	5.09
	10	217.5	4.19×10^{-1}	6.47	3.02
	15	212.5	7.21×10^{-1}	5.36	2.03
	20	204.1	9.10×10^{-1}	6.03	1.43
PEN/CNT 0.5	2.5	235.4	2.92×10^{-3}	6.15	9.59
	5	228.2	1.57×10^{-1}	5.11	5.85
	10	218.3	5.16×10^{-1}	5.07	3.06
	15	213.2	7.11×10^{-1}	5.15	2.36
	20	205.1	8.23×10^{-1}	5.32	1.89
PEN/CNT 1.0	2.5	235.8	4.58×10^{-3}	5.80	9.68
	5	229.2	1.34×10^{-1}	5.48	5.92
	10	220.0	6.31×10^{-1}	4.79	3.12
	15	216.8	7.59×10^{-1}	5.12	2.50
	20	209.0	8.19×10^{-1}	5.52	1.96
PEN/CNT 2.0	2.5	238.0	2.31×10^{-3}	6.42	9.97
	5	229.7	1.26×10^{-1}	5.57	6.09
	10	222.1	5.16×10^{-1}	5.42	3.46
	15	218.6	7.44×10^{-1}	4.89	2.52
	20	211.6	8.21×10^{-1}	4.60	2.07

In addition, for a given cooling rate, the T_p values for the PEN/CNT nanocomposite are higher than that for pure PEN, while the $t_{1/2}$ values are lower than that for pure PEN. Thus, the incorporation of the CNT into the PEN matrix increases the crystallization rate of PEN. In order to investigate the effect of CNT content on the $t_{1/2}$ values for the PEN/CNT nanocomposites, the variation and normalization of the $t_{1/2}$ values, with respect to those for pure PEN, are shown in Figure 9. The lowering of the $t_{1/2}$ values for the PEN/CNT nanocomposites with the introduction of the CNT was more pronounced at lower CNT content. This result suggests that a lower CNT content is more effective for enhancing the crystallization of the PEN matrix.

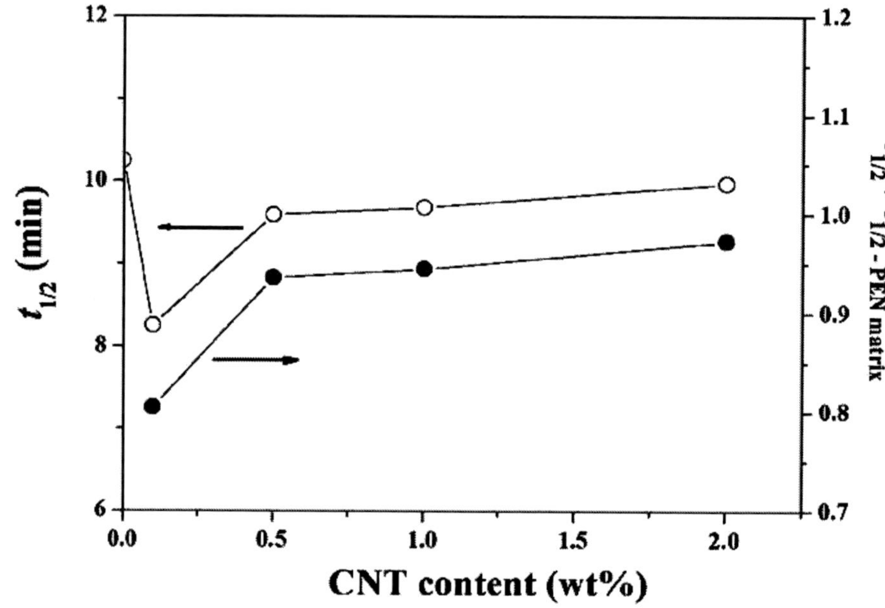

Figure 9. Effect of CNT content on the half-time of crystallization ($t_{1/2}$) for the PEN/CNT nanocomposites.

3.4. Nucleation Activity and Activation Energy for Non-Isothermal Crystallization

Non-isothermal crystallization kinetics can be analyzed by using the extension of the Avrami theory [68-70] and proposed by Ozawa [71, 72]. This analysis accounts for the effect of cooling rate on crystallization from the melt by replacing the time variable in the Avrami equation with a variable cooling rate term, that is, by replacing t in the equation (3) with T/a as shown in the equation (4):

$$1 - X(T) = \exp(-Z_t t^n) \tag{3}$$

$$1 - X(T) = \exp[-\frac{K(T)}{a^m}] \quad (4)$$

where $X(T)$ is the relative degree of crystallinity; n is the Avrami exponent depending on the mechanisme of the nucleation and crystal growth; Z_t is the crystallization rate parameter involving both nucleation and growth rate parameters; a is the cooling rate; m is the Ozawa exponent depending on the dimension of crystal growth, and $K(T)$ is the cooling function related to the overall crystallization rates. Parameters such as Z_t and n have an explicit physical meaning relating to isothermal crystallization, but in non-isothermal crystallization their physical meaning does not have the same significance, due to the constant change in temperature, which influences both nucleation and crystal growth.

The Ozawa plot of *log*[-*ln*[1-*X*(*T*)]] versus *log a* for pure PEN and the PEN/CNT nanocomposites, taking the double logarithmic form of the equation (4) are shown in Figure 10. The *X*(*T*) values calculated at different temperatures decreased with increasing cooling rate at a given temperature. Some curvature in the plot was observed, indicating that the Ozawa exponent is not consistent with temperature during non-isothermal crystallization, and this makes it difficult to estimate the cooling function, *K*(*T*) related to the overall crystallization rate. This result may arise due to inaccurate assumption in Ozawa's theory, and he has a disregard for the secondary crystallization and the dependence of the fold length on temperature [73-75]: slow secondary crystallization could lower the determined values of the Ozawa exponent, and if changes in fold length of polymer chains depend on the temperature during dynamic crystallization, the fold length should be considered in the deviation of the theory [73]. Therefore, we deduce that the Ozawa analysis does not effectively describe the non-isothermal crystallization of the PEN/CNT nanocomposites.

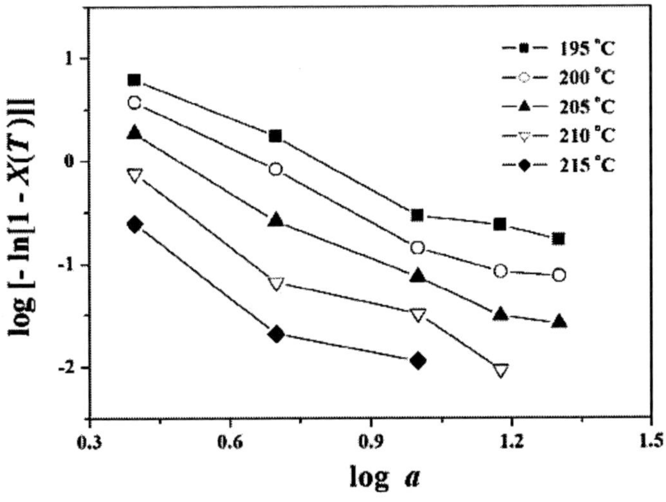

(a) PEN

Figure 10. Ozawa plots of (a) PEN and (b) the PEN/CNT 0.5 nanocomposites during non-isothermal crystallization.

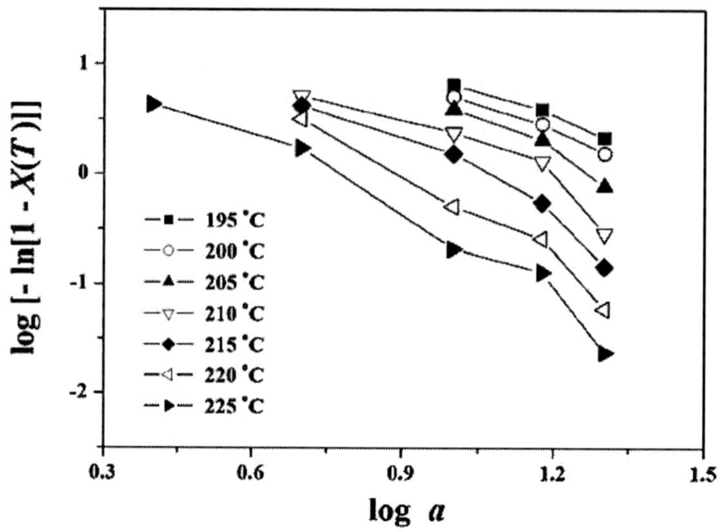

(b) PEN/CNT 0.5

Figure 10 (Continued)

Considering the non-isothermal character of the process investigated, Jeziorny [76] suggested that the rate parameter, Z_t should be corrected as follows:

$$\log Z_c = \frac{\log Z_t}{a} \tag{5}$$

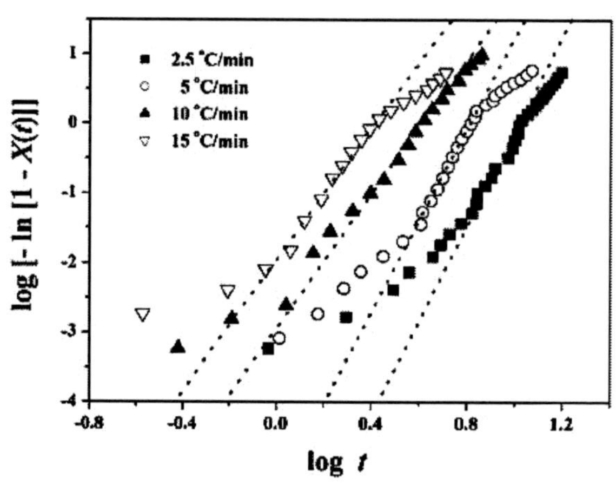

(a) PEN

Figure 11. Avrami plots of (a) PEN and (b) the PEN/CNT 0.5 nanocomposites during non-isothermal crystallization.

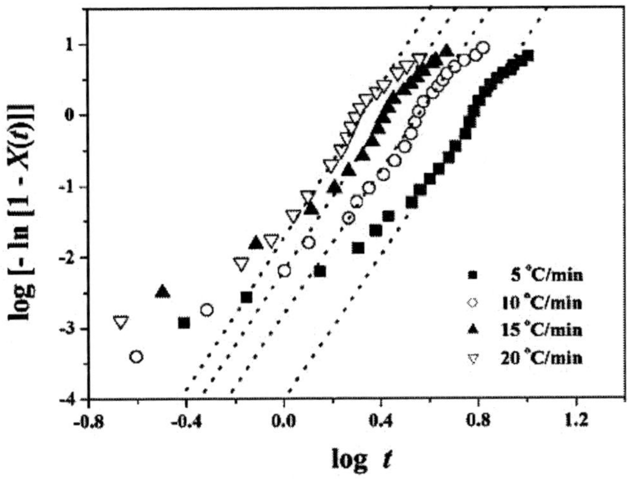

(b) PEN/CNT 0.5

Figure 11 (Continued)

The values of the Avrami exponent (n) and the rate parameter (Z_t) can be determined from the slope and intercept of the plot of $log[-ln\{1\text{-}X(T)\}]$ versus $log\ t$. The plots of $log[-ln\{1\text{-}X(T)\}]$ versus $log\ t$ for PEN and the PEN/CNT 0.5 nanocomposites are shown in Figure 11. It can be seen that the plots exhibit a poor linear relationship, consisting of three linear regions, indicating that the modified Avrami analysis does not effectively describe non-isothermal crystallization of PEN/CNT nanocomposites. The kinetic data in the central linear region were selected to estimate the Avrami parameter for non-isothermal crystallization of pure PEN and the PEN/CNT nanocomposites. The values of the Avrami exponent (n) and the crystallization rate parameter (Z_c) are shown in Table 4. The Avrami exponent (n) was in the range of 3.89~5.81 for pure PEN, and 4.60~6.47 for the PEN/CNT nanocomposites, depending on cooling rate. The dependence of the crystallization kinetics on temperature is a complex function, and many theoretical models based on the Avrami equation have been developed [75-78]. As shown in Table 4, the PEN/CNT nanocomposites exhibited the values of n higher than four. This result indicates that the non-isothermal crystallization mechanism of the PEN/CNT nanocomposites is very complicated, suggesting that the CNT significantly influences the mechanism of nucleation and crystal growth of the PEN/CNT nanocomposites. In addition, the values of Z_c and $t_{1/2}$ decreased with increasing cooling rate.

In order to describe the non-isothermal crystallization process more effectively for comparison, Liu et al. [75] suggested a convenient procedure for characterizing non-isothermal crystallization kinetics by combining the Avrami and Ozawa equations based on the assumption that the degree of crystallinity was correlated to the cooling rate and crystallization time. Therefore, their relationship for non-isothermal crystallization can be derived by combining the equations (3) and (4) as follows:

$$\log Z_t + n \log t = \log K(T) - m \log a \qquad (6)$$

$$\log a = \log F(T) - b \log t \tag{7}$$

where the kinetic parameter, $F(T) = [K(T)/Z_t]^{1/m}$ represents the value of the cooling rate chosen at unit crystallization time when the systems have a defined degree of crystallinity; *a* is the cooling rate, and *b* is the ratio of the Avrami exponent (*n*) to the Ozawa exponent (*m*). From the equation (7), the plots of *log a* versus *log t* at a certain degree of crystallinity are shown in Figure 12, and they exhibit a good linear relationship, suggesting that this analysis may be more effective in describing non-isothermal crystallization kinetics of the PEN/CNT nanocomposites. The values of log *F*(*T*) and *b* were determined from the slope and intercept of the plots are shown in Table 5. The value of *F*(*T*) increased with increasing relative degree of crystallinity, indicating that at unit crystallization time, a higher relative degree of crystallinity was obtained with a higher cooling rate. The value of *b*, i.e., the ratio of the Avrami exponent to the Ozawa exponent, varied from 1.126 to 1.358 and from 1.125 to 1.362, for the PEN and the PEN/CNT nanocomposites, respectively. This result suggests that the presence of the CNT as strong nucleating agents influences the non-isothermal crystallization process involving the type of nucleation and crystal growth for the PEN/CNT nanocomposites.

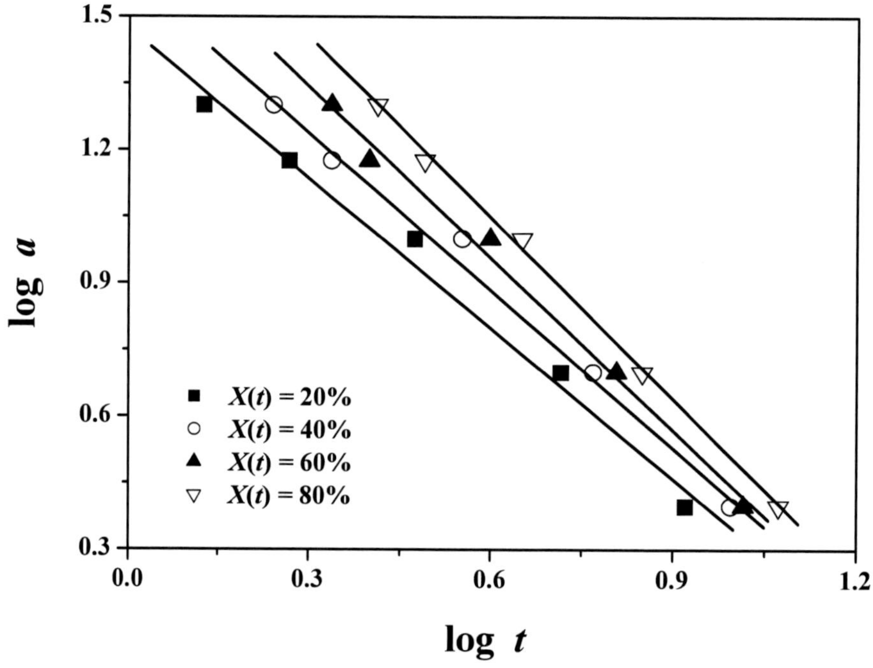

(a) PEN

Figure 12. Plots of log a versus log t from the combined Avrami and Ozawa equations at different relative degree of crystallinity for (a) PEN and (b) the PEN/CNT 0.5 nanocomposites.

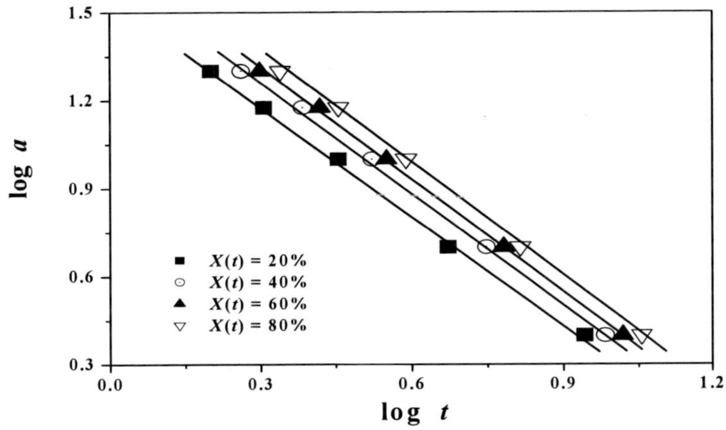

(b) PEN/CNT 0.5

Figure 12. (Continued)

Table 5. Values of *b* and *F*(*T*) for the PEN and the PEN/CNT nanocomposite obtained from the combined Avrami and Ozawa equation

Materials	*X(T)*	log *F(T)*	*b*	E_a (kJ/mol)
PEN	20% 40% 60% 80%	1.478 1.598 1.729 1.858	1.126 1.181 1.291 1.358	138.2
PEN/CNT 0.1	20% 40% 60% 80%	1.410 1.498 1.563 1.631	1.125 1.176 1.224 1.253	128.2
PEN/CNT 0.5	20% 40% 60% 80%	1.552 1.651 1.695 1.747	1.235 1.270 1.272 1.279	136.2
PEN/CNT 1.0	20% 40% 60% 80%	1.551 1.661 1.725 1.812	1.323 1.305 1.327 1.362	155.9
PEN/CNT 2.0	20% 40% 60% 80%	1.538 1.671 1.717 1.783	1.233 1.289 1.292 1.318	159.7

For the dynamic crystallization kinetics of polymer melts in the presence of nucleation agents, Dobreva et al. [79, 80] suggested a simple method for calculating the nucleation activity of different substrates. It is known that the nucleation activity (ϕ) is a factor by which the work of thee-dimensional nucleation decreases with the addition of a foreign substrate [79-82]. If the foreign substrate is extremely active, the nucleation activity approaches zero, while for inert particles, it is unity. For nucleation from melts near their melting temperature, the cooling rate can be represented as follows:

$$\log a = A - \frac{B}{2.3\Delta T_p^2} \tag{8}$$

where a is the cooling rate; A is a constant; ΔT_p is the degree of supercooling, i.e., $\Delta T_p = T_m - T_p$, T_p is the temperature corresponding to the peak temperature of DSC crystallization, and $B = \varpi\sigma^3 V_m^2/3kT_m\Delta S_m^2 n$ is a parameter related to three-dimensional nucleation (V_m is the molar volume of the crystallizing substance; ΔS_m is the entropy of melting; ω is a geometrical factor; σ is the specific surface energy; n is the Kolmogorov-Avrami exponent, and k is the Boltzmann constant). The nucleation activity can be calculated from the following equation [79-82]:

$$\phi = \frac{B^*}{B^0} \tag{9}$$

where B^0 and B^* are the values of B for homogeneous and heterogeneous nucleations, respectively. The values of B^0 and B^* can be obtained from the slope of the plot of *log a* versus $1/\Delta T_p^2$, and thus, the nucleation activity can be calculated from the equation (9) by using these values. The plots of log a versus $1/\Delta T_p^2$ for pure PEN and the PEN/CNT nanocomposites are shown in Figure 13. The values for the nucleation activity (ϕ) of the PEN/CNT nanocomposites were calculated as 0.29, 0.26, 0.27, and 0.23, respectively. This result demonstrates that the CNT can act as an excellent nucleating agent for the PEN/CNT nanocomposites. Kim et al. [64] reported that in the case of the PEN/silica nanocomposite, the ϕ value for silica nanoparticles was estimated to be 0.71 in the PEN matrix. For the montmorillonite/nylon 1212 nanocomposite, the ϕ value for the clay was calculated as 0.71 in the nylon 1212 matrix [62, 63]. In addition, Alonso et al. [82] reported that for the talc/isotactic polypropylene nanocomposite system, the ϕ value for the untreated talc was approximately 0.56, while that for the talc modified with silane coupling agent was estimated to be approximately 0.45 in the isotactic polypropylene matrix. From the above results, it can be deduced that the CNT in the PEN/CNT nanocomposites exhibit much higher nucleation activity than any other nanoreinforcing filler reported to date.

The activation energy for non-isothermal crystallization can be derived from the combination of cooling rate and crystallization peak temperature, and Kissinger [83, 84] suggested a method for calculating the activation energy for non-isothermal crystallization as follows:

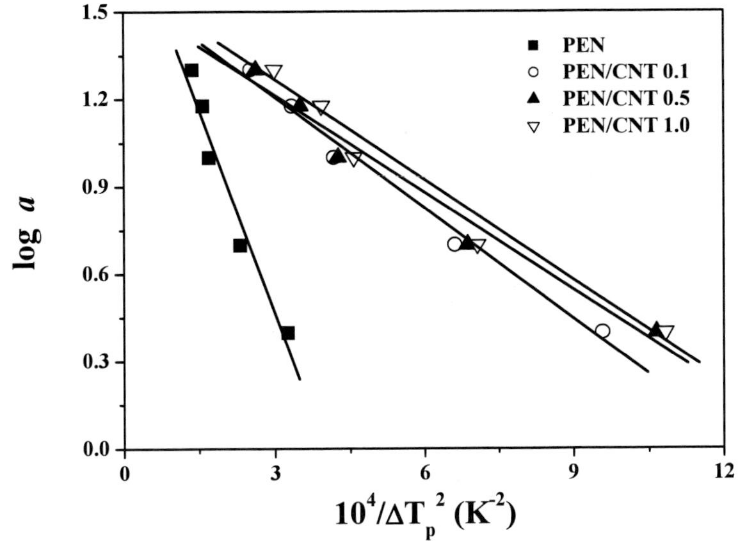

Figure 13. Plots of log *a* versus $1/\Delta T_p^2$ for the PEN/CNT nanocomposites with the CNT content.

$$\frac{d[\ln(\Phi / T_p^2)]}{d(1/T_p)} = -\frac{\Delta E_a}{R} \tag{10}$$

where R is the universal gas constant; T_p is the crystallization peak temperature; Φ is the cooling rate, and ΔE_a is the crystallization activation energy. The activation energies of the non-isothermal crystallization for pure PEN and the PEN/CNT nanocomposites were obtained from the slope of the plot of $ln\ (\Phi/T_p^2)$ versus $1/T_p$, according to the equation (10) and the results are shown in Table 5. The activation energy of crystallization for the PEN/CNT nanocomposites containing lower CNT content (≤ 0.5 wt%) was lower than that of pure PEN, while the activation energy increased with further increasing CNT content compared with that of pure PEN. The variation in the activation energy for the non-isothermal crystallization of the PEN/CNT nanocomposites may be explained by changes in the crystallization mechanism and the free energy of nucleation with the degree of supercooling [85]. In this PEN/CNT nanocomposite system, the CNT seem to perform two functions in the PEN matrix. One is that the CNT can act as nucleating agents and may accelerate the non-isothermal crystallization of PEN/CNT nanocomposites, which was confirmed by the kinetic parameters determined for non-isothermal crystallization and crystallization half-time. CNT may also adsorb the PEN molecular segments and restrict the movement of chain segments [58], thereby making crystallization difficult. The PEN molecular segments require more energy to rearrange, resulting in an increment in the activation energy of non-isothermal melt crystallization. The incorporation of the CNT into the PEN matrix accelerates the non-isothermal crystallization of the PEN/CNT nanocomposites. Furthermore, this enhancing effect of the CNT is more pronounced at lower CNT content (≤ 0.5 wt%), as described in the result showing a lower activation energy and crystallization half-time for non-isothermal crystallization compared with the PEN matrix.

4. Influenece of CNT on Physical Properties of PEN Nanocomposies

4.1. Dynamic Mechanical Thermal Analysis

The storage modulus (E') and loss modulus (E'') of the PEN/CNT nanocomposites are shown in Figure 14. As the molecular motion within a polymer changes, the modulus of a polymer varied with temperature. The storage modulus of a polymer decreases rapidly, whereas the loss modulus and tan δ undergoes a maximum when a polymer is heated though the glass transition region. The apparent glass transition region was revealed by a rapid decrease in the storage modulus of the PEN/CNT nanocomposites, and this temperature corresponding to the T_g of the PEN matrix was not affected by the addition of CNT. The incorporation of a very small quantity of the CNT significantly improved the storage and loss moduli of the PEN/CNT nanocomposites. This behavior of the PEN/CNT nanocomposites was attributed to the physical interactions between the CNT and the PEN matrix due to the high surface area of the CNT as well as the stiffening effect of the CNT in the PEN matrix; this effect being more pronounced at lower CNT content.

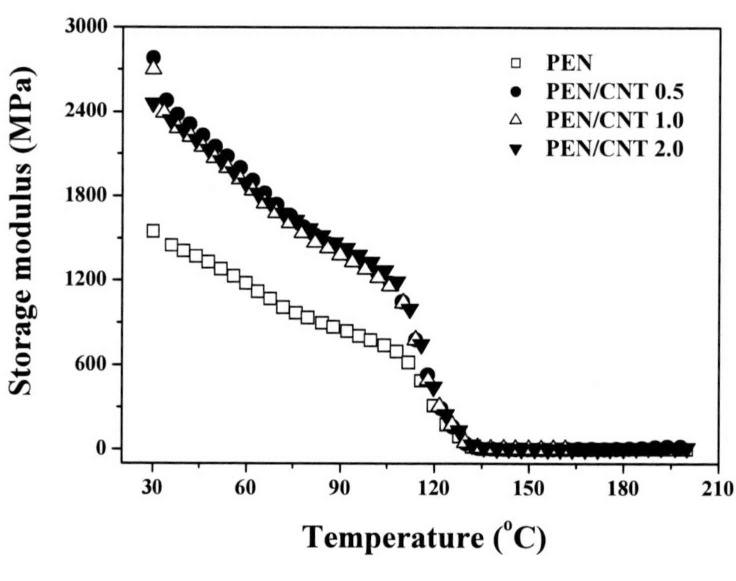

(a)

Figure 14. (a) Storage modulus and (b) loss modulus of the PEN/CNT nanocomposites with the CNT content as a function of temperature

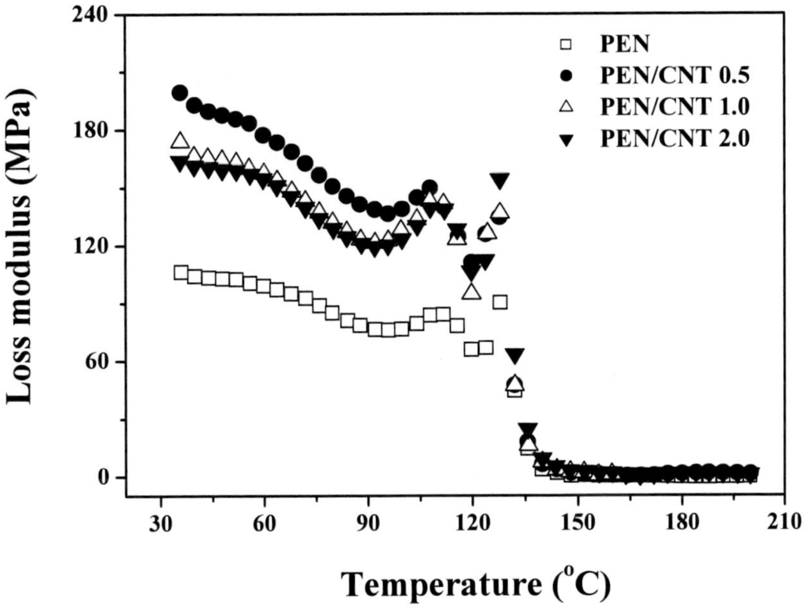

(b)

Figure 14. (Conintued)

The PEN/CNT nanocomposites exhibited a plateau region above 130 °C, implying that after a critical temperature, the storage modulus of the PEN/CNT nanocomposites was not significantly affected by the incorporation of the CNT and they exhibited a strong dependence of the storage modulus on the polymer matrix [86, 87]. Wu and Liu [87] reported that at the temperature above 140 °C, it has almost reached the softening point of the PEN matrix, which strongly reduce the elastic response of the PEN/layered silicate nanocomposites. The E'' values of the PEN/CNT nanocomposites increased with the introduction of the CNT, while they decreased with further increasing CNT content above 1 wt%. This behavior may be explained by the fact that the increasing degree of agglomeration of the nanotubes at higher CNT content results in less energy dissipating in the PEN/CNT nanocomposites under viscoelastic deformation [88]. The tan δ peak of the PEN/CNT nanocomposites as a function of temperature is shown in Figure 15. The position of the tan δ peak of the PEN/CNT nanocomposites was not significantly affected by the incorporation of CNT, while the peak height decreased.

4.2. Rheological Behavior

The complex viscosity ($|\eta^*|$) of pure PEN and the PEN/CNT nanocomposites at 285 oC as a function of frequency are shown in Figure 16(a). The $|\eta^*|$ of the PEN/CNT nanocomposites decreased with increasing frequency, indicating a non-Newtonian behavior over the frequency range investigated. The shear thinning behavior observed in the PEN/CNT nanocomposites may be attributed to the orientation of the rigid molecular chains in the nanocomposites during the applied shear force. The effect of the CNT on the |η*| of the PEN/CNT nanocomposites is more significant at low frequency compared with high frequency, and this effect was reduced with increasing frequency because of the strong shear thinning behavior induced by incorporating CNT

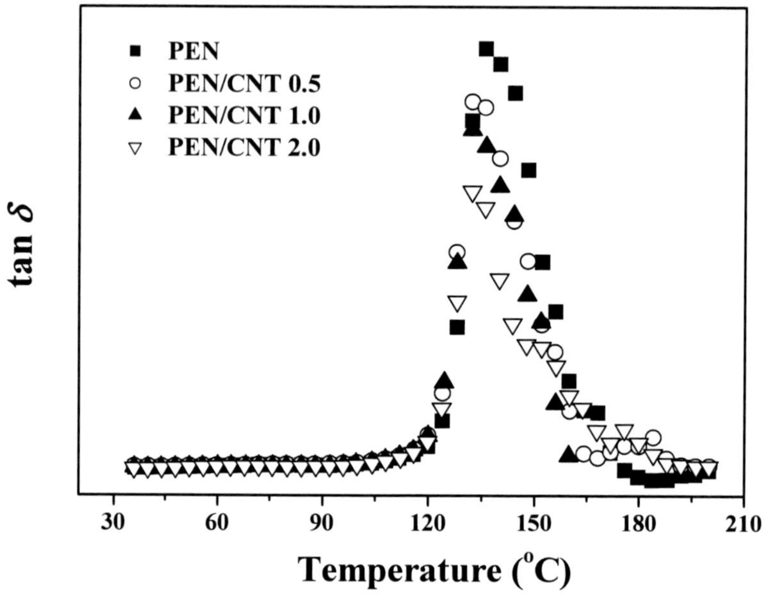

Figure 15. Variations of tan δ peak of the PEN/CNT nanocomposites with the CNT content as a function of temper.

In addition, the irregular decrease in the $|\eta^*|$ with increasing frequency indicates pseudo-plastic characteristics of the PEN/CNT nanocomposites due to random orientation and entangled molecules in this nanocomposite system. The frequency dependence of the $|\eta^*|$ of the PEN/CNT nanocomposites measured at various temperatures is shown in Figure 16(b). The $|\eta^*|$ of the PEN/CNT nanocomposites was decreased with increasing temperature. At low shear force region, the temperature had little effect on the $|\eta^*|$ of the PEN/CNT nanocomposites. However, the rheological properties of the PEN/CNT nanocomposites were affected by the temperature at high shear force region, and the $|\eta^*|$ significantly decreased with increasing temperature.

The shear thinning exponents (n) for the PEN/CNT nanocomposites can be obtained from the relationship of $|\eta^*| \approx \omega^n$ [89, 90], and their result are shown in Table 6. In this study, the shear thinning exponents of the PEN/CNT nanocomposites were estimated by fitting a straight line to the data at low frequency, because some curvature was observed in the plot of log $|\eta^*|$ versus log ω.

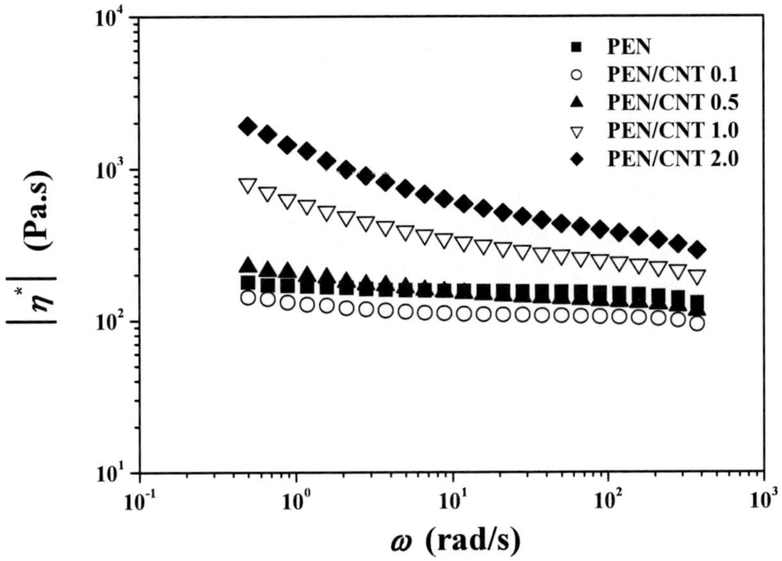

(a)

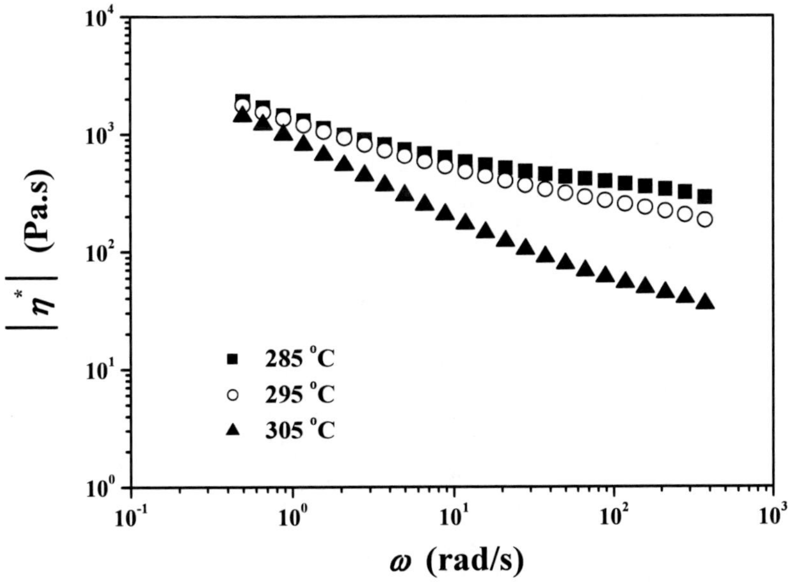

(b)

Figure 16. Variations of complex viscosity ($|\eta^*|$) of (a) the PEN/CNT nanocomposites with the CNT content measured at 285 °C and (b) the PEN/CNT 2.0 nanocomposites at different temperatures, with the frequency.

The n values of the PEN/CNT nanocomposites decreased with increasing CNT content, indicating that shear thinning behavior of the PEN/CNT nanocomposites significantly depended on CNT content. As shown in Table 6, the activation energy for the flow of the PEN/CNT nanocomposites increased with increasing CNT content. This result suggests that the incorporation of CNT leads to more rigid and stiffer polymer chains in the PEN/CNT nanocomposites, resulting in the increase in the activation energy for the flow process.

Table 6. Variations of shear thinning exponent and activation energy of the PEN/CNT nanocomposites with the CNT content.

Materials	Shear thinning exponent (n)	Activation energy, E_a (kJ/mol)
PEN	-0.13	59.40 kJ/mol
PEN/CNT 0.1	-0.11	58.40 kJ/mol
PEN/CNT 0.5	-0.17	66.52 kJ/mol
PEN/CNT 1.0	-0.24	85.23 kJ/mol
PEN/CNT 2.0	-0.34	87.23 kJ/mol

The variations of the $|\eta^*|$ of the PEN/CNT nanocomposites with the CNT content at different frequencies is shown in Figure 17.

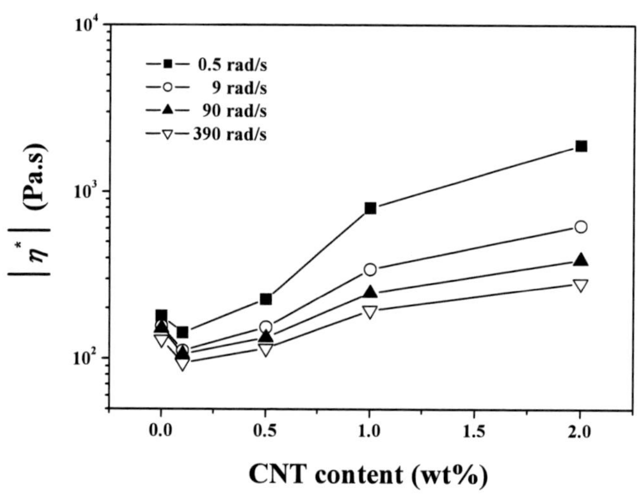

Figure 17. Variations of complex viscosities ($|\eta^*|$) of the PEN/CNT nanocomposites at different frequencies as a function of CNT content.

It can be seen that the $|\eta^*|$ of the PEN/CNT nanocomposites increased with increasing CNT content over the frequency ranges investigated. In addition, the extent of increase in the $|\eta^*|$ with increasing CNT content was more pronounced at low frequency compared with that at high frequency. Interestingly, it can be observed that the incorporation of a very small quantity (0.1 wt%) of CNT into the PEN matrix slightly decreased the complex viscosity of the PEN/CNT nanocomposites. This phenomenon may be attributed to the formation of viscous surface layers around the dispersed nanotubes leading to an increase in the free volume in this nanocomposite system, making it easier for flow to occur [91]. However, with further increase in CNT content, the $|\eta^*|$ of the PEN/CNT nanocomposites increased, and this may be attributed to the increase in physical interactions between the PEN matrix and the CNT with high aspect ratio and large surface area. The increase in the $|\eta^*|$ of the PEN/CNT nanocomposites with the CNT was closely related to the large increase in the storage modulus, which will be described in the following section.

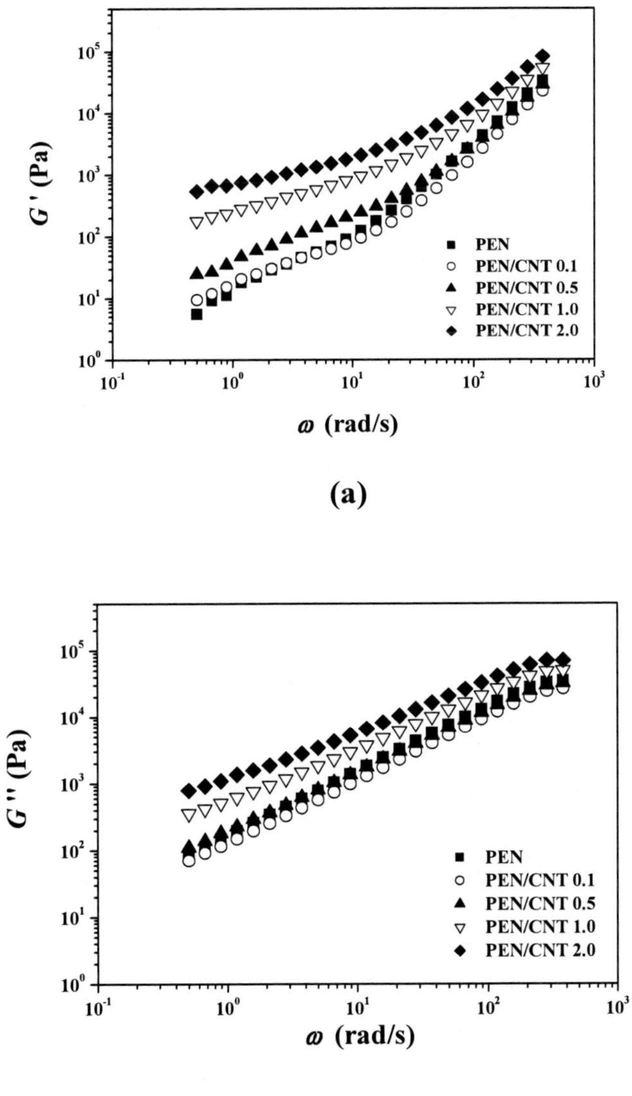

Figure 18. (a) Storage modulus (G′) and (b) loss modulus (G″) of the PEN/CNT nanocomposites with the CNT content as a function of frequency.

The storage modulus (G') and loss modulus (G'') of the PEN/CNT nanocomposites as a function of frequency are shown in Figure 18. The values of G' and G'' of the PEN/CNT nanocomposites increased with increasing frequency and CNT content, this increment being more significant at low frequency. This rheological response is similar to the relaxation behavior of typical filled-polymer composite systems [92, 93].

It is known that polymer chains are fully relaxed and exhibit characteristic homopolymer-like terminal flow behavior, resulting in that the flow curves of polymers being expressed by the power law $G' \propto \omega^2$ and $G'' \propto \omega$ [93-95]. Krisnamoorti and Giannelis [96] reported that the slopes of $G'(\omega)$ and $G''(\omega)$ for polymer/layered silicate nanocomposites were much smaller than 2 and 1, respectively, which are the values expected for linear homodispersed polymer melts. They suggested that large deviations in the presence of a small quantity of layered silicate were caused by the formation of a network structure in the molten state. The slopes of the terminal zone of G' and G' for the PEN/CNT nanocomposites are presented in Table 7. This result indicated the non-terminal behavior with the power-law dependence for G' and G' of the PEN/CNT nanocomposites: the flow curves of the PEN/CNT nanocomposites can be expressed by a power law of $G' \propto \omega^{1.19\sim0.72}$ and $G'' \propto \omega^{0.95\sim0.73}$, respectively. Similar non-terminal low-frequency rheological behavior has been observed in ordered block copolymers and smectic liquid-crystalline small molecules [97, 98]. The decrease in the slopes of G' and G' for the PEN/CNT nanocomposites with increasing CNT content was explained by the fact that the nanotube-nanotube or the nanotube-polymer interactions could lead to the formation of interconnected or network-like structures of the CNT, resulting in the pseudo-solid-like behavior of the PEN/CNT nanocomposites. In the PEN/CNT nanocomposites, the large deviations observed at high CNT content may be explained by the CNT agglomerates which could act as large particles in terms of the dominant nanotube-nanotube interactions in the PEN/CNT nanocomposites matrix with increasing CNT content. For CNT/epoxy nanocomposite systems, it has been reported that the difference in the terminal zone slope was closely related to the internal structure of the CNT/expoxy nanocomposites which is affected by particle-particle interaction of CNT in the polymer matrix [99]. In addition, the slopes of G' and G'' for the PEN/CNT nanocomposites decrease with increasing temperature, implying that the temperature can also affect the relaxation behavior of polymer chains in the PEN/CNT nanocomposites.

Table 7. Terminal Zone Slopes of $G'(\omega)$ and $G''(\omega)$ for the PEN/CNT nanocomposites

	Slope of $G'(\omega)$			Slope of $G''(\omega)$		
Materials	285 °C	295 °C	305 °C	285 °C	295 °C	305 °C
PEN	1.26	0.94	0.81	0.95	0.88	0.69
PEN/CNT 0.1	1.19	0.82	0.70	0.94	0.87	0.61
PEN/CNT 0.5	1.02	0.77	0.62	0.89	0.84	0.58
PEN/CNT 1.0	0.81	0.65	0.40	0.80	0.75	0.51
PEN/CNT 2.0	0.72	0.50	0.32	0.73	0.64	0.35

The variation of storage modulus (G') and loss modulus (G'') of the PEN/CNT nanocomposites with the CNT content at different frequencies is shown in Figure 19

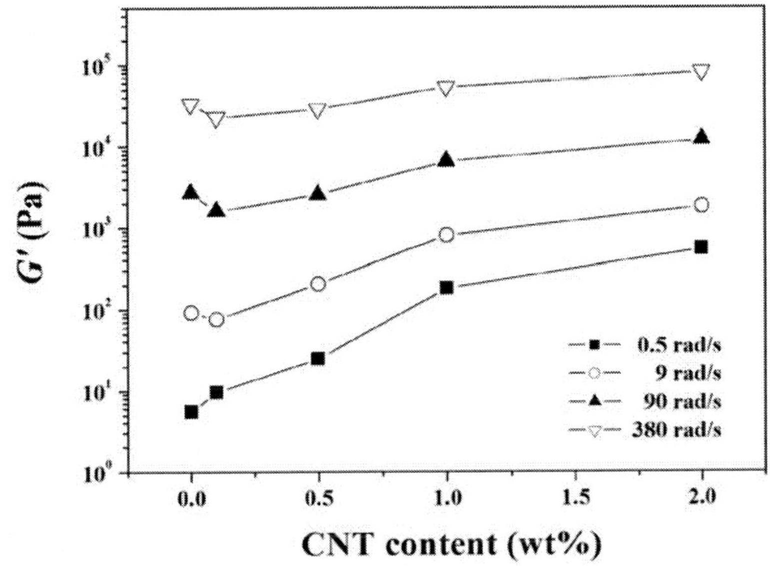

(a)

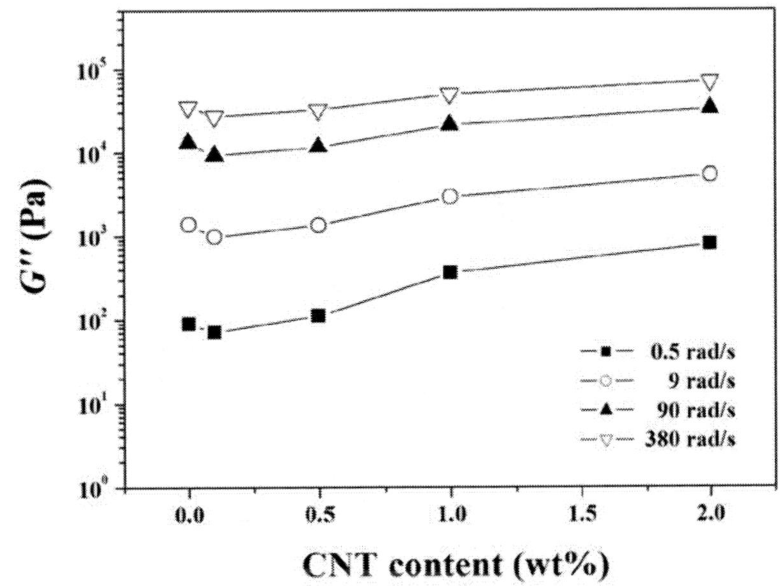

(b)

Figure 19. Variations of (a) storage modulus (G') and (b) loss modulus (G'') of the PEN/CNT nanocomposites at different frequencies, as a function of CNT content.

It can be seen that the incorporation of a very small quantity of the CNT (0.1 wt%) into the PEN matrix resulted in a decrease in the values of G' for the PEN/CNT nanocomposites over the frequency range investigated, except for the results measured at very low frequency. This phenomenon was similar to the tendency observed in the variations of the complex viscosity for the PEN/CNT nanocomposites with the CNT content. As CNT content increased, the physical interactions between the nanotubes could lead to the formation of interconnected or network-like structure of the CNT in the polymer matrix [29]. The extent of the increase in G' of the PEN/CNT nanocomposites is higher that that of G'' over the frequency range investigated. In addition, the storage modulus and loss modulus of the PEN/CNT nanocomposites were significantly improved relative to pure PEN, particularly at low frequency.

The variation of tan δ for the PEN/CNT nanocomposites with the CNT content as a function of frequency is shown in Figure 20. Shear deformation leads to the partial orientation of the molecules in polymer chains, resulting in the decrease in tan δ of the PEN/CNT nanocomposites with increasing frequency. It can be seen that the loss tangent maximum of the PEN/CNT nanocomposites shifted to higher frequency with increasing CNT content, implying the formation of dense interconnected or network structures in the polymer nanocomposites [31]. Hsiao et al. [100] reported that in the case of the polymer nanocomposites consisting of ethylene-propylene copolymers (EP) and polyhedral oligomeric silsesquioxane (POSS) molecules, the maximum in tan δ curves shifts towards a higher frequency as the POSS concentration increased, indicating the densification of physically crosslinked network in the polymer nanocomposites.

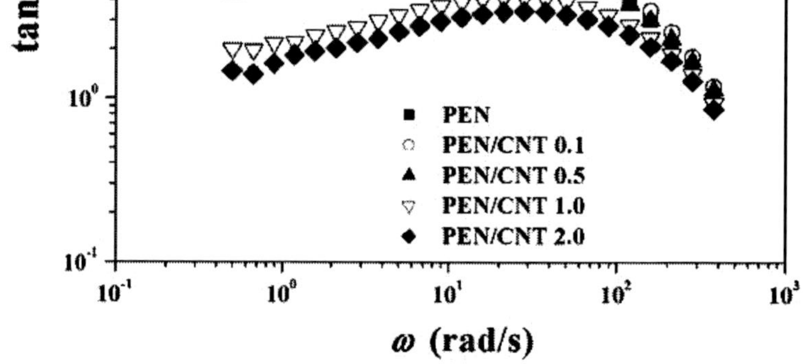

Figure 20. Variation of tan δ for the PEN/CNT nanocomposites with the CNT content as a function of frequency.

The plots of phase angle (δ) versus the absolute value of complex modulus ($|G^*|$) for the PEN/CNT nanocomposites, which is known as the Van Gulp-Palmen plot [101-103], are shown in Figure 21. A significant change in the phase angle occurred on the incorporation of the CNT, and the decrease in the phase angle with decreasing the complex modulus indicated the enhancement of the elastic behavior. The PEN/CNT nanocomposites exhibited smaller δ values at lower $|G^*|$ values as compared to pure PEN, indicating that the incorporation of the CNT into the PEN matrix enhanced the elastic behavior of the PEN/CNT nanocomposites and this enhancing effect was more pronounced at higher CNT content

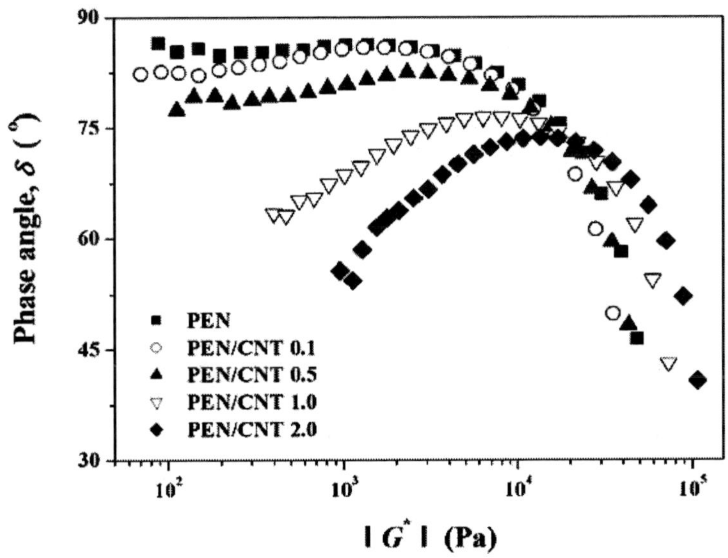

Figure 21. Plots of phase angle versus complex modulus of the PEN/CNT nanocomposites measured at 285 °C.

As described in the previous section, the storage modulus and loss modulus of the PEN/CNT nanocomposites were higher than that of pure PEN, and increased with increasing CNT content. In general, the Cole-Cole plot, a logarithmic plot of G' versus G'' in dynamic rheology, provides a master curve with a slope of 2 for isotropic and homogeneous polymer melts irrespective of temperatures [104]. The extent of deviation from the slope of 2 has been commonly utilized in judging the heterogeneity of polymeric systems, because the slope of the plot of log G' versus log G'' decreases as the heterogeneity increases. The plots of log G' versus log G'' for the PEN/CNT nanocomposites with the CNT content are shown in Figure 22. The PEN/CNT nanocomposites do not provide a perfect single master curve, and exhibit the shifting and the change of the slope of the plot with the the introduction of the CNT. The slopes in the terminal regime of PEN/CNT nanocomposites were less than 2, implying that the PEN/CNT nanocomposites underwent some chain conformational changes with the incorporation of the CNT. As shown in Figure 22, the storage modulus of the PEN/CNT nanocomposite at a given loss modulus increased with increasing CNT content. In addition, the slope of the plot of log G' versus log G'' for the PEN/CNT nanocomposites decreased with increasing CNT content. Therefore, it can be deduced that incorporation of the CNT has a significant effect on the microstructures for the PEN/CNT nanocomposites. For

CNT/polycarbonate (PC) composites prepared via direct melt compounding, Pötschke et al. [29] suggested that the shift and the change in slope of the plot of G' versus G'' for CNT/PC composites indicated a significant change in the microstructure of the polymer composites with the addition of the CNT.

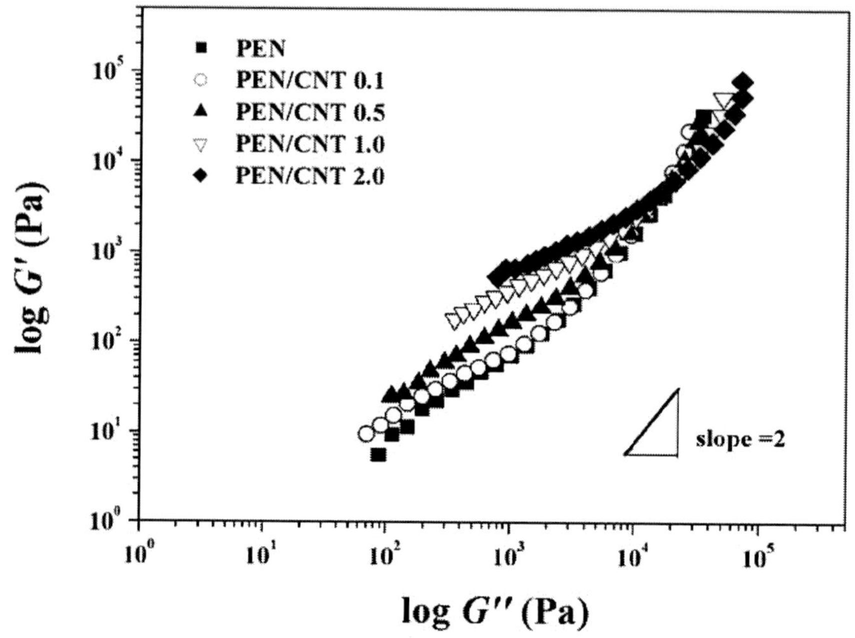

Figure 22. The plots of log G' versus log G'' for the PEN/CNT nanocomposites with the CNT content.

The relaxation time of the PEN/CNT nanocomposites under dynamic shear in the polymeric systems that involved the pseudo-structures can be calculated from the following equation [105]:

$$J' = \frac{G'}{\left(|\eta^*|\omega\right)^2} = \frac{\lambda}{|\eta^*|} \tag{11}$$

where J' and η^* are the compliance and the complex viscosity, respectively. It is expected that the presence of some molecular order or physical structure leads to a much longer relaxation time [105]. As shown in Figure 23, the relaxation time of the PEN/CNT nanocomposites decreased with increasing frequency. In addition, the relaxation time of the PEN/CNT nanocomposites increased with increasing CNT content. As CNT content increases, the nanotube-nanotube interactions will be dominant in the nanocomposite systems, and lead to the formation of interconnected structures of CNT in the PEN/CNT nanocomposites, which affects the relaxation behavior, this effect being more pronounced at higher concentrations.

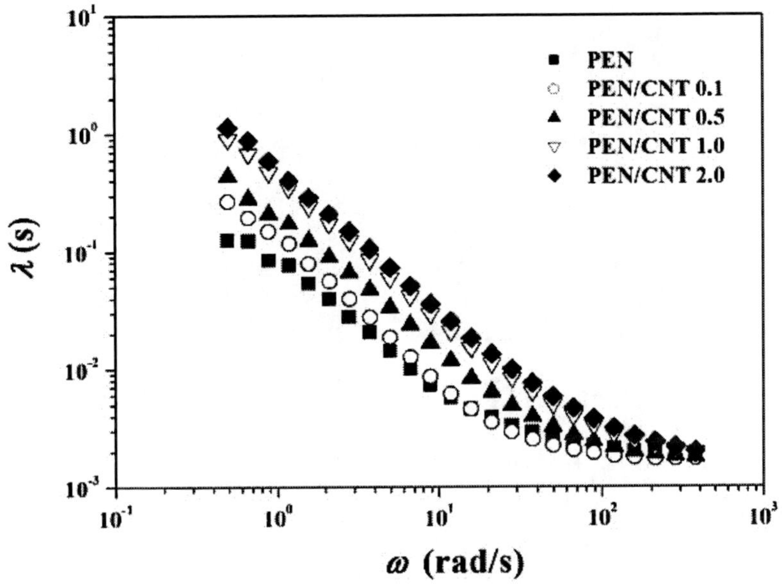

Figure 23. Variation of the relaxation time of the PEN/CNT nanocomposites with the CNT content.

4.3. Mechanical Properties and Thermal Stability

Representative SEM images of pristine CNT and the PEN/CNT 2.0 nanocomposites are shown in Figure 24. The CNT exhibits highly curved and random coiled features, which may be attributed to hydrogen bonding and van der Waals attractions between the individual nanotubes [57, 58]. As shown in Figure 24(b), the CNT were randomly dispersed in the PEN matrix, and the PEN/CNT nanocomposites exhibited the interconnected or network-like structures of the CNT in the PEN matrix. The CNT with small size, high aspect ratio, and large surface area, is often subjected to self-agglomeration or bundle formation at higher concentration, and thus easily form interconnected or network-like structures in the molten polymer matrix. TEM images of the PEN/CNT nanocomposites are shown in Figure 25. It can be seen that the CNT were randomly dispersed in the PEN matrix, with some entanglements or bundles of the CNT. On a larger scale, however, the CNT were uniformly dispersed in the PEN matrix, despite some agglomerated CNT structures. In addition, the CNT were randomly oriented and formed interconnecting structures [106].

As shown in Figure 25(b), less uniformly dispersed and more entangled bundles of the CNT were formed in the PEN matrix at higher CNT content, which may be the cause of the stress concentration phenomenon.

As a result, the tensile strength of the PEN/CNT nanocomposites was not increased significantly at high CNT content, as expected, compared with that at low content, resulting from some agglomerated structures of randomly dispersed CNT in the PEN/CNT nanocomposites

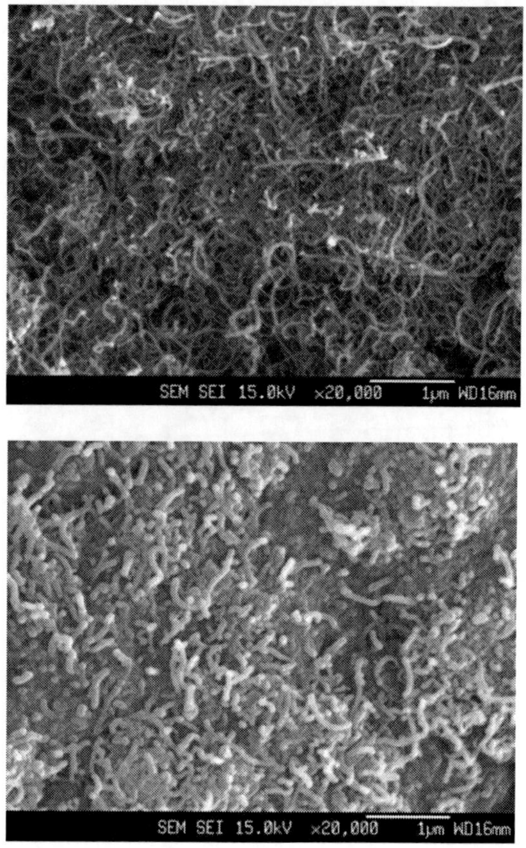

(a) **(b)**

Figure 24. SEM images of (a) pristine CNT and (b) the PEN/CNT 2.0 nanocomposites.

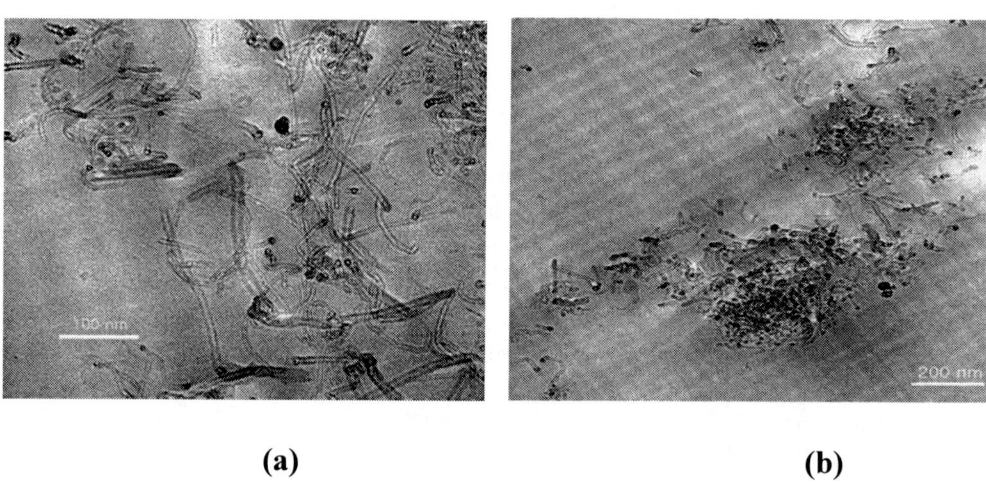

(a) **(b)**

Figure 25. TEM images of the PEN/CNT nanocomposites at (a) 0.1 wt% and (b) 2.0 wt% CNT content.

Therefore, for the fabrication of the PEN/CNT nanocomposites with further enhanced mechanical properties, it should be required for improving the interfacial adhesion between the CNT and the PEN matrix as well as the uniform dispersion of the CNT in the PEN matrix.

The variation of tensile strength and tensile modulus of the PEN/CNT nanocomposites with the CNT content is shown in Figure 26. As CNT content increased, the tensile strength and tensile modulus of the PEN/CNT nanocomposites significantly improved due to the nanoreinforcing effect of the CNT with high aspect ratios, and this enhancing effect was more significant with lower CNT content. However, the CNT often tend to bundle together because intrinsic van der Waals attractions between the individual tubes in combination with high aspect ratio and surface area of the nanotubes, leading to some agglomeration, thus prevents efficient load transfer to the polymer matrix [19, 57].

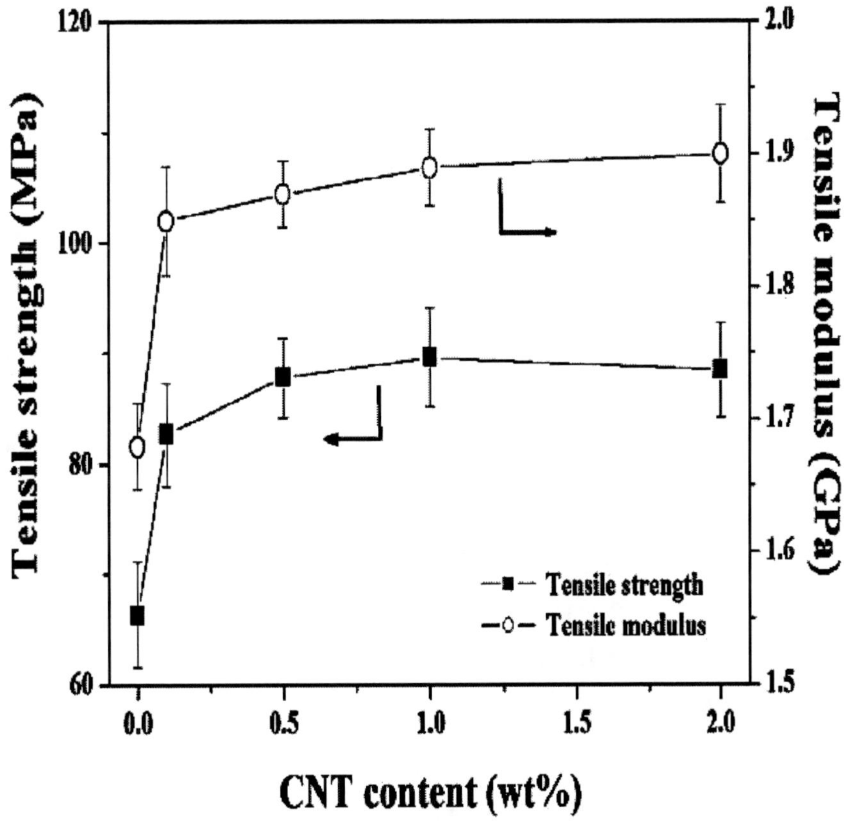

Figure 26. Variation of the tensile strength and tensile modulus of the PEN/CNT nanocomposites with the CNT content.

Thermal stability of polymer nanocomposites plays a significant role in determining the processing and industrial applications because it can affect the limit of working temperatures and dimensional stability for targeted applications. Results from the TGA thermograms of the PEN/CNT nanocomposites with the CNT content at a heating rate of 10 °C·min^{-1} under a nitrogen atmosphere are shown in Table 8. The thermal decomposition temperatures, including T_i, T_{10}, T_{60}, and T_{dm}, and the residual yields of the PEN/CNT nanocomposites increased with increasing CNT content, indicating that the incorporation of the CNT into the PEN matrix could retard the thermal decomposition of the PEN/CNT nanocomposites with higher residual yields. The presence of the CNT could stabilize the PEN matrix, resulting in the enhancement of the thermal stability of the PEN/CNT nanocomposites. The improvement in thermal stability of the PEN/CNT nanocomposites was attributed to the physical barrier effect induced by the CNT that prevent the transport of volatile decomposed products in the nanocomposites and that retard the thermal decomposition of the PEN/CNT nanocomposites [59, 60]. Kashiwagi et al. [107, 108] reported that CNT layers exhibited good barrier effect on the thermal degradation process, and they can not only insulate polymers but also reduce the weight-loss rate of thermal degradation products, resulting in the significant improvement in the thermal stability of the polymer nanocomposites. Therefore, it can be deduced that the CNT is beneficial to act as effective thermal decomposition-resistant nanoreinforcing fillers in the PEN matrix, thus making it possible to significantly improve the thermal stability of the PEN/CNT nanocomposites even with the introduction of a very small quantity of CNT.

Table 8. Thermal stability of the PEN/CNT nanocomposites with the CNT content

Materials	T_i [a] (°C)	T_{10} [b] (°C)	T_{60} [b] (°C)	T_{dm} [c] (°C)	W_{800} [d] (%)
PEN	389.2	412.4	447.4	436.8	14.8
PEN/CNT 0.5	390.6	417.2	457.1	438.7	22.7
PEN/CNT 1.0	397.7	420.4	461.4	439.8	26.9
PEN/CNT 2.0	398.2	422.2	467.9	440.9	30.3

[a] Initial thermal decomposition temperature in TGA thermograms at a heating rate of 10 °C/min.
[b] Thermal decomposition temperatures at 10% and 60% of the weight-loss, respectively.
[c] Thermal decomposition temperature at the maximum rate of the weight-loss.
[d] Residual yield in TGA thermograms at 800 °C under N_2.

5. Crystallization, Melting Behavior, and Mechanical Properties of CNT and PEN Nanocomposites

5.1. Isothermal Crystallization and Melting Behavior

The thermal properties of the PEN/CNT nanocomposites with various CNT contents are shown in Table 9. The incorporation of the CNT has a little effect on the glass transition temperature (T_g) and the melting temperature (T_m) of the PEN/CNT nanocomposites. As the CNT content increased, the cold-crystallization temperature (T_{cc}) of the PEN/CNT nanocomposites decreased, and the melt-crystallization temperature (T_{mc}) increased. These results, together with the fact that the PEN/CNT nanocomposites have a lower degree of supercooling ($\Delta T = T_m - T_{mc}$) for crystallization with increasing CNT content, suggest that the CNT can effectively act as nucleating agents in the PEN/CNT nanocomposites. The CNT promoted the formation of heterogeneous nuclei with lower energy consumption necessary to reach critical stability for crystal growth [61]. Therefore, the incorporated CNT had a strong nucleation effect on the PEN matrix with the enhancement of the PEN crystallization through heterogeneous nucleation.

Table 9. Thermal Behavior of the PEN/CNT nanocomposites with the CNT content

Materials	T_g (°C)	T_{cc} [a] (°C)	T_m (°C)	T_{mc} [b] (°C)	ΔT [c] (°C)
PEN	116.8	194.0	263.0	195.7	67.3
PEN/CNT 0.1	116.7	186.7	264.5	217.5	47.0
PEN/CNT 0.5	116.0	186.1	263.6	219.6	44.0
PEN/CNT 1.0	116.2	184.4	263.9	222.1	41.8
PEN/CNT 2.0	116.1	183.3	262.6	222.7	39.9

[a] Cold-crystallization temperature measured from DSC traces at a heating rate of 10 °C/min.
[b] Melt-crystallization temperature measured from DSC traces at a cooling rate of 10 °C/min.
[c] Degree of supercooling, $\Delta T = T_m - T_{mc}$.

Isothermal crystallization exotherms of the PEN/CNT nanocomposites with the CNT content and crystallization temperature are shown in Figure 27. As the crystallization temperature increased, the exothermic peaks shifted to longer time, implying that the crystallization rate decreased with increasing crystallization temperature. In addition, the exothermic peaks shifted toward shorter time region with increasing CNT content, indicating faster crystallization rates with increasing CNT content.

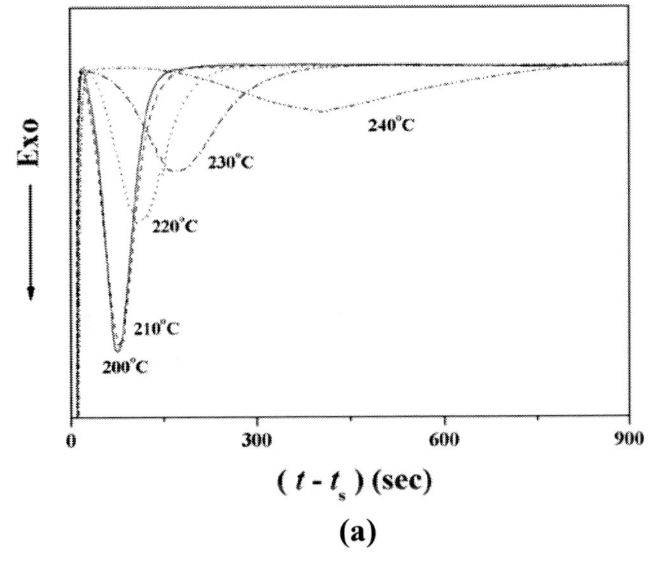

Figure 27. Isothermal crystallization exotherms of (a) the PEN/CNT 0.1 nanocomposites with the crystallization temperature and (b) the PEN/CNT nanocomposites at 220 °C with the CNT content.

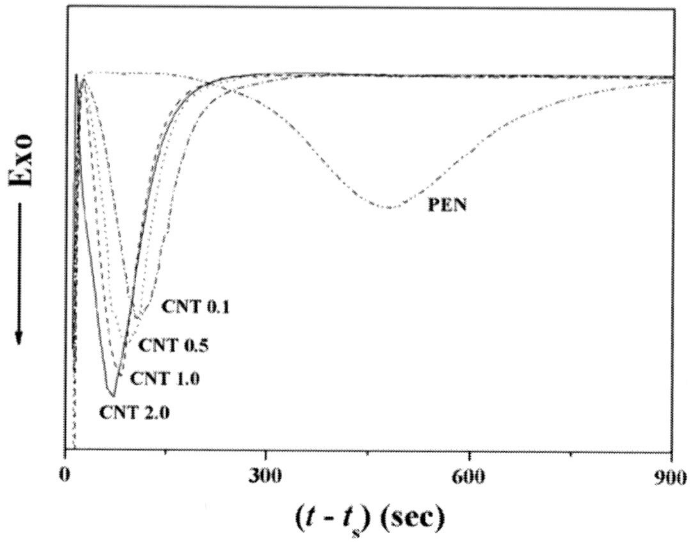

(b)

Figure 27. (Continued)

The relative degree of the crystallinity (X_t) at various crystallization times can be calculated from the ratio of the area of the exothermic peak up to time t divided by that of the total exotherms of the crystallization, and all isotherms exhibited a sigmoidal dependence on time. The time for completing the crystallization was reduced with increasing CNT content, and the relative degree of the crystallinity for the PEN/CNT nanocomposites was higher than that of pure PEN. The crystallization rate of the PEN/CNT nanocomposites was decreased with the isothermal crystallization temperature. The isothermal crystallization kinetics of the PEN/CNT nanocomposites was analyzed using the Avrami equation [68-70]:

$$1 - X_t = \exp[-k(t - t_s)^n] \tag{12}$$

where k is the rate constant; t is the crystallization time; t_s is the initial time of crystallization, and n is the Avrami exponent that related to the nucleation mechanism and crystal growth. As the crystallization temperature increased, the Avrami plots of the PEN/CNT nanocomposites shifted toward longer time, indicating that the crystallization rate decreased with increasing crystallization temperature. The values of n and k were calculated from the slope and the intercept of the plots of $\log[-\ln(1 - X_t)]$ versus $\log (t - t_s)$. The dependences of the kinetic parameters on the crystallization temperature and CNT content are shown in Table 10.

The crystallization rate constant (k) of the PEN/CNT nanocomposites increased with the introduction of the CNT, indicating the acceleration of the PEN crystallization by the introduction of the CNT. In addition, the k values of the PEN/CNT nanocomposites were very sensitive to the crystallization temperature, and they decreased in all specimens.

Table 10. Kinetic parameters of the PEN/CNT nanocomposites with the CNT content during the isothermal crystallization

Materials	T_c (°C)	n	k (min^{-1})
PEN	200	3.75	9.14×10^{-3}
	210	3.51	3.98×10^{-3}
	220	3.06	2.08×10^{3}
	230	2.19	1.07×10^{-3}
	240	1.58	8.47×10^{-4}
PEN/CNT 0.1	200	3.33	5.31×10^{-1}
	210	3.09	4.44×10^{-1}
	220	2.85	1.38×10^{-1}
	230	2.82	3.62×10^{-2}
	240	1.87	2.19×10^{-2}
	250	1.89	2.82×10^{-3}
PEN/CNT 0.5	210	2.17	6.63×10^{-1}
	220	2.16	3.56×10^{-1}
	230	1.96	1.26×10^{-1}
	240	1.82	6.16×10^{-2}
	250	1.61	7.53×10^{-3}
PEN/CNT 1.0	210	1.64	9.15×10^{-1}
	220	1.59	8.08×10^{-1}
	230	1.56	2.87×10^{-1}
	240	1.49	5.52×10^{-2}
	250	1.48	9.06×10^{-3}
PEN/CNT 2.0	210	1.50	1.23×10^{0}
	220	1.59	9.49×10^{-1}
	230	1.54	3.72×10^{-1}
	240	1.46	7.69×10^{-2}
	250	1.41	1.79×10^{-2}

The crystallization and the growth mechanism of the PEN were affected by the presence of the CNT in the polymer nanocomposites, and the variation of the n values for the PEN/CNT nanocomposites with the CNT content indicates different nucleation and crystalline growth process. In general, the value of the Avrami exponent (n) close to 3 implies the athermal nucleation process followed by a three-dimensional heterogeneous crystal growth [68-70]. The significant increase in the crystallization rate with the addition of the CNT is attributed to the enhancement of the heterogeneous nucleation in the presence of the CNT. The n values of the PEN/CNT nanocomposites were lower than that of pure PEN, and this implies that on the incorporation of the CNT, crystal growth may not occur in the three dimensions at an equal rate [87], leading to a low n values for the PEN/CNT nanocomposites. Thus, the incorporation of the CNT leads to the heterogeneous nucleation induced by the change in the crystal growth process from three-dimensional crystal growth to mixed spherulite growth for the PEN/CNT nanocomposites [87]. In addition, the k values increased with the increasing CNT content, indicating a significant increase in the heterogeneous nucleation for the PEN/CNT nanocomposites. The higher nucleation density of the dispersed CNT has the heterogeneous nucleation effect on the PEN crystallization. The crystal growth of the PEN/CNT nanocomposites was occurred much fast by incorporating the CNT with high nucleation density and large surface area, leading to the higher possibility of the impingement of the spherulites in the limited space. This result relates to the fact that the nucleation parameter, the folded surface energy, and the work of chain folding were decreased with the introduction of the CNT, which is elaborated during the following section for secondary nucleation theory. The half-time of crystallization ($t_{1/2}$) can be estimated from the equation (13), and the lower $t_{1/2}$ implies the higher crystallization rate.

$$t_{1/2} = \left(\frac{\ln 2}{k} \right)^{1/n} \tag{13}$$

The dependence of the crystallization temperature on the half-time of crystallization for the PEN/CNT nanocomposites is shown in Figure 28.

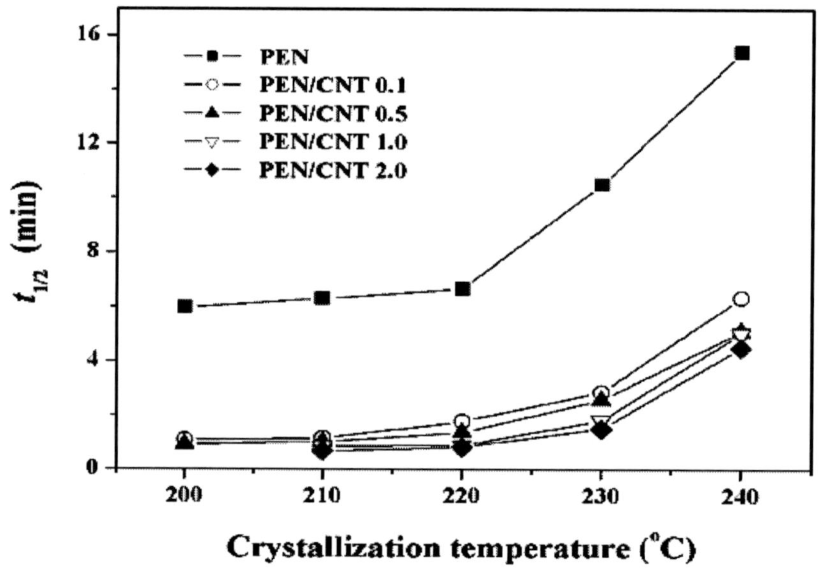

Figure 28. Dependence of crystallization temperature on the half-time of crystallization ($t_{1/2}$) for the PEN/CNT nanocomposites with various CNT contents.

The PEN/CNT nanocomposites exhibited lower $t_{1/2}$ values than that of PEN matrix. The isothermal crystallization rate of the PEN/CNT nanocomposites conducted by the $t_{1/2}$ value was decreased with increasing crystallization temperature and increased with increasing CNT content, implying that the acceleration of the PEN crystallization process by incorporating the CNT. Thus, it can be deduced that the strong nucleation effect of the CNT enhances the crystallization of the PEN through more heterogeneous nucleation, leading to higher crystallization rate. Assuming that crystallization process is thermally activated, the crystallization rate parameter (k) can be expressed by the following equation [87, 109, 110]:

$$\frac{1}{n}(\ln k) = \ln k_0 - \frac{\Delta E}{RT} \tag{14}$$

where k_0 is a temperature independent pre-exponential factor; R is the universal gas constant, and ΔE is the total activation energy for the crystallization, which consists of the transport activation energy (ΔE^*) referring to the activation energy required to transport molecular segments across the phase boundary to the crystallization surface and the nucleation activation energy (ΔF) referring to the free energy of formation of the critical sized crystal nuclei at T_c [87, 110]. As shown in Figure 29, the activation energy for crystallization can be determined from the slope of the plot of $(1/n)\ln k$ versus $1/T_c$. There is a dependence of the ΔE value of the PEN/CNT nanocomposites on the CNT content: the ΔE values for the PEN, the PEN/CNT 0.1, the PEN/CNT 1.0, and the PEN/CNT 2.0 nanocomposites were 98.9, 92.6, 120.2, and 160.9 kJ·mol^{-1}, respectively. In the case of the PEN/CNT nanocomposites, the incorporation of a very small quantity of the CNT into the PEN matrix may induce more heterogeneous nucleation, causing a lower ΔE values relative to the PEN. However, at higher CNT content, the transportation of polymer chains in the PEN/CNT nanocomposites may be hindered by the presence of more CNT that was induced more steric hindrance [87, 110], leading to a higher ΔE value.

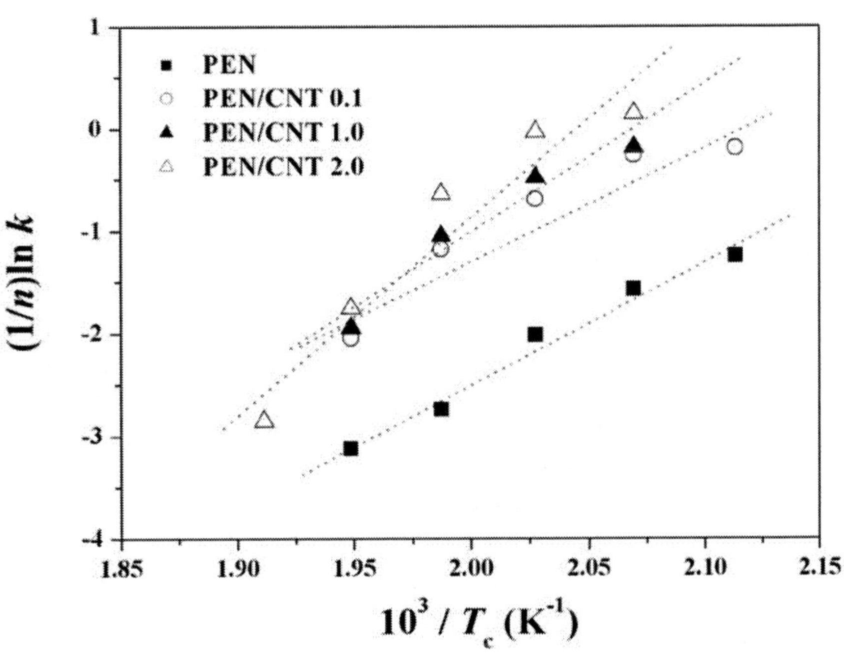

Figure 29. Variations of activation energy for crystallization of the PEN/CNT nanocomposites with various CNT contents.

DSC heating thermograms of pure PEN and the PEN/CNT nanocomposites after the completion of the isothermal crystallization are shown in Figure 30.

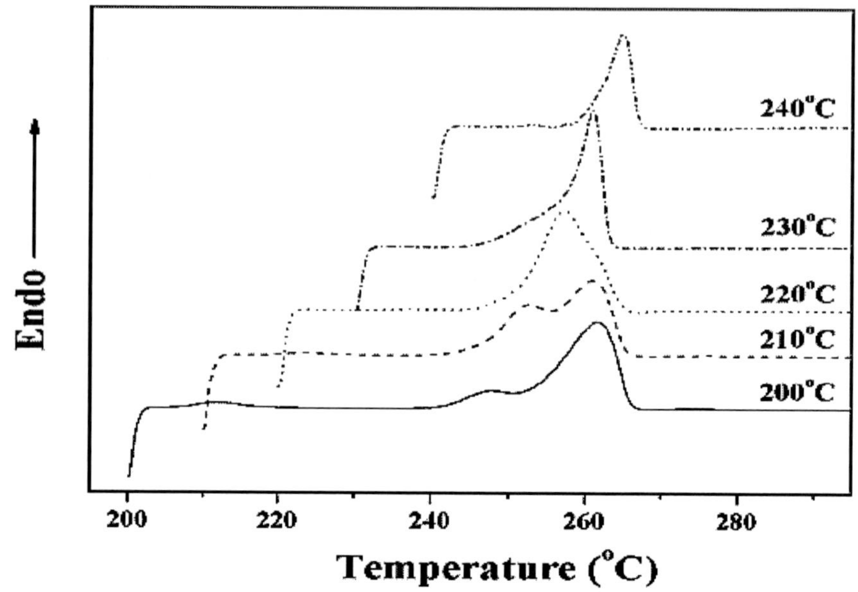

(a) PEN

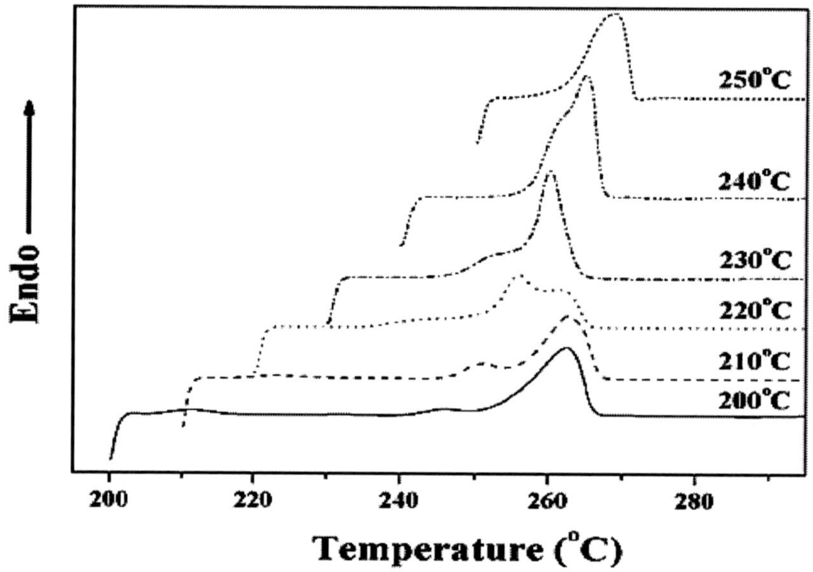

(b) PEN/CNT 0.1

Figure 30. DSC heating thermograms of (a) PEN and (b) the PEN/CNT 0.1 nanocomposites crystallized at various crystallization temperatures.

Both pure PEN and the PEN/CNT nanocomposites exhibited multiple melting peaks in the DSC heating thermograms. The small and low melting peaks observed at approximately 10~15 °C above the isothermal crystallization temperature was attributed to the melting of small crystallites formed between main lamellar populations, the middle melting peaks corresponded to the melting of the primary crystals grown at the isothermal crystallization temperature, and the high melting peaks was attributed to the melting of crystallites produced by the melting-recrystallization process. As the crystallization temperature increased, low temperature melting peaks gradually shifted toward high temperature region, and eventually merged into a single peak at higher crystallization temperature. This result indicated that the crystallization temperature influences the melting behavior of the PEN/CNT nanocomposites. In general, the multiple melting behaviors of polymers or polymer composites can be influenced by various factors, including the change in morphology, the orientation effects, the presence of more than one crystal modification, and the melting-recrystallization-remelting processes occurring during DSC scans [111, 112, 113-117]. The increase in the perfection of the crystalline structures with crystallization temperature was confirmed by the fact that the low and middle melting temperatures shifted toward higher temperature regions with higher crystallization temperature. The original imperfect thin crystals or lamellae in the polymers could melt and recrystallize to form the crystals with higher perfection during DSC scans [117]. The equilibrium melting temperatures of the PEN/CNT nanocomposites with the CNT content can be estimated from the theoretical relation suggested by Hoffman and Weeks [118], and their results are shown in Table 11. The values of the T_m^0 for the PEN/CNT nanocomposites decreased with increasing CNT content. Similar observation has been reported that the decrease in the T_m^0 value of the PEN/layered silicate nanocomposites with increasing the content of layered silicate was probably due to the presence of more heterogeneous nucleation to reduce the perfection of the PEN crystallites in the polymer nanocomposites [87, 110]. Thus, the decrease in the T_m^0 value of the PEN/CNT nanocomposites may be attributed to the formation of less perfect crystallites in the PEN/CNT nanocomposites because of the heterogeneous nucleation effect of the CNT, compared to the PEN matrix.

Table 11. Values of T_m^0, K_g, σ_e, and q of the PEN/CNT nanocomposites with the CNT content

Materials	T_m^0 (°C)	K_g ($K^2 \times 10^{-5}$)	σ_e (Jm^{-2})	q (J)
PEN	285.4	1.54	8.28×10^{-2}	6.10×10^{-20}
PEN/CNT 0.1	283.5	1.46	7.90×10^{-2}	5.82×10^{-20}
PEN/CNT 0.5	279.1	1.19	6.49×10^{-2}	4.78×10^{-20}
PEN/CNT 1.0	275.0	1.04	5.69×10^{-2}	4.19×10^{-20}
PEN/CNT 2.0	273.6	0.93	5.14×10^{-2}	3.79×10^{-20}

The dependence of the linear growth rate on the temperature can be described by Laurizen-Hoffman theory for secondary nucleation [119]. According to the nucleation theory, the crystal growth rate (G) can be represented as follows:

$$G = G_0 \exp\left[\frac{-U^*}{R(T_c - T_\infty)}\right] \exp\left[\frac{-K_g}{T_c(\Delta T)f}\right] \qquad (15)$$

where G_0 is the front factor; U^* is the activation energy for segment diffusion to the site of crystallization; R is the universal gas constant; T_c is the crystallization temperature; T_∞ is the hypothetical temperature below which all viscous flow ceases; K_g is the nucleation parameter; $\Delta T = T_m^0 - T_c$ is the degree of supercooling; T_m^0 is the equilibrium melting temperature, and f is the correction factor given as $2T_c/(T_m^0 + T_c)$. The plot of $\ln G + [U^*/R(T_c - T_\infty)]$ versus $1/T_c(\Delta T)f$ for the PEN/CNT nanocomposites are shown in Figure 31. As the crystallization kinetics are governed by the nucleation term, the growth rates are insensitive to the values of U^* and T_∞ employed in the equation (15) [119]. In this study, universal values of $U^* = 6300$ J/mol and $T_\infty = T_g - 30$ K were used in all of the calculations [119-122]. With these assumptions, the values of K_g can be determined the slope of the plot shown in Figure 31. The slope of the plots of $\ln G + [U^*/R(T_c - T_\infty)]$ versus $1/T_c(\Delta T)f$ for the PEN/CNT nanocomposites decreased with increasing CNT content. For a secondary nucleation, the values of K_g contains the contribution from the surface free energy, and it can be expressed by the following equations:

$$K_g = \frac{n\sigma\sigma_e b_0 T_m^0}{\Delta h_f k_B} \tag{16}$$

where σ is the lateral surface energy; σ_e is the fold surface free energy; b_0 is the layer thickness in the direction normal to the growth plane, $\Delta h_f = \Delta H_f \times \rho_c$ is the heat of fusion per unit volume, and k_B is the Boltzmann constant. In this study, the heat of fusion per unit volume (Δh_f) was determined by using the crystal density ($\rho_c = 1.407$ g/cm^3) and the heat of fusion per unit mass ($\Delta H_f = 190$ J/g) [123]. In general, the values of n are equal to 4 for regime II growth and 2 for regime I or III growth, respectively [119]. The values of σ_e for the PEN and the PEN/CNT nanocomposites can be estimated using the obtained K_g values according to the following empirical relations for the lateral surface free energy [124]:

$$\sigma = \alpha(\Delta h_f)(a_0 b_0)^{1/2} \tag{17}$$

where α was derived empirically to be 0.11 by analogy with the well-known behavior of hydrocarbons [125], and a_0 and b_0 are the monomolecular width and layer thickness, respectively. In this study, the values of $a_0 = 0.651$ nm and $b_0 = 0.566$ nm were used in all calculations [123]. The values of K_g and σ_e for the PEN/CNT nanocomposites estimated from the equations (15)-(17) are shown in Table 11. The values of K_g and σ_e for the PEN/CNT nanocomposites decreased with increasing CNT content. The work of chain folding (q) is one of the parameters related to the molecular structure, and its value is apparently proportional to chain stiffness [119, 126, 127]. The work of chain folding per molecular fold that is required for bending the polymer chain in the appropriate configuration, can be defined as follows [128]:

$$q = 2a_0 b_0 \sigma_e \tag{18}$$

From the equation (18), it can be deduced that the value of q is apparently proportional to the σ_e value. The q values of PEN/CNT nanocomposites with the CNT content are shown in Table 11. The q value of pure PEN was estimated to be 6.10×10^{-20} J per molecular chain fold, while that of the PEN/CNT 2.0 nanocomposite was estimated to be 3.79×10^{-20} J per molecular chain fold. The values of σ_e and q of the PEN/CNT nanocomposites decreased with increasing CNT content, implying that the incorporation of the CNT has a significant effect on the work of chain folding, resulting in the lowering the works required in folding the macromolecular chains in the PEN/CNT nanocomposites. Karayannidis and coworkers [129] reported that the presence of silica nanoparticles decreased the work required for creating a new surface, resulting in higher crystallization rate. Therefore, it can be deduced that the incorporation of the CNT as nanoreinforcing fillers into the PEN matrix increases the nucleation and crystallization of the PEN/CNT nanocomposites, because the CNT can effectively as heterogeneous nuclei in the nucleation of crystallization with higher crystallization rate.

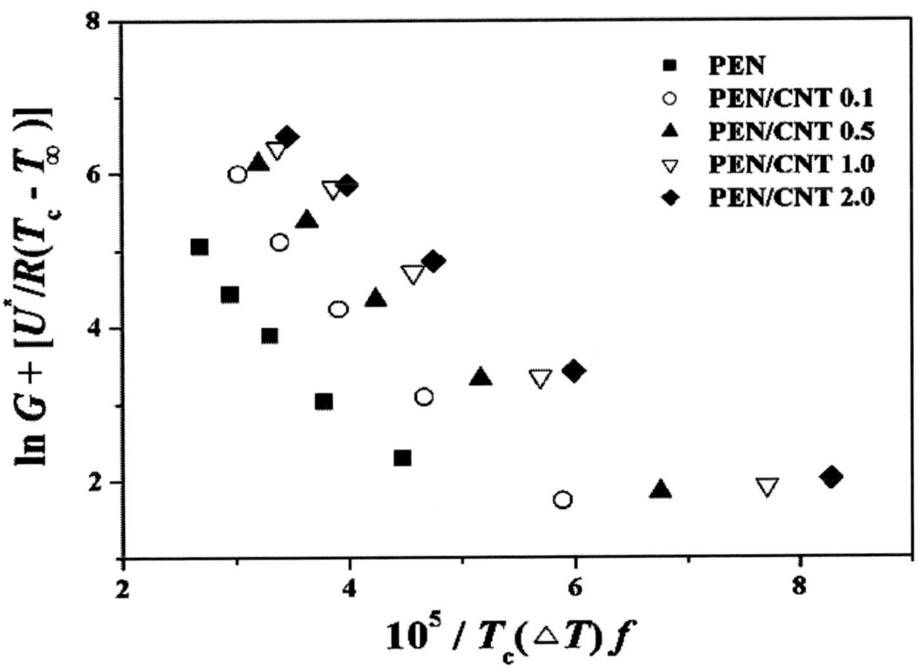

Figure 31. The plots of ln $G + [U^*/R(T_c\text{-}T_\infty)]$ versus $1/T_c(\Delta T)f$ for the PEN/CNT nanocomposites with various CNT contents.

WAXD patterns of the PEN/CNT nanocomposites crystallized at 240 °C with the CNT content are shown in Figure 32. It is known that PEN crystallizes two different crystal modifications such as α and β forms, depending on the crystallization conditions [123, 130, 131]. The unit cell of the α form, determined by Mencik [130], was triclinic with a = 0.651 nm, b = 0.575 nm, c = 1.32 nm, α = 81.33°, β = 144°, and γ = 100°, and a single chain passes through each unit cell of the α form. The β form was suggested by Zachmann et al. [131] to be triclinic unit cell with a = 0.926 nm, b = 1.559 nm, c = 1.273 nm, α = 121.6°, β = 95.57°, and γ = 122.52°, and four chains pass through each unit cell of the β form. For the PEN, characteristic peaks corresponding to the α form were observed at 2θ of 15.5, 23.2, and 26.8°, respectively, indicating the (010), (100), and (-110) reflections [149]. This result suggested that the α form population was dominated crystalline phase for the PEN. However, the PEN/CNT nanocomposites exhibited additional peaks at 2θ of 16.2, 18.4, 24.2, and 25.7°, respectively, as well as the reflections of the α form. These characteristic peaks indicated the (-1-11), (020), (-202), and (2-42) reflections corresponding to the β form [149]. As shown in Figure 32, the diffraction intensities of the characteristic peaks corresponding to the β-form population were increased with increasing CNT content, suggesting that the growth of the β-form crystals in the PEN/CNT nanocomposites was preferentially favored with the introduction of the CNT, and this effect being more significant at higher CNT content. Therefore, it can be deduced that the incorporation of more CNT facilitates the formation of the β-form in the PEN/CNT nanocomposites.

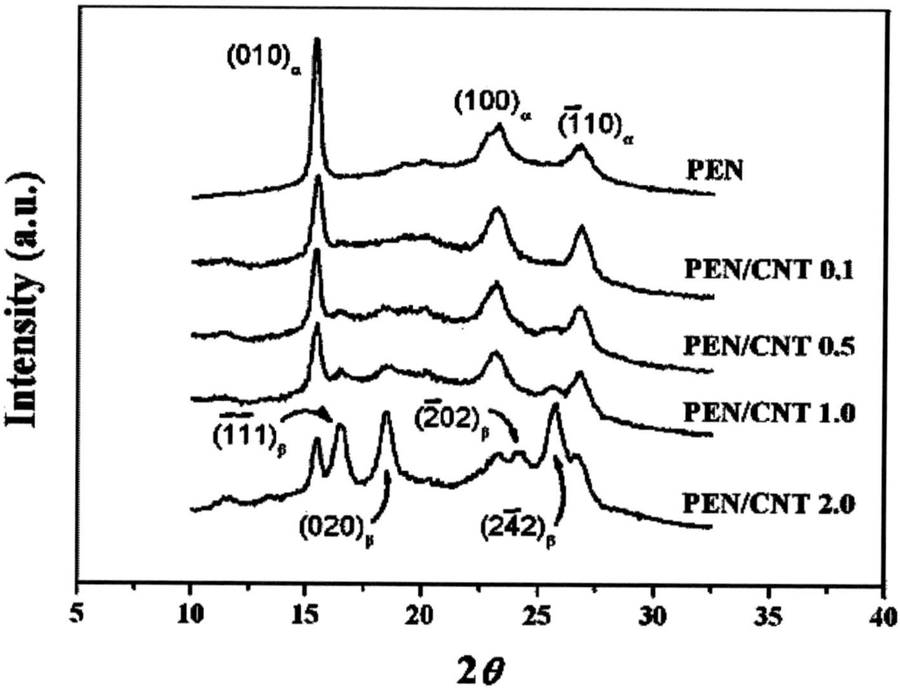

Figure 32. WAXD patterns of the PEN/CNT nanocomposites crystallized at 240 °C.

5.2. Mechanical Properties and Theoretical Approach

The tensile strength, the tensile modulus, and the elongation at break of the PEN/CNT nanocomposites with the CNT content are shown in Figure 33. There is significant dependence of the mechanical properties of the PEN/CNT nanocomposites on the CNT content. The incorporation of CNT improved the mechanical properties of the PEN/CNT nanocomposites because of the reinforcement effect of the CNT with high aspect ratio and their uniform dispersion in the PEN matrix, and this improvement being more significant at lower CNT content.

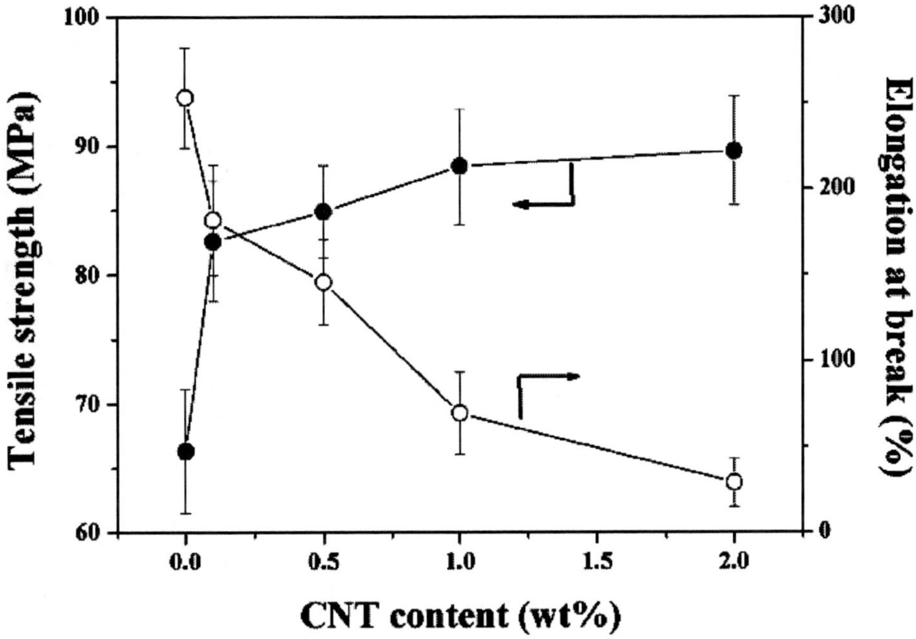

(a)

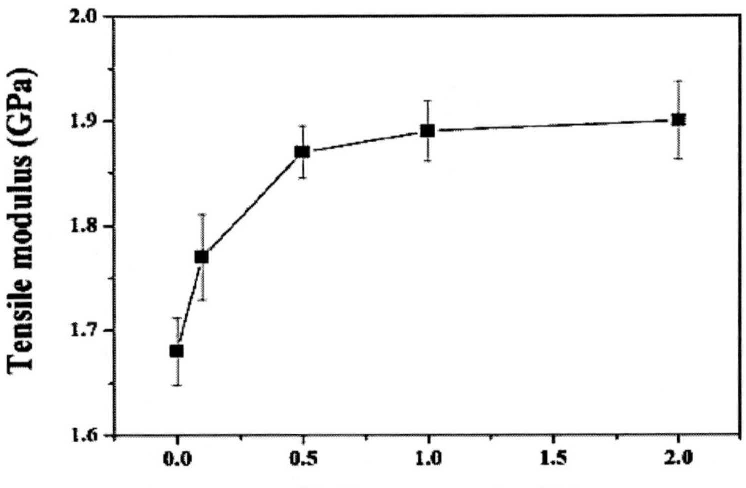

(b)

Figure 33. Variations of the mechanical properties of the PEN/CNT nanocomposites as a function of CNT content.

The morphologies of pristine CNT and the PEN/CNT nanocomposites are shown in Figure 34. The CNT have the diameter of 10-30 nm and the length of several micrometers, indicating that their aspect ratio reaches > 1000. The highly curved and random coiled features of the CNT were attributed to hydrogen bonding and van der Waals attractive interactions between the individual nanotubes [57].

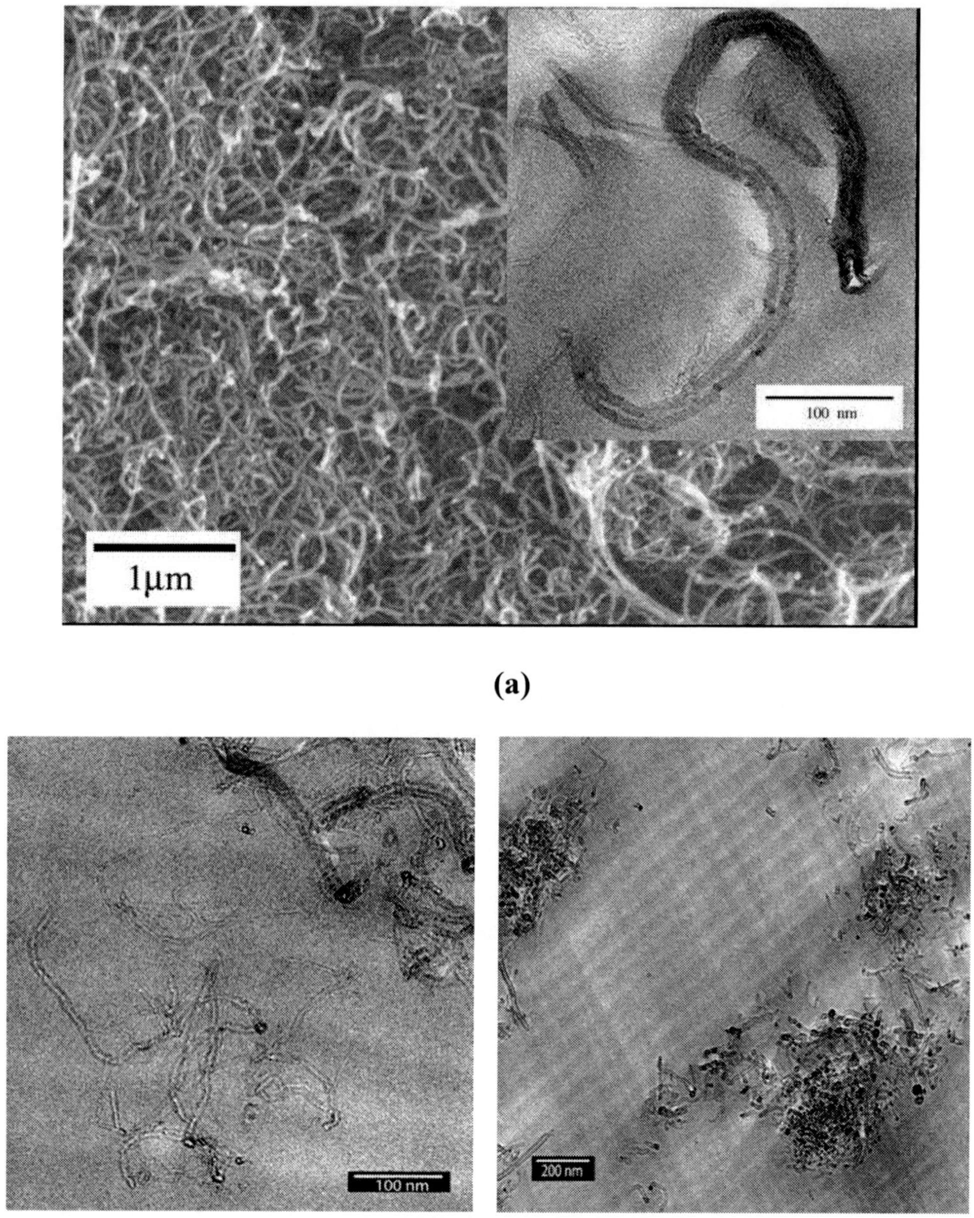

(b) (c)

Figure 34. Morphologies of (a) pristine CNT and the PEN/CNT nanocomposites containing (b) 0.1 wt% and (c) 2.0 wt% of CNT content.

As shown in Figure 34(b), the incorporated CNT was, on a large scale, uniformly dispersed in the PEN matrix, despite some entangled CNT or bundles. However, at higher CNT content, the mechanical properties of the PEN/CNT nanocomposites was not significantly increased, as expected, when compared with that at low CNT content. In general, the CNT often tend to bundle together and to form some agglomeration because of the intrinsic van der Waals attractions between the individual nanotubes, leading to the obstruction of efficient load transfer from the reinforcing phase to the matrix phase [17-19, 57]. Thus, for the PEN/CNT nanocomposites containing higher CNT content, less uniformly dispersed and highly entangled CNT in the PEN matrix may prevent efficient load transfer to the polymer matrix, which was confirmed by the morphological observation of the PEN/CNT 2.0 nanocomposites as shown in Figure 34(c). In addition, the elongation at break of the PEN/CNT nanocomposites decreased gradually with increasing CNT content, indicating that the PEN/CNT nanocomposites became somewhat brittle as compared to pure PEN because of the increased stiffness of the PEN/CNT nanocomposites and the micro-voids formed around the nanotubes during the tensile testing

For the characterizing the effect of the CNT on the composite modulus of the PEN/CNT nanocomposites, it is also instructive to compare the experimental results with the values predicted form the theoretical models. By assuming the PEN/CNT nanocomposites as randomly oriented discontinuous fiber lamina, the composite modulus (E_C) can be determined from the following equation [150, 151]:

$$E_C = \left[\left(\frac{3}{8}\right)\left(\frac{1+2(l_{CNT}/d_{CNT})\eta_L V_{CNT}}{1-\eta_L V_{CNT}}\right)+\left(\frac{5}{8}\right)\left(\frac{1+2\eta_T V_{CNT}}{1-\eta_T V_{CNT}}\right)\right]E_{PEN} \quad (19)$$

$$\eta_L = \frac{(E_{CNT}/E_{PEN})-1}{(E_{CNT}/E_{PEN})+2(l_{CNT}/d_{CNT})}$$

$$\eta_T = \frac{(E_{CNT}/E_{PEN})-1}{(E_{CNT}/E_{PEN})+2}$$

where E_{PEN} and E_{CNT} represent the modulus of PEN and CNT, respectively; l_{CNT}/d_{CNT} is the ratio of length to diameter for the nanotubes, and V_{CNT} is the volume fraction of the CNT in the polymer nanocomposites. To fit the equation (19) to the experimental results for the PEN/CNT nanocomposites, the weight fraction was transformed to the volume fraction, taking the densities of PEN (1.407 g/cm^3) [149] and perfectly graphitized CNT (2.16 g/cm^3) [22, 44]. The theoretical values of the composite modulus can be estimated assuming the aspect ratio of ~1000 and E_{CNT} of ~450 GPa for the CNT. The E_{CNT} values used in this study represents a mid-range value in the modulus ranges of the CNT previously measured [22, 44]. The theoretically predicted values and the experimental data for the composite modulus of the PEN/CNT nanocomposites are compared in Figure 35. At low CNT content (0.1 wt%), the E_C value of the PEN/CNT nanocomposites was fitted well with the theoretically predicted value. However, the PEN/CNT nanocomposites exhibited much lower E_C value at higher CNT content when compared with theoretically predicted E_C value, and the deviation was larger with increasing CNT content. As the CNT content increases, CNT often tends to bundles together because of intrinsic van der Waals attraction between the individual nanotubes in combination with high aspect ratio and large surface area of the CNT, leading to some agglomeration, and thus prevent an efficient load transfer to the polymer matrix [17, 19].

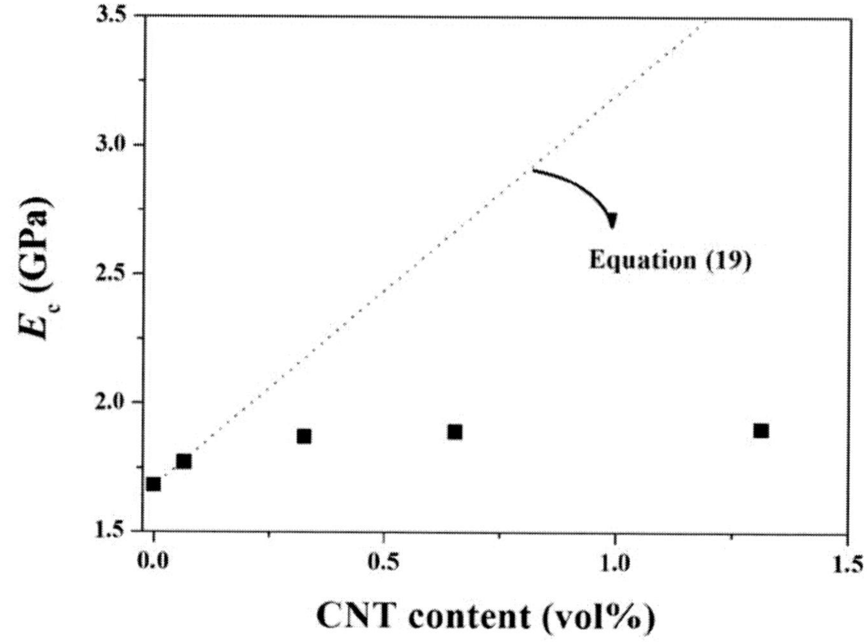

Figure 35. Variations of the mechanical properties of the PEN/CNT nanocomposites as a function of CNT content.

As a consequence, the mechanical properties of the PEN/CNT nanocomposites was not significantly increased at higher CNT content, as expected, when compared with that at lower content. Sokolov and coworkers [150] reported that for PP/SWNT composite fibers, the experimental values of the composite modulus at 1 wt% of SWNT concentration were in good agreement with the theoretical calculation, while the negative deviation was greater with increasing SWNT concentration up to 5 wt%, which was attributed to the poor distribution of SWNT that became even more heterogeneous in the PP matrix with increasing SWNT concentration. Thus, to achieve further enhanced mechanical properties of polymer nanocomposites, the improvement in the dispersion of the CNT in the polymer matrix and the enhancement of the interfacial adhesion between the CNT and the polymer matrix through functionalization of the CNT should be required.

6. Thermal Stability and Degradation Behavior of PEN/CNT Nanocomposites

6.1. Dynamic Mechanical Thermal Properties

The dynamic mechanical thermal properties of the PEN/CNT nanocomposites as a function of temperature are shown in Figure 36. There is significant dependence of the storage modulus (E') and the loss tangent (tan δ) of the PEN/CNT nanocomposites on the presence of the CNT. As shown in Figure 36, the *E′* value decreased with increasing temperature, and simultaneously the tan δ went through the maximum. The apparent glass transition region was revealed by a rapid decrease in the storage modulus of the PEN/CNT nanocomposites, and this temperature corresponding to the glass transition of the PEN matrix. The incorporation of the CNT significantly increased the storage modulus of the PEN/CNT nanocomposites. For instance, the *E′* value at 35 °C was increased from 1.45 GPa for pure PEN to 2.34 GPa for the PEN/CNT 2.0 nanocomposites. This increase in the storage modulus of the PEN/CNT nanocomposites may be attributed to some physical interactions between the PEN matrix and the CNT with high aspect ratio and large surface area and the stiffening effect of the CNT in the PEN matrix, providing the possibility to allow an efficient load transfer in the PEN/CNT nanocomposites. After a critical temperature, the storage modulus of the PEN/CNT nanocomposites was not significantly affected by the CNT, and they exhibited a strong dependence of the storage modulus on the PEN matrix [30], which was reflected by a plateau region above approximately 135 °C. In addition, the peak position of tan δ for the PEN/CNT nanocomposites was not affected by the presence of the CNT, while the peak height decreased.

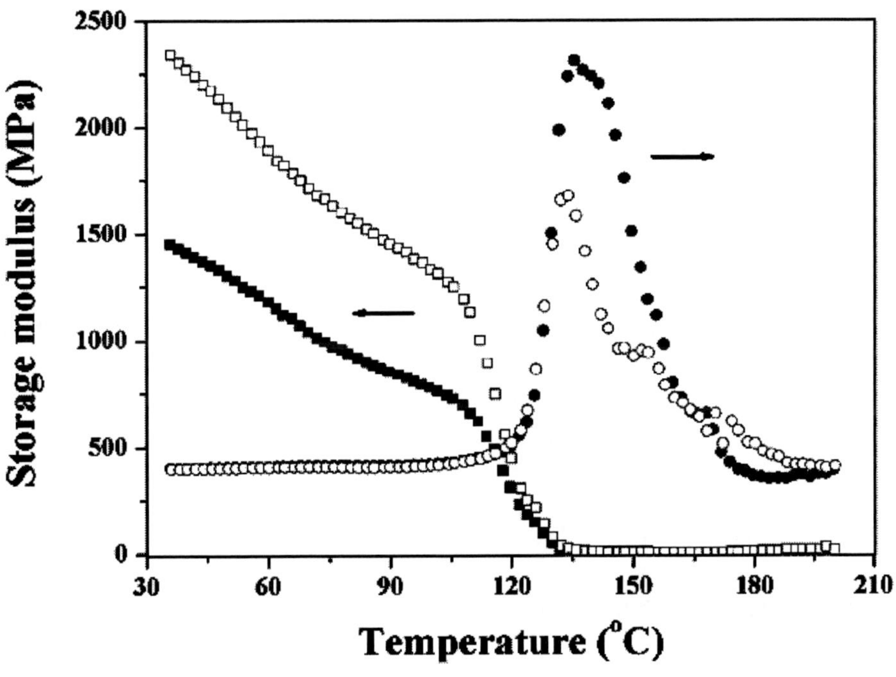

Figure 36. Dynamic mechanical thermal properties of PEN (filled symbols) and the PEN/CNT nanocomposites (unfilled symbols).

6.2. Thermal Stability

Thermal stability of polymers or polymer composites determines the upper-limit of the working temperature and the environmental conditions for use, which are related to their thermal degradation temperature and degradation rate [132, 133]. The TGA thermograms of pure PEN and the PEN/CNT 2.0 nanocomposites at 20 °C·min^{-1} under nitrogen and air are shown in Figure 37.

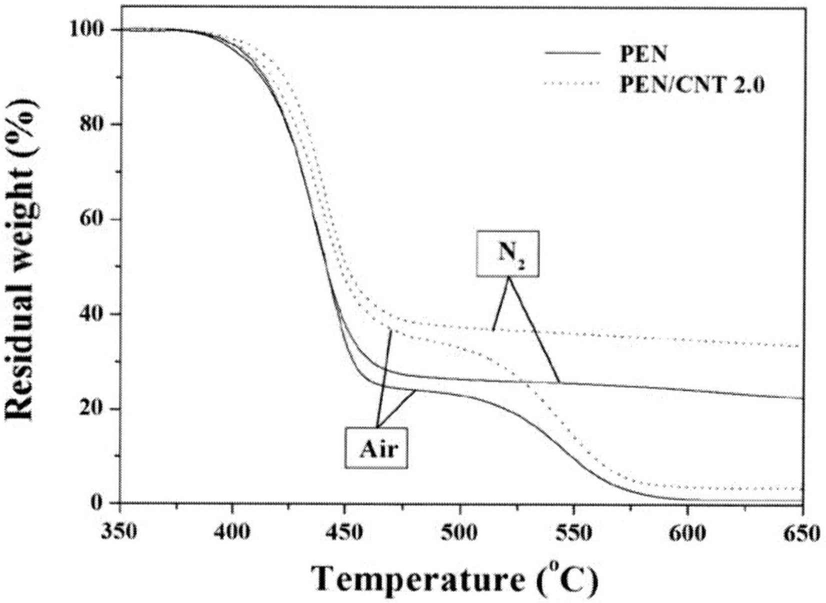

Figure 37. TGA curves of PEN and the PEN/CNT 2.0 nanocomposites at a heating rate of 20 °C·min^{-1} under nitrogen and air.

The TGA curves of the thermal degradation for the PEN/CNT nanocomposites exhibited only one weight-loss step during thermal degradation under nitrogen, which was attributed to the random scission of the PEN macromolecular chains [151]. The patterns of TGA curves for the PEN/CNT nanocomposites are similar to that of the PEN matrix, either under nitrogen or air, indicating that the features of the weigh-loss for the thermal degradation of the PEN/CNT nanocomposites may mostly stem from the PEN matrix

Under air atmosphere, the PEN/CNT nanocomposites exhibited two weight-loss stages during thermal degradation. The first weight-loss step during thermal degradation was attributed to the thermal decomposition of the PEN main chains from higher molecular weight macromolecules to small fragments due to the random scission of polymer chains [151], while the second weight-loss step resulted from the thermo-oxidative decomposition of the decomposed smaller chain fragments into volatile decomposed products under air [152-154]. The thermal degradation temperatures of the PEN/CNT nanocomposites were higher than those of pure PEN, indicating that the incorporation of CNT increased the thermal stability of the polymer nanocomposites. TGA thermograms of the PEN/CNT nanocomposites with heating rates under nitrogen and air are shown in Figure 38. TGA curves of the PEN/CNT nanocomposites shifted towards to higher temperature regions with increasing heating rates. This result may be explained by the fact that polymer molecules did not have enough time to exhaust the heat with increasing heating rate, leading to slower degradation rate and higher degradation temperature due to slow diffusion of heat.

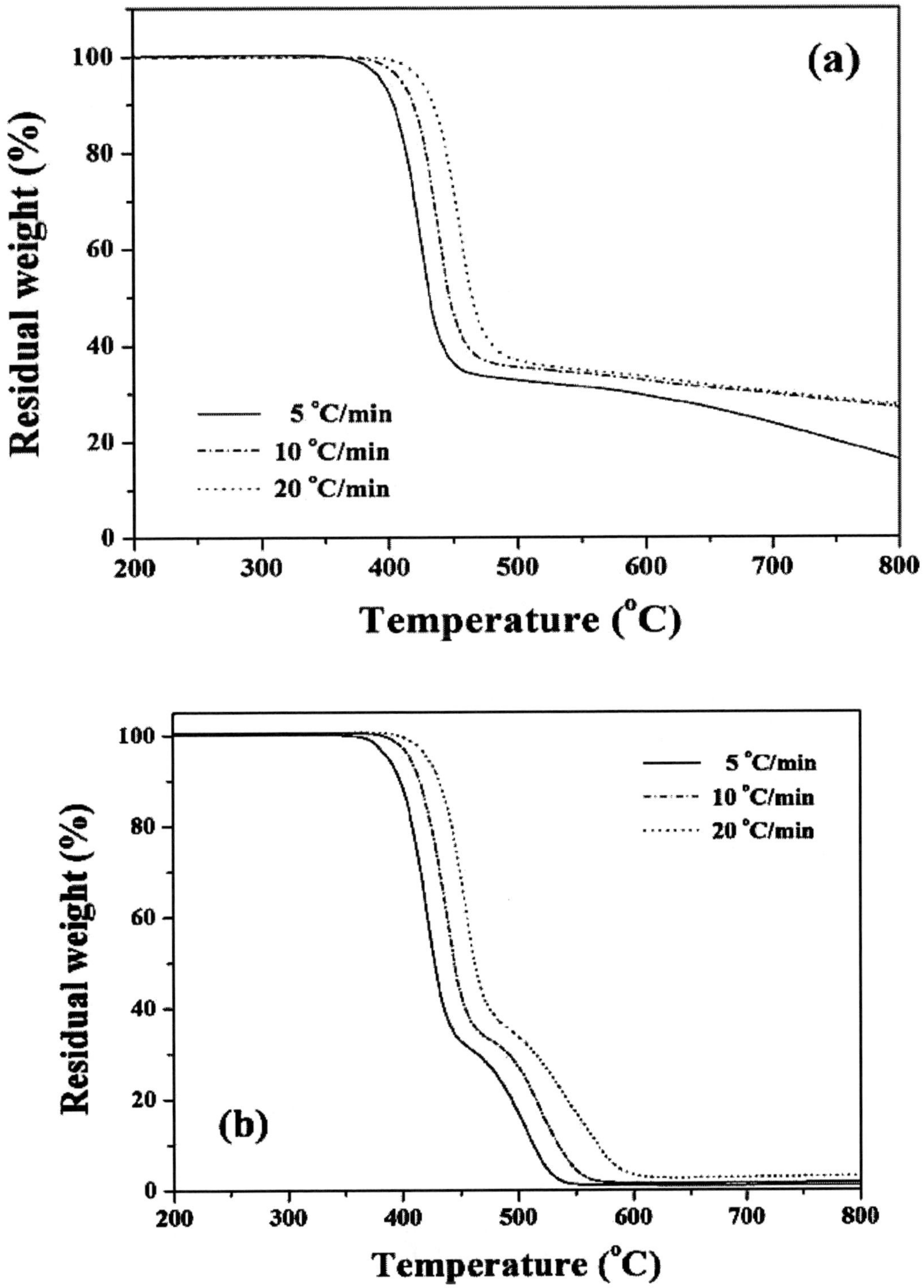

Figure 38. TGA curves of the PEN/CNT 2.0 nanocomposites with the heating rate under (a) nitrogen and (b) air.

TGA thermograms of the PEN/CNT nanocomposites with the CNT content at 10 °C·min^{-1} under nitrogen and the variations of the thermal degradation temperatures of the PEN/CNT nanocomposites with the CNT content under air are shown in Figure 39(a) and (b), respectively. The PEN/CNT nanocomposites exhibited the slower rate of the weight-loss and higher thermal degradation temperatures than that of the PEN matrix, and they tended to increase with increasing CNT content.

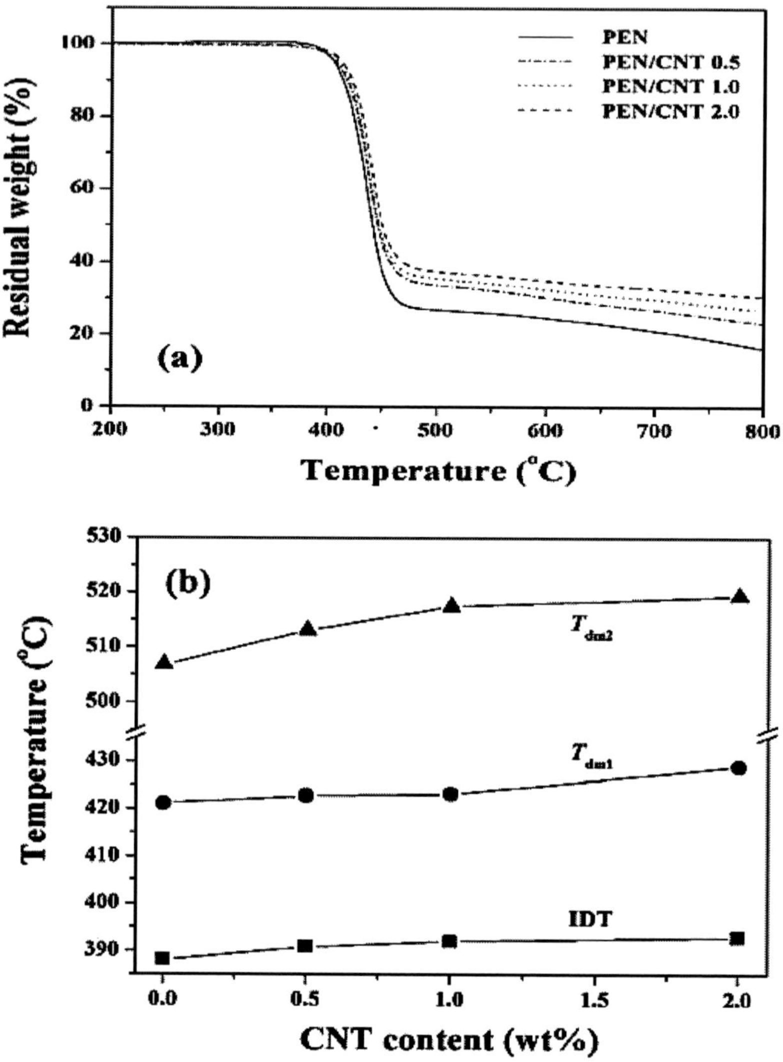

Figure 39. (a) TGA curves of PEN/CNT nanocomposites with the CNT content under nitrogen and (b) the variations of thermal degradation temperatures of PEN/CNT nanocomposites with the CNT content under air.

The increase in the thermal degradation temperatures of the PEN/CNT nanocomposites with the incorporation of the CNT may be explained by the decrease in the evolution rate of volatile degradation products in the polymer nanocomposites. The thermal degradation temperatures and degradation kinetic parameter are in common use to estimate the thermal stability of a polymer. In this study, for determining the thermal stability factor of polymer nanocomposites based on the Doyle's proposition [134], the integral procedure decomposition temperature (*IPDT*) can be calculated from the TGA thermograms using the following equation [135].

$$IPDT = A \cdot K(T_f - T_i) + T_i \quad (20)$$

where A is the area ratio of total experimental curve divided by total TGA thermogram; K is the coefficient of A; T_i is the initial experimental temperature (30 °C), and T_f the final experimental temperature (800 °C). As shown in Table 12, the thermal stability factors, including *IDT*, T_{dm}, A, K, and *IPDT*, of the PEN/CNT nanocomposites increased with increasing CNT content. In particular, the values of *IPDT* significantly increased from 686.5 °C for pure PEN to 982.9 °C for the PEN/CNT 2.0 nanocomposites. During the thermal degradation process, high thermal stability of CNT resulted in the retardation of the thermal volatilization of the PEN matrix, suggesting that the incorporation of a small quantity of CNT significantly improves the thermal stability of the PEN/CNT nanocomposites and CNT is beneficial to act as the thermal degradation resistant nanoreinforcing fillers in the PEN matrix. In addition, the residual yields of the PEN/CNT nanocomposites increased with increasing CNT content

Table 12. Effect of CNT on the thermal stability of the PEN/CNT nanocomposites

Materials	*IDT* [a] (°C)	T_{dm} [b] (°C)	A	K	*IPDT* [c] (°C)	W_R [d] (%)
PEN	389.2	436.8	0.6360	1.3404	686.5	14.8
PEN/CNT 0.5	390.6	438.7	0.6632	1.5312	812.1	22.7
PEN/CNT 1.0	397.7	439.8	0.6774	1.6605	896.2	26.9
PEN/CNT 2.0	398.2	440.9	0.6903	1.7925	982.9	30.3

[a] The initial degradation temperature.
[b] The degradation temperature of maximum rate of the weight-loss.
[c] The integral procedure degradation temperature.
[d] The residual yield at 800 °C in TGA thermograms.

In this PEN/CNT nanocomposite system, the dispersed CNT layers acting as physical barriers to thermal degradation in the polymer nanocomposites, may prevent volatile decomposed products or small molecules produced during thermal degradation from diffusing out of the PEN/CNT nanocomposites, leading to the retardation of their thermal degradation by the higher protecting and slow-down effect of CNT layers against the thermal degradation in the polymer nanocomposites [35, 136]. Kashiwagi et al. [107, 108] reported that CNT layers good barrier effect on the gases such as oxygen and nitrogen, and they can not only insulate polymers but also reduce the weight-loss rate of degradation products, resulting in the improvement in the thermal stability of the polymer nanocomposites. Thus, the retardation of thermal degradation of the PEN/CNT nanocomposites with CNT content may be attributed to the physical barrier effect promoted by the dispersed CNT layers acting as an effective thermal insulator in the PEN/CNT nanocomposites. As s result, the thermal stability of the PEN/CNT nanocomposites was significantly improved with the incorporation of a small quantity of the CNT. This improvement effect of the PEN/CNT nanocomposites with the incorporation of the CNT will be also elaborated during the following discussion on the activation energy for the thermal degradation of the PEN/CNT nanocomposites.

6.3. Thermal Degradation Kinetics

In the thermal degradation process of a polymer, the degree of thermal degradation or conversion (α) can be typically calculated as follows [137]:

$$\alpha = \frac{W_0 - W_t}{W_0 - W_f} \tag{21}$$

where W_t, W_0, and W_f are the actual, initial, and final weights of the samples, respectively. Assuming that the rate of conversion is a near function of temperature-dependent rate constant, $K(T)$ and a emperature-independent/ eight loss-dependent function, $F(\alpha)$, the rate of degradation can be expressed as follows:

$$\frac{d\alpha}{dt} = K(T)F(\alpha) \tag{22}$$

where t and T are the reaction time and temperature, respectively, and $F(\alpha) = (1 - \alpha)^n$ depends on the mechanism of thermal degradation reaction. The function $K(T)$ can be usually described by Arrehenius equation as follows:

$$K(T) = A\exp\left(\frac{-E_a}{RT}\right) \tag{23}$$

where A is the pre-exponential factor; E_a is the activation energy, and R is the universal gas constant. At constant heating rate, β (=$d\alpha/dT$), the basic equation for the thermal degradation kinetics can be expressed in terms of combination of the equations (22) and (23) as follows:

$$\frac{d\alpha}{dT} = \left(\frac{A}{\beta}\right)\exp\left(\frac{-E_a}{RT}\right)F(\alpha) \tag{24}$$

Using this equation, theoretical approaches based on differential mode and integral mode, were applied for calculation of thermal degradation kinetic parameters [138-141].

The Flynn-Wall-Ozawa method [142, 143] has been used to determine the activation energy from dynamic tests by plotting the logarithm of heating rate as a function of the inverse of the temperature, at different conversions. Being integrated with the initial condition of $\alpha = 0$ at $T = T_0$, the equation (24) can be arranged as follows:

$$F(\alpha) = \left(\frac{A}{\beta}\right)\int_0^T \exp\left(\frac{-E_a}{RT}\right)dT \tag{25}$$

Then, using Doyle's approximation [144], the equation (25) can be simplified as follows:

$$\log\beta = \log\left(\frac{AE_a}{F(\alpha)R}\right) - 2.315 - 0.4567\left(\frac{E_a}{RT}\right) \tag{26}$$

where β is the heating rate; α is the weight loss; A is the pre-exponential factor; R is the universal gas constant; T is the absolute temperature, and E_a is the activation energy of the thermal degradation reaction. The values of the activation energy (E_a) for a specific weight loss can be calculated from the slope of the plot of log β versus $1/T$ at different heating rates.

The Kissinger's method [138] has been used to estimate the activation energy of the thermal degradation, involving the maximum temperature (T_{dm}) of the first derivative weight loss curves in TGA measurements at a constant heating rate, and the resulting equation can be expressed as follows:

$$\ln\left(\frac{\beta}{T_{dm}^{2}}\right) = \left[\ln\left(\frac{ZR}{E_a}\right) - \ln F(\alpha)\right] - \left(\frac{E_a}{RT_{dm}}\right) \tag{27}$$

where β is the heating rate; T_{dm} is the absolute temperature at the maximum rate of the thermal degradation; Z is the frequency factor; R is the universal gas constant, and E_a is the activation energy of the thermal degradation reaction. The value of the activation energy (E_a) can be calculated from the slope of the plot of $\ln(\beta/T_{dm}^2)$ versus $1/T_{dm}$.

Based on the data obtained from the TGA thermograms at different heating rates, the plot of log β versus $1/T$ of the PEN/CNT nanocomposites for a given conversion under nitrogen and air are shown in Figures 40 and 41, respectively. The Flynn-Wall-Ozawa plot of the PEN/CNT nanocomposites exhibited a good linear relationship at a constant conversion, indicating that the Flynn-Wall-Ozawa analysis was effective in describing the thermal degradation kinetics of the PEN/CNT nanocomposites in terms of the experimental TGA data.

In addition, the thermal degradation of the PEN/CNT nanocomposites may vary with the degree of conversion for thermal degradation under nitrogen and air

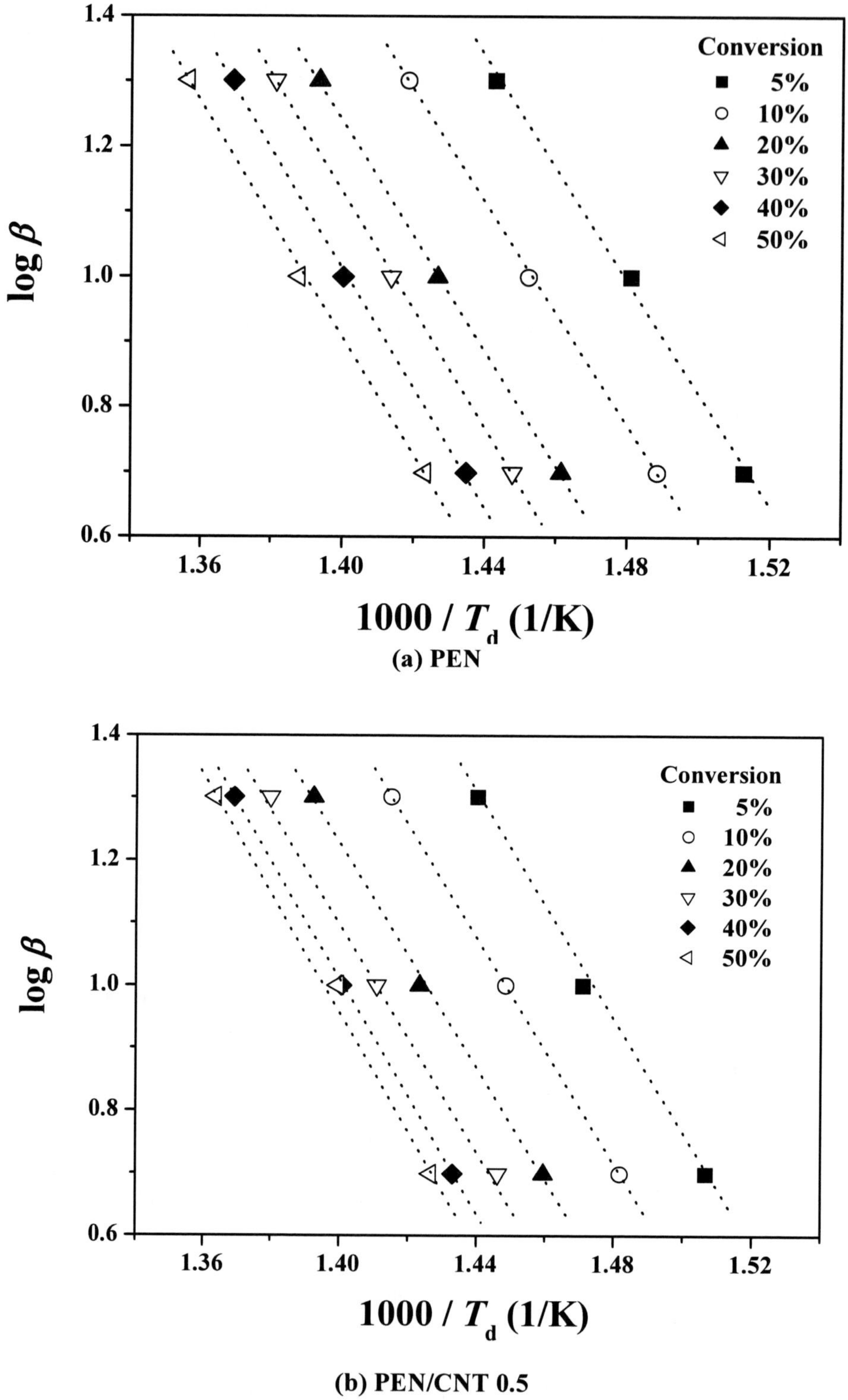

Figure 40. Flynn-Wall-Ozawa plots of (a) PEN, (b) PEN/CNT 0.5, (c) PEN/CNT 1.0, and (d) PEN/CNT 2.0 nanocomposites under nitrogen.

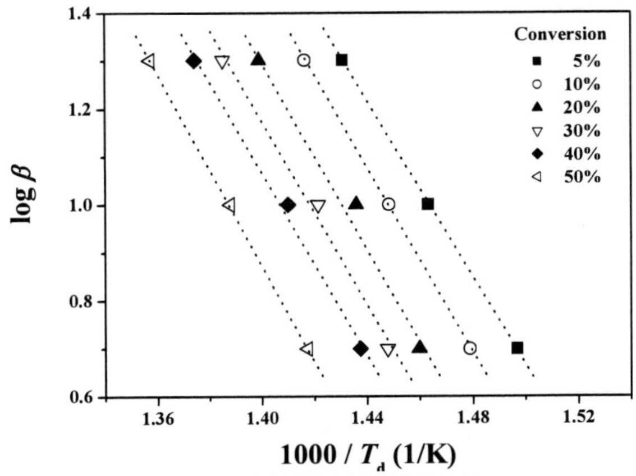

(c) PEN/CNT 1.0

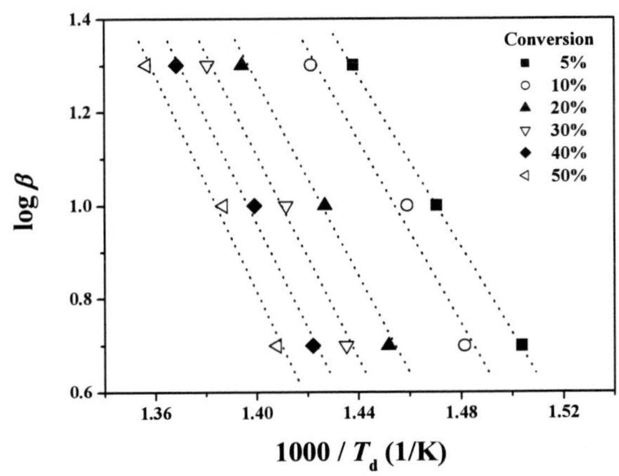

(d) PEN/CNT 2.0

Figure 40. (Continued).

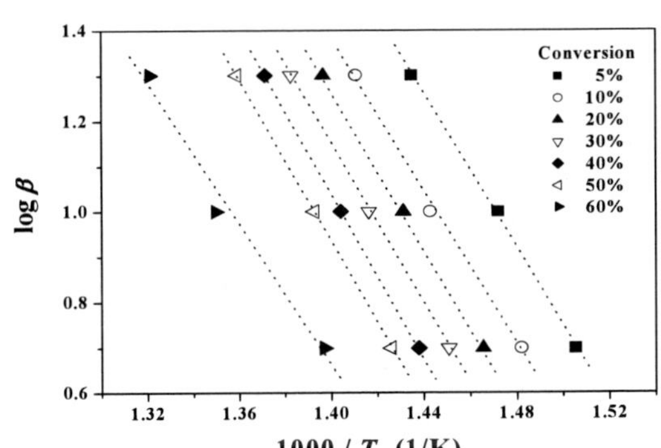

(a) PEN

Figure 41. Flynn-Wall-Ozawa plots of (a) PEN, (b) PEN/CNT 0.5, (c) PEN/CNT 1.0, and (d) PEN/CNT 2.0 nanocomposites under air.

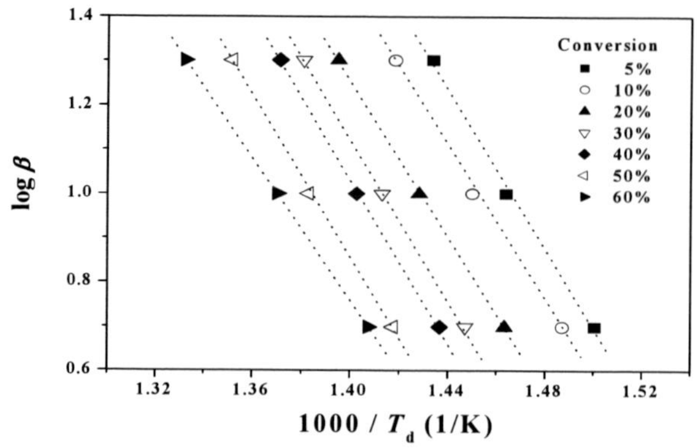

(b) PEN/CNT 0.5

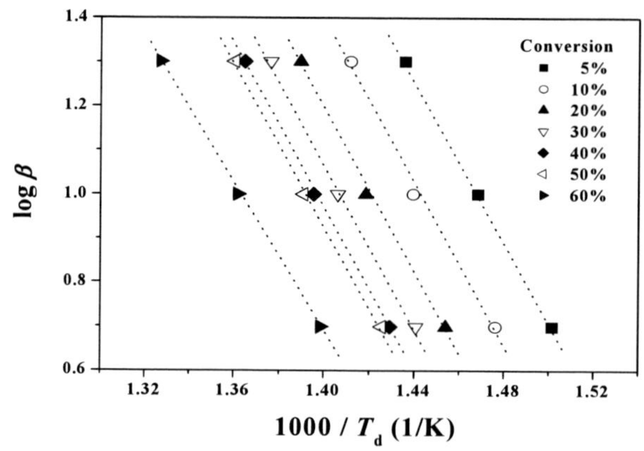

(c) PEN/CNT 1.0

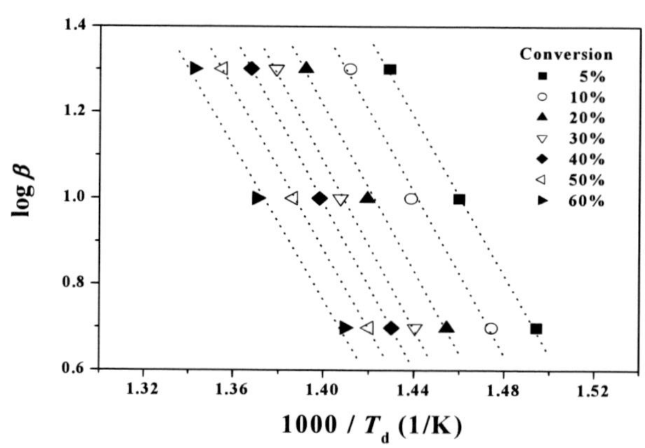

(d) PEN/CNT 2.0

Figure 41. (Continued)

The variation of the activation energy for thermal degradation (Ea) of the PEN/CNT nanocomposites under nitrogen and air estimated from the Flynn-Wall-Ozawa analysis are presented in Tables 13 and 14, respectively.

Table 13. Activation energy (E_a) and correlation coefficient (r^2) of the PEN/CNT nanocomposites with the CNT content under nitrogen from the Flynn-Wall Ozawa method

	PEN		PEN/CNT 0.5		PEN/CNT 1.0		PEN/CNT 2.0	
Conversion (%)	E_a (kJ/mol)	r^2	E_a (kJ/mol)	r^2	E_a (kJ/mol)	r^2	E_a (kJ/mol)	r^2
5	156.6	0.9985	164.1	0.9991	166.5	0.9999	167.0	0.9999
10	155.9	0.9998	163.6	0.9999	175.5	0.9999	179.2	0.9892
20	161.2	0.9999	162.7	0.9989	177.1	0.9925	189.7	0.9971
30	164.3	0.9998	164.6	0.9993	172.9	0.9960	199.4	0.9969
40	166.4	0.9995	170.8	0.9999	172.3	0.9972	203.6	0.9964
50	163.5	0.9995	172.0	0.9967	182.0	0.9999	211.5	0.9947
Average	161.3		166.3		174.4		191.7	

Table 14. Activation energy (E_a) and correlation coefficient (r) of the PEN/CNT nanocomposites with the CNT content under air from the Flynn-Wall Ozawa method

	PEN		PEN/CNT 0.5		PEN/CNT 1.0		PEN/CNT 2.0	
Conversion (%)	E_a (kJ/mol)	r^2	E_a (kJ/mol)	r^2	E_a (kJ/mol)	r^2	E_a (kJ/mol)	r^2
5	155.3	0.9995	163.0	0.9985	165.9	0.9999	166.7	0.9995
10	153.4	0.9983	158.3	0.9989	167.9	0.9968	172.7	0.9968
20	158.7	0.9999	160.1	0.9998	169.3	0.9983	173.8	0.9975
30	161.2	0.9999	165.0	0.9998	169.0	0.9991	176.9	0.9992
40	164.6	0.9999	166.7	0.9997	169.4	0.9996	175.6	0.9999
50	163.6	0.9999	166.2	0.9995	167.8	0.9993	168.3	0.9997
60	141.8	0.9902	145.6	0.9999	153.3	0.9999	161.9	0.9956
Average	156.9		160.7		166.1		170.8	

The Ea values of the PEN/CNT nanocomposites were higher than pure PEN, and they tended to increase with the CNT content. In addition, the Ea values of the PEN/CNT nanocomposites in air were lower than that in nitrogen, indicating that the thermal degradation mechanisms of the PEN/CNT nanocomposites in both gases were different, because the active effect of oxygen under air lead to the occurrence of the thermo-oxidative reactions and the increase in the degradation rate, thus accelerating the thermal degradation for the PEN/CNT nanocomposites. As the previously described, the improvement in the thermal stability of the

PEN/CNT nanocomposites induced by the incorporation of the CNT may be explained by the function of the CNT as an effective physical barrier to retard the thermal degradation of volatile components and to prevent the diffusion out of the decomposed polymeric molecules in the PEN/CNT nanocomposites [145, 146]. Based on the Flynn-Wall-Ozawa analysis, it can be deduced that the Ea values of the PEN/CNT nanocomposites calculated from Flynn-Wall-Ozawa method exhibited good reliance on describing the thermal degradation kinetics of their nanocomposites, which was confirmed by the fact that the values of the correlation coefficient (r2) were greater than 0.99.

The Kissinger plots of the PEN/CNT nanocomposites under nitrogen and air are shown in Figure 42. The slopes of the plots of ln (β/T_{dm}^2) or ln β versus of $1/T_{dm}$ for the PEN/CNT nanocomposites were increased with the addition of the CNT, indicating the higher E_a values of the PEN/CNT nanocomposites relative to pure PEN. As shown in Table 15, the E_a values of the PEN/CNT nanocomposites increased with increasing CNT content. For instance, the E_a values were increased from 158.6 kJ/mol for pure PEN to 194.1 kJ/mol for the PEN/CNT 2.0 nanocomposites. The thermal degradation of the PEN/CNT nanocomposites was retarded due to the physical barrier effect of CNT in the PEN matrix, leading to higher E_a values with increasing CNT content. The PEN/CNT nanocomposites exhibited higher E_a values under nitrogen than that under air. This result may be explained by the fact that the active effect of oxygen under air lead to the occurrence of the thermo-oxidative reactions and increase the degradation rate, thus accelerating the thermal degradation for the PEN/CNT nanocomposites. In addition, based on the Kissinger analysis, the values of correlation coefficient (r^2) were greater than 0.99, indicating good reliance on describing the thermal degradation of the PEN/CNT nanocomposites.

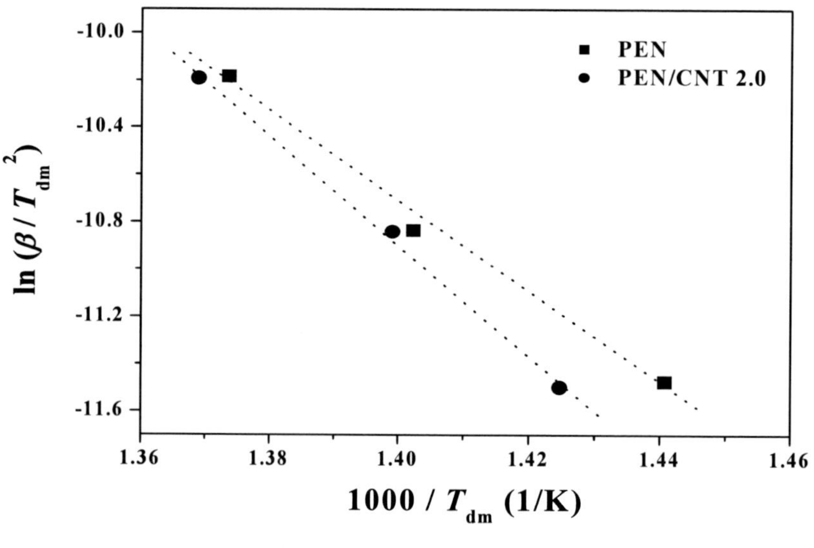

(a)

Figure 42. Kissinger plots of PEN and PEN/CNT 2.0 nanocomposites under (a) nitrogen and (b) air.

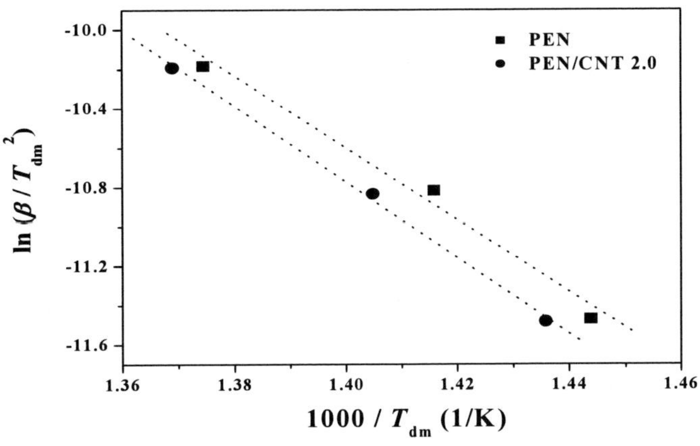

(b)

Figure 42. (Continued)

Table 15. Activation energy (E_a) and correlation coefficient (r) of the PEN/CNT nanocomposites with the CNT content under nitrogen and air from the Kissinger method

Materials	Nitrogen		Air	
	E_a (kJ/mol)	r^2	E_a (kJ/mol)	r^2
PEN	158.6	0.9960	152.2	0.9929
PEN/CNT 0.5	159.4	0.9984	160.2	0.9989
PEN/CNT 1.0	167.0	0.9973	160.5	0.9936
PEN/CNT 2.0	194.1	0.9987	163.1	0.9937

6.4. Interconnencted Network-Like Structures of MWCNT

The rheological properties of the PEN/CNT nanocomposites with the CNT content are shown in Figure 43. The complex viscosities ($|\eta^*|$) of pure PEN and the PEN/CNT nanocomposites were decreased with increasing frequency. However, at lower frequency region, the curve shape of the $|\eta^*|$ for pure PEN was almost flat, indicating the independence of the complex viscosity on the frequency. The incorporation of the CNT significantly increased the $|\eta^*|$ and the storage modulus (G') of the PEN/CNT nanocomposites, and this effect being more pronounced at lower frequency region. According to the relationship of $|\eta^*| \approx \omega^n$, reported by Wagner and Reisinger [90], the shear thinning exponent (n) of the PEN/CNT nanocomposites can be determined in the lower frequency region. In this polymer nanocomposite system, the n values were -0.13 for pure PEN and -0.34 for the PEN/CNT 2.0

nanocomposites, respectively. In addition, the G' of the PEN/CNT nanocomposites significantly increased by incorporating the CNT, particularly at lower frequency region. In general, these variations of the complex viscosity and the storage modulus in the dynamic oscillatory measurement imply the formation of the interconnected or network-like structures of the incorporated fillers, resulting from the percolation of those [104, 147]. At higher CNT content, the nanotubes-nanotubes interactions due to their high aspect ratio and large surface area were more dominated in the polymer nanocomposites, and lead to the formation of the interconnected or network-like structures of the nanotubes in the polymer matrix.

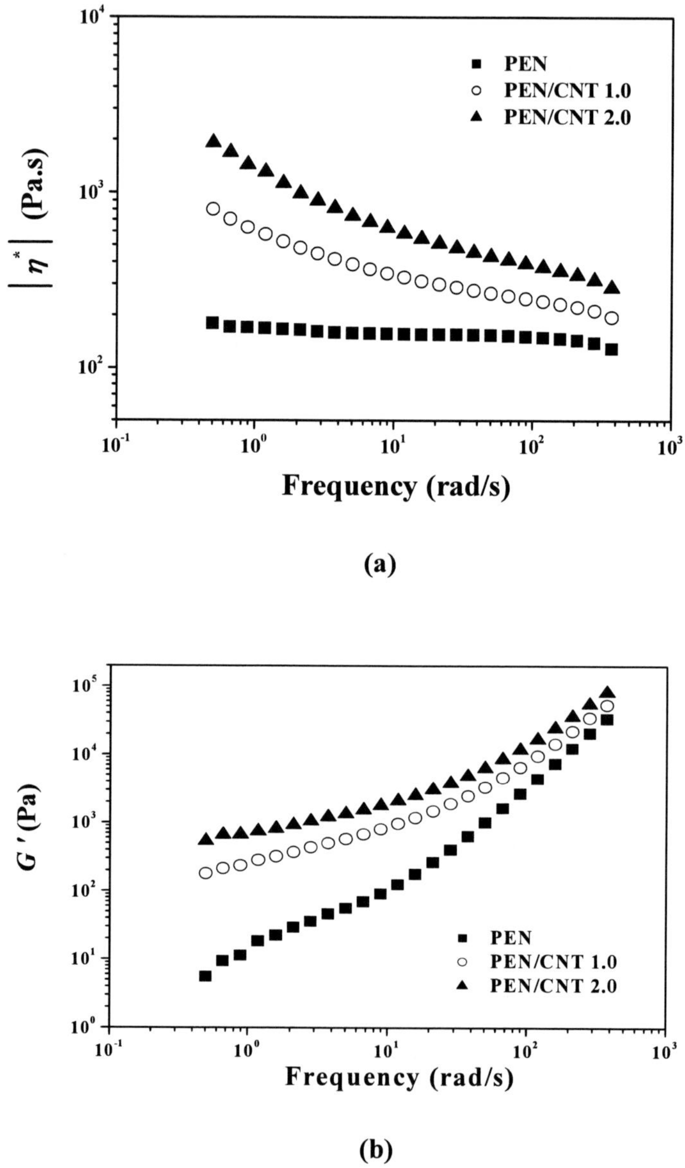

Figure 43. (a) Complex viscosity ($|\eta^*|$) and (b) storage modulus (G') of the PEN/CNT nanocomposites with the CNT content.

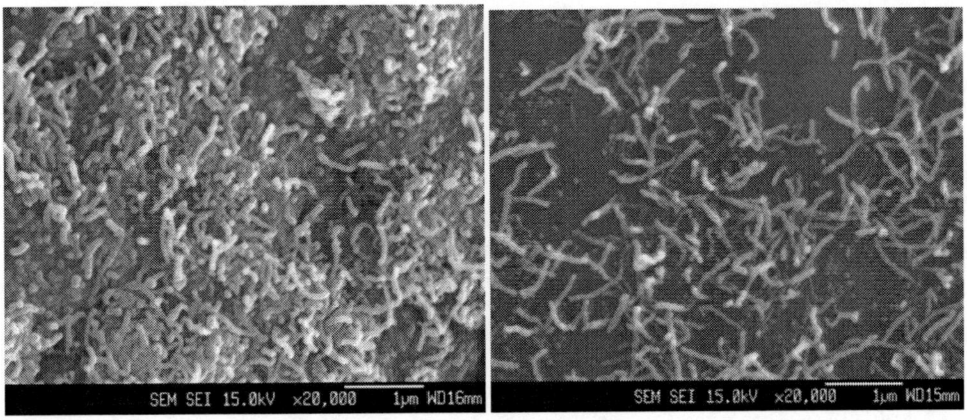

Figure 44. SEM images of (a) the PEN/CNT 2.0 nanocomposites and (b) the residue of the PEN/CNT nanocomposites after thermal degradation.

The morphologies of the PEN/CNT 2.0 nanocomposites and the residue of that after thermal degradation are shown in Figure 44. From the SEM image of the fractured PEN/CNT 2.0 nanocomposites, it can be seen that the CNT was randomly dispersed in the PEN matrix, and the interconnected or network-like structures of the CNT were observed in the PEN matrix. The CNT with small size, high aspect ratio, and large surface area are often subjected to self-agglomeration or bundle formation at higher content, and thus easily form the interconnected or network-like structures in the molten polymer matrix. As shown in Figure 44(b), the CNT kept the interconnected network-like structures in the PEN matrix after thermal degradation, despite some collapse or loss of their forms. It can be deduced that this morphological feature of the CNT in the polymer matrix may be also a possible reason for the improvement in the thermal stability of the PEN/CNT nanocomposites, because the interconnected network structures of the CNT in the PEN matrix resulted in the good physical barrier effect on the thermal degradation of the PEN/CNT nanocomposites by retarding the thermal degradation and by preventing the diffusion-out of volatile decomposed products in the nanocomposites. Schartel et al. [148] reported that the increased viscosity of the polyamide/multiwall nanotube nanocomposites and the fiber-network character of the incorporated nanofillers are the dominant mechanisms influencing fire retardancy, suggesting that the interconnected network structures of the nanotubes stabilized the melt in the pyrolysis zone and residue. In this PEN/CNT nanocomposite system, the incorporation of the CNT into the PEN matrix significantly improves the thermal stability of the polymer nanocomposites and reduced their weight-loss rates, resulting in the increase in the activation energy for thermal degradation. The enhancing effect of the CNT on the thermal degradation of the PEN/CNT nanocomposites may be also explained by the high thermal resistance of the CNT that increased the energy required for the thermal degradation of the polymer matrix and by the function of the interconnected network-like structures of the CNT as thermal insulating layers on the polymer matrix [148].

7. SUMMARY

We prepared aromatic polyester nanocomposites based on a very small quantity of CNT and PEN nanocomposites using a melt blending in a twin-screw extruder to create advanced materials for possible practical applications in various industrial fields. There is significant dependence of non-isothermal crystallization of the PEN/CNT nanocomposites on the CNT content and cooling rate. The incorporation of the CNT accelerates the mechanism of nucleation and crystal growth of PEN, this effect being more significant at lower CNT content. Combined Avrami and Ozawa analysis was found to be effective in describing the non-isothermal crystallization of the PEN/CNT nanocomposites. The CNT in the PEN/CNT nanocomposites exhibits much higher nucleation activity than any nanoreinforcing filler. The incorporation of the CNT improved the storage modulus (G') and loss modulus (G'') of the PEN/CNT nanocomposites, with this effect also being more pronounced at lower CNT content. A gradual decrease in the terminal zone slope of G' for the PEN/CNT nanocomposites with increasing CNT content may be explained by the fact that the nanotube-nanotube interactions will be dominant at higher CNT content, and lead to the formation of the interconnected or network-like structures of the CNT in the polymer nanocomposites. The decrease in the slope of the plot of log G' versus log G'' for the PEN/CNT nanocomposites with increasing CNT content suggests that the microstructures of PEN/CNT nanocomposites is affected by the presence of the CNT. There is significant dependence of the crystallization and melting behavior for the PEN/CNT nanocomposites on the CNT content and crystallization temperature, and the equilibrium temperatures decreased with increasing CNT content. The presence of more CNT may facilitate the formation of the β-form in the PEN/CNT nanocomposites. The values of the nucleation parameter (K_g) and fold surface free energy (σ_e) for the PEN/CNT nanocomposites decreased with increasing CNT content, indicating that the incorporation of the CNT lowers the work required in folding macromolecular chains in the PEN/CNT nanocomposites. In addition, the incorporation of a very small quantity of the CNT significantly improved the mechanical properties of the PEN/CNT nanocomposites, and this reinforcing effect being more significant at lower CNT content. There was significant dependence of the thermal degradation of the PEN/CNT nanocomposites on the presence and the concentration of the CNT. The incorporation of a very small quantity of the CNT into the PEN matrix greatly improved the thermal stability of the PEN/CNT nanocomposites. The variations of the activation energy for thermal degradation confirmed that thermal stability of the PEN/CNT nanocomposites was improved with the introduction of the CNT. The morphological observations revealed the formation of the interconnected or network-like structures of the CNT in the PEN matrix. In the PEN/CNT nanocomposite systems, the interconnected network-like structures of the CNT in the PEN matrix plays a significant role on the improvement in the thermal stability by acting as an efficient physical barrier to retard the thermal degradation of the PEN and to prevent the transport of volatile decomposed products in the PEN/CNT nanocomposites during thermal degradation.

REFERENCES

[1] Iijima, S. *Nature* 1991, vol 354, p 56.
[2] Bethune, D. S.; Kiang, C. H.; de Vries, M. S.; Gorman, G.; Savoy, R.; Vazquez, J.; Beyers, R. *Nature* 1993, vol 363, p 605.
[3] Lambert, J. M.; Ajayan, P. M.; Bernier, P.; Planeix, J. M.; Brotons, V.; Coq, B.; Castaing, *J. Chem. Phys. Lett.* 1994, vol 226, p364.
[4] Journet, C.; Maser, W. K.; Bernier, P.; Loiseau, A.; Chapelle, M. L.; Lefrant, S.; Deniard, P.; Lee, R.; Fisher, J. E. *Nature,* 1997, vol 388, p 756.
[5] Thess, A.; Lee, R.; Nikolaev, P.; Dai, H.; Petit, P.; Robert, J.; Xu, C.; Lee, Y. H.; Kim, S. G.; Rinzler, A. G.; Colbert, D. T.; Scuseria, G. E.; Tomanek, D.; Fisher, J. E.; Smalley, R. E. *Science,* 1996, vol 273, p 483.
[6] Guo, T.; Nikolaev, P.; Rinzler, A. G.; Tomanek, D.; Colbert, D. T.; Smalley, R. E. *J. Phys. Chem.*, 1995; vol 99, p 10694.
[7] Yudasaka, M.; Komatsu, T.; Ichihashi, T.; Iijima, S. *Chem. Phys. Lett.,* 1997, vol 278, p 102.
[8] Bronikowski, M. J.; Willis, P. A.; Colbert, D. T.; Smith, K. A.; Smalley, R. E. *J. Vac. Sci. Technol.* A 2001, vol 19, p 1800.
[9] Journet C.; Bernier, P. Appl. Phys. A: *Mater. Sci. Process.* 1998, vol 67, p 1.
[10] Liang, Q.; Gao, L. Z.; Li, Q.; Tang, S. H.; Liu, B. C.; Yu, Z. L. *Carbon* 2001, vol 39, p 897.
[11] Chen, Y.; Shaw, D. T.; Guo, L. *Appl. Phys. Lett. 2000*, vol 76, p 2469.
[12] Sen, R.; Govindaraj, A.; Rao, C. N. R. *Chem. Phys. Lett.* 1997, vol 267, p 276.
[13] Li, D. C.; Dai, L.; Huang, S.; Mau, A. W. H.; Wang, Z. L. *Chem. Phys. Lett.* 2000, vol 316, p 349.
[14] Chen, Q.; Dai, L. *Appl. Phys. Lett.* 2000, vol 76, p 2719.
[15] Hsu, W. K.; Hare, J. P.; Terrones, M.; Kroto, H. W.; Walton, D. R. M. *Nature* 1995, vol 377, p 687.
[16] Laplaze, D.; Bernier, P.; Maser, W. K.; Flamant, G.; Guillard, T.; Loiseau, A. *Carbon* 1998, vol 36, p 685.
[17] Ebbesen, T. W. *Ann. Rev. Mater. Sci.* 1994, vol 24, p 235.
[18] Iijima, S.; Ichihashi, T. *Nature* 1993, vol 363, p 603.
[19] Dresselhaus, M. S.; Dresselhaus, G.; Avouris, P. H. *Carbon Nanotubes: Synthesis, Structure, Properties, and Applications*, Springer, Berlin, 2001.
[20] Thostenson, E. T.; Ren, Z.; Chou, T. W. *Compos. Sci. Technol.* 2001, vol 61, p 1899.
[21] Goze, C.; Bernier, P.; Henrard, L.; Vaccarini, L.; Hernandez, E.; Rubio, A. *Synth. Metals* 1999, vol 103, p 2500.
[22] Wong, E. W.; Sheehan, P. E.; Lieber, C. M. *Science* 1997, vol 277, p 1971.
[23] Yao, Z.; Zhu, C. C.; Cheng, M.; Liu, *J. Comput. Mater. Sci.* 2001, vol 22, p 180.
[24] Yu, M. F.; Files, B. S.; Arepalli, S.; Ruoff, R. S. *Phys. Rev. Lett.* 2000, vol 84, p 5552.
[25] Antonucci, V.; Hsiao, K. T.; Advani, S. G. Advanced Polymeric Materials, Shonaike, G. O.; Advani, S. G. Eds., CRC Press, New York, 2003, Chap. 11.
[26] Haggenmuller, R.; Conmas, H. H.; Rinzler, A. G.; Fischer, J. E.; Winey, K. I. *Chem. Phys. Lett.* 2000, vol 330, p 219.

[27] Jia, Z. J.; Wang, Z. Y.; Liang, J.; Wei, B. Q.; Wu, D. H. *Mater. Sci. Eng*. A, 1999, vol 271, p 395.
[28] Jin, J. X.; Pramoda, K. P.; Goh, S. H.; Xu, G. Q. Mater. Res. *Bull.* 2002, vol 37, p 271.
[29] Pötschke, P.; Fornes, T. D.; Paul, D. R. *Polymer* 2002, vol 43, p 3247.
[30] Kim, J. Y.; Park, H. S.; Kim, S. H. *Polymer* 2006, vol 47, p 1379.
[31] Kim, J. Y.; Kim, S. H. J. Polym. Sci. Part B: *Polym. Phys.* 2006, vol 44, p 1062.
[32] Kim, J. Y.; Park, H. S.; Kim, S. H. *J. Appl. Polym. Sci*., 2007, vol 103, p 1450.
[33] Kim, J. Y.; Han, S. I.; Kim, S. H. *Polym. Eng. Sci.* 2007, vol 47, p 1715.
[34] Kim, J. Y.; Han, S. I.; Hong, S. *Polymer* 2008, vol 49, p 3335.
[35] Kim, J. Y.; Park, H. S.; Kim, S. H. *J. Polym. Sci. Part B: Polym. Phys.*2008, submitted.
[36] Park, J. G.; Kang, S. G. *Tech Trend Report* 2002, KISTI.
[37] Geng, H. Z.; Rosen, R.; Zheng, B.; Shimoda, H.; Fleming, L.; Liu, J.; Zhou, O. *Adv. Mater.* 2002, vol 14, p 1387.
[38] Grady, B. P.; Pompeo, F.; Shambaugh, R. L.; Resasco, D. E. *J. Phys. Chem.* B 2002, vol 106, p 5852.
[39] Kumar, S.; Doshi, H.; Srinivasarao, M.; Park, J. O.; Schiraldi, D. A. *Polymer* 2001, vol 43, p 1701.
[40] Thostenson, E. T.; Chou, T. W. J. Phys. D: *Appl. Phys.* 2002, vol 35, p L77.
[41] Lozano, K.; Barrera, E. V. *J. Appl. Polym. Sci.* 2001, vol 79, p 125.
[42] Ajayan, P. M.; Schadler, L. S.; Giannaris, C.; Rubio, *A. Adv. Mater*. 2000, vol 12, p 750.
[43] Gong, X.; Liu, J.; Baskaran, S.; Voise, R. D.; Young, J. S. *Chem. Mater.* 2000, vol 12, p 1049.
[44] Qian, D.; Dickey, E. C.; Andrew, R.; Rantell, T. *Appl. Phys. Lett.* 2000, vol 76, p 2868.
[45] Lourie, O.; Wagner, H. D. *Compos. Sci. Technol.* 1999, vol 59, p 975.
[46] Fornes, T. D.; Yoon, P. J.; Keskkula, H.; Paul, D. R. *Polymer* 2001, vol 42, p 9929.
[47] Kumar, S.; Doshi, H.; Srinivasrao, M.; Park, J. O.; Schiraldi, D. A. *Polymer* 2002, vol 43, p 1701.
[48] Mitchell, C. A.; Krishnamoorti, R. *Polymer* 2005, vol 46, p 8796.
[49] Bhattacharyya, A. R.; Sreekumar, T. V.; Liu, T.; Kumar, S.; Ericson, L. M.; Hauge, R. H.; Smalley, R. E. *Polymer* 2003, vol 44, p 2373.
[50] Zeng, H.; Gao, C.; Wang, Y.; Watts, P. C. P.; Konh, H.; Cui, X.; Yan, D. *Polymer* 2006, vol 47, p 113.
[51] Phang, Y.; Ma, J.; Shen, L.; Liu, T.; Zhang, W. D. *Polym. Int.* 2006, vol 55, p 71.
[52] Radhakrishnan, T. S. *J. Appl. Polym. Sci.* 1999, vol 73, p 441.
[53] Li, L. Q.; Guan, C. X.; Zhang, A. Q.; Chen, D. H.; Qing, Z. B. *Polym. Degrad. Stab.* 2004, vol 84, p 369.
[54] Yang, S.; Castilleja, J. R.; Barrera, E. V.; Lozano, K. *Polym. Degrad. Stab*. 2004, vol 83, p 383.
[55] Sengupta, R.; Sabharwal, S.; Bhowmick, A. K.; Chaki, T. K. *Polym. Degrad. Stab.* 2006, vol 91, p 1311.
[56] Gao, F.; Beyer, G.; Yuan, Q. *Polym. Degrad. Stab.* 2005, vol 89, p 559.
[57] Ebbesen, T. Carbon Nanotubes: Preparation and Properties, CRC Press, New York, 1997.
[58] Li, S. N.; Li, Z. M.; Yang, M. B.; Hu, Z. Q.; Xu, X. B.; Huang, R. *Mater. Lett.* 2001, vol 58, p 3967.

[59] Zanetti, M.; Camino, G.; Reichert, P.; Mulhaupt, R. Macromol. *Rapid Commun.* 1997, vol 22, p 176.
[60] Troitskii, B. B.; Troitskaya, L. S.; Yakhnov, A. S.; Lopatin, M. O.; Novikova, M. A. *Eur. Polym. J.* 1997, vol 33, p 1587.
[61] Vincent, B. F. *Calorimetry and Thermal Analysis of Polymers*, *Springer*, Berlin, 2001.
[62] Xu, W.; Ge, M.; He, P. *J. Polym. Sci. Part B: Polym. Phys.* 2002, vol 40, p 408.
[63] Ma, J.; Zhang, S.; Qi, Z.; Li, G.; Hu, Y. *J. Appl. Polym. Sci.* 2002, vol 83, p 1978.
[64] Kim, S. H.; Ahn, S. H.; Hirai, T. *Polymer* 2003, vol 44, p 5625.
[65] Gopakumar, T. G.; Lee, J. A.; Kontopoulou, M.; Parent, *J. S. Polymer* 2002, vol 43, p 5483.
[66] Cheng, S. Z. D.; Wunderlich, B. *Macromolecules* 1988, vol 21, p 789.
[67] Lopez, C.; Wilkes, G. L. *Polymer* 1989, vol 30, p 882.
[68] Avrami, J. *Chem. Phys.* 1939, vol 7, p 1103.
[69] Avrami, *J. Chem. Phys.* 1940, vol 8, p 212.
[70] Avrami, *J. Chem. Phys.* 1941, vol 9, p 177.
[71] Ozawa, T. *Polymer* 1971, vol 12, p 150.
[72] Ozawa, T. *J. Thermal Analysis* 1976, vol 9, p 369.
[73] Di Lorenzo, M. L.; Silvestre, C. *Prog. Polym. Sci.* 1999, vol 24, p 917.
[74] Eder, M.; Wlochowicz, A. *Polymer* 1983, vol 24, p 1593.
[75] Liu, T.; Mo, Z.; Wang, S.; Zhang, H. *Polym. Eng. Sci.* 1997, vol 37, p 568.
[76] Jeziorny, *Polymer* 1978, vol 19, p 1142.
[77] Ravindranath, K.; Jog, J. P. *J. Appl. Polym. Sci.* 1993, vol 49, p 1395.
[78] Lambrigger, M. *Polym. Eng. Sci.* 1998, vol 38, p 610.
[79] Dobreva; Gutzow, I. J. *J. Non-Cryst. Solids* 1993, vol 162, p 1.
[80] Dobreva; Gutzow, I. J. *J. Non-Cryst. Solids* 1993, vol 162, p 13.
[81] Kim, J. Y.; Kang, S. W.; Kim, S. H.; Kim, B. C.; Shim, K. B.; Lee, J. G. *Macromol. Res.* 2005, vol 13, p 19.
[82] Alonso, M.; Velasco, J. I.; De Saja, J. A. *Eur. Polym. J.* 1997, vol 33, p 255.
[83] Kissingerm, H. E. *J. Res. Natl. Stand.* 1956, vol 57, p 217.
[84] Kissinger, H. E. *J. Thermal Analysis* 1957, vol 9, p 369.
[85] Wunderlich, *Macromolecular Physics,* Academic Press, New York, 1976. vol. 2.
[86] Ray, S. S.; Okamoto, M. *Prog. Polym. Sci.* 2003, vol 28, p 1539.
[87] Wu, T. M.; Liu, C. Y. *Polymer* 2005, vol 46, p 5621.
[88] Gojny, F. H.; Schulte, K. *Compos. Sci. Technol.* 2004, vol 64, p 2303.
[89] Abdel-Goad, M.; Pötschke, P. *J. Non-Newtonian Fluid Mech.* 2005, vol 128, p 2.
[90] Wagener, R.; Reisinger, T. J. G. *Polymer* 2003, vol 44, p 7513.
[91] Kulichikhin, V. G.; Shumskii, V. F.; Semakov, A. Rheology and Processing of Liquid Crystal Polymers, Acierno, D.; Collyer, A. A. Eds., Chapman and Hall, New York, 1996, Chap. 5.
[92] Enikolopyan, N. S.; Fridman, M. L.; Stalnova, I. U.; Popov, V. L. *Adv. Polym. Sci.* 1990, vol 96, p 1.
[93] Krishnamoorti, R.; Vaia, R. A.; Giannelis, E. P. *Chem. Mater.* 1996, vol 8, p 1728.
[94] Ferry, J. Viscoelastic Properties of Polymers, Wiley, New York, 1980.
[95] Okamoto, M. *Encyclopedia of Nanoscience and Nanotechnology,* Nalwa, H. S. Ed., American Scientific Publisher, LA, 2004, Vol. 8, pp 791-843.
[96] Krisnamoorti, R.; Giannelis, E. P. *Macromolecules* 1997, vol 30, p 4097.

[97] Rosedalev, J. H.; Bates, F. S. *Macromolecules* 1990, vol 23, p 2329.
[98] Larson, R. G.; Winey, K. I.; Patel, S. S.; Watanabe, H.; Bruinsma, R. *Rheol. Acta* 1993, vol 32, p 245.
[99] Song, Y. S.; Youn, J. R. *Carbon* 2005, vol 43, p 1378.
[100] Fu, X.; Gelfer, M. Y.; Hsiao, B. S.; Phillips, S.; Viers, B.; Blanski, R.; Ruth, P. *Polymer* 2003, vol 44, p 1499.
[101] Van Gulp, M.; Palmen, *J. Rheol. Bull.* 1998, vol 67, p 5.
[102] Trinkle, S.; Friedrich, C. *Rheol. Acta* 2002, vol 41, p 103.
[103] Honerkamp, J.; Weese, *J. Rheol. Acta* 1993, vol 32, p 57.
[104] Ottenbrite, R. M.; Utracki, L. A.; Inoue, S. *Current Topics in Polymer Science*, Carl Hanser, Munich, 1991.
[105] Wissbrun, K. F.; Griffin, A. C. *J. Polym. Sci. Polym. Phys. Ed.* 1982, vol 20, p 1895.
[106] Park, S. J.; Kim, J. S. *Carbon* 2001, vol 39, p 2011.
[107] Kashiwagi, T.; Grulke, E.; Hilding, J.; Harris, R.; Awad, W.; Douglas, *J. Macromol. Rapid Commun.* 2002, vol 23, p 761.
[108] Kashiwagi, T.; Shields, J. R.; Harris, R. H.; Davis, R. D. *J. Appl. Polym. Sci.* 2003, vol 87, p 1541.
[109] Lu, X. F.; Hay, J. *N. Polymer* 2001, vol 42, p 9423.
[110] Wu, T. M.; Chen, E. C. J. *Polym. Sci. Part B: Polym. Phys*. 2006, vol 44, p 598.
[111] Kim, J. Y.; Seo, E. S.; Kim, S. H.; Kikutani, T. *Macromol. Res.* 2003, vol 11, p 62.
[112] Kim, J. Y.; Kim, S. H.; Kang, S. W.; Chang, J. H.; Ahn, S. H. *Macromol. Res.* 2006, vol 14, p 146.
[113] Cheng, S. Z. D.; Wu, Z. Q.; Wunderlich, B. *Macromolecules* 1987, vol 20, p 2802.
[114] Kim, H. G.; Robertson, R. E. *J. Polym. Sci. Part B: Polym. Phys.* 1998, vol 36, p 1757.
[115] Blundell, J. *Polymer* 1987, vol 37, p 1167.
[116] Medellin-Rodriguez, J.; Phillips, P. J.; Lin, J. S.; Avila-Orta, C. A. *J. Polym. Sci. Part B: Polym. Phys.* 1998, vol 36, p 763.
[117] Lee, Y.; Porter, R. S. *Macromolecules* 1987, vol 20, p 1336.
[118] Hoffman, J. D.; Week, J. J. *J. Chem. Phys.* 1962, vol 37, p 1723.
[119] Hoffman, J. D.; Davis, G. T.; Lauritzen, J. I. *Treatise on Solid Sstate Chemistry: Crystalline and Non-crystalline Solids*, Hannay, N. B. Ed., Plenum Press, New York, 1976, Vol. 3.
[120] Park, J. W. Kim, D. K.; Im, S. S. *Polym. Int.* 2002, vol 51, p 239.
[121] Lee, W. D.; Yoo, E. S.; Im, S. S. *Polymer* 2003, vol 44, p 6617.
[122] Han, S. I.; Kang, S. W.; Kim, B. S.; Im, S. S. Adv. Funct. *Mater.* 2005, vol 15, p 367.
[123] Buchner, S.; Wiswe, D.; Zachmann, H. G. *Polymer* 1989, vol 30, p 480.
[124] Lauritzen, J.; Hoffmann, J. D. *J. Appl. Phys.* 1973, vol 44, p 4340.
[125] Daubeny, R.; De, P.; Bunn, C. W. *Proc. R Soc. London A* 1954, vol 226, p 531.
[126] Lovinger, A. J.; Davis, D. D.; Padden, F. J. *Polymer* 1985, vol 26, p 1595.
[127] Xing, P.; Ai, X.; Dong, L.; Feng, Z. *Macromolecules* 1998, vol 31, p 6898.
[128] Lauritzen, J. I.; Hoffman, J. D. *J. Res. Natl. Bur. Stand. Sect.* 1960, vol 64A, p 73.
[129] Papageorgiou, Z.; Achilias, D. S.; Bikiaris, D. N.; Karayannidis, G. P. *Thermochim. Acta* 2005, vol 427, p 117.
[130] Mencik, Z. *Chem. Prum.* 1967, vol 17, p 78.
[131] Zachmann, G.; Wiswe, D.; Gehrke, R.; Riekel, C. *Makromol. Chem.* 1985, vol 12, p 175.

[132] Salim, M.; Seferis, J. C. *J. Appl. Polym. Sci.* 1993, vol 47, p 847.
[133] Nair, C. P. R.; Bindu, R. L.; Joseph, V. C. *J. Polym. Sci. Part A: Polym. Chem.* 1995, vol 33, p 621.
[134] Doyle, C. D. *Anal. Chem.* 1961, vol 33, p 77.
[135] Park, S. J.; Cho, M. S. J. *Mater. Sci.* 2000, vol 35, p 3525.
[136] Gilman, W.; Jackson, C. L.; Morgan, A. B.; Harris, R.; Manias, E.; Giannelis, E. P. Wuthenow, W.; Hilton, D.; Phillips, S. H. *Chem. Mater.* 2000, vol 12, p 1866.
[137] Vyazovkin, S. *Int. Rev. Phys. Chem.* 2000, vol 19, p 45.
[138] Kissinger, H. E.; J. *Anal. Chem.* 1957, vol 29, p 1072.
[139] Ozawa, T.; *J. Thermal Anal.* 1970, vol 2, p 301.
[140] Liu, Y. L.; Hsiue, G. H.; Lan, C. W.; Chiu, Y. S. *Polym. Deg. Stab.* 1997, vol 56, p 291.
[141] Hsiue, G. H.; Liu, Y. L.; Liao, H. H. *J. Polym. Sci. Part A: Polym. Chem.* 2001, vol 39, p 986.
[142] Ozawa, T. Bull. *Chem. Soc. Jpn.* 1965, vol 38, p 1881.
[143] Flynn, H.; Wall, L. A. *Polym. Lett.* 1966, vol 4, p 323.
[144] Doyle, C. D. *J. Appl. Polym. Sci.* 1962, vol 6, p 639.
[145] Macan, J.; Ivankovic, H.; Ivankovic, M.; Mencer, H. J. *J. Appl. Polym. Sci.* 2004, vol 92, p 498.
[146] Macan, J.; Brnardic, I.; Orlic, S.; Ivankovic, H.; Ivankovic, M. *Polym. Deg. Stab.* 2006, vol 91, p 122.
[147] Shenoy, A. V.; Rheology of Filled Polymer Systems, Kluwer Academic Publishers, London, 1999.
[148] Schartel, B.; Pötschke, P.; Knoll, U.; Abdel-Goad, M. Eur. *Polym. J.* 2005, vol 41, p 1061.
[149] Ito, M.; Kikutani, T. Handbook of Thermoplastic Polyesters, Fakirov, S. Ed., Wiley-VCH, weinheim, 2002, Chap. 10.
[150] Chang, T. E.; Jensen, J. R.; Kisliuk, A.; Pipes, R. B.; Pyrz, R.; Sokolov, A. P. *Polymer* 2005, vol 46, p 439.
[151] Holland, B. J. ; Hay, J. N. *Polymer* 2002, vol 43, p 1835.
[152] Botelho, G.; Queiro, A.; Gijsman, P. *Polym. Degrad. Stab.* 2000, vol 70, p 299.
[153] Samperi, F.; Puglisi, C.; Alicata, R.; Montaudo, G. *Polym. Degrad. Stab.* 2004, vol 83, p 3.
[154] Li, X. G. ; Huang, M. R. ; Bai, H. *Angrew. Makromol. Chem.* 1998, vol 256, p 9.

In: Polyester: Properties, Preparation and Applications
Editors: Hina Yamashita, Yui Nakano
ISBN: 978-1-60456-753-3

Chapter 3

Recent Developments in Modification of Cyanate Ester Resins

A. Fainleib* and O. Grigoryeva
[1]Institute of Macromolecular Chemistry of the National Academy of Ukraine, 02160 Kyiv, Ukraine

1. Introduction

Cyanate Ester Resins (CER) offer a variety of excellent thermal and good mechanical properties, which commend them for use in high performance technology (e.g. as matrices for composites for high-speed electronic circuitry and transportation). The product of CER curing process polycyanurates (PCN) are synthesized by a catalytic high temperature polycyclotrimerization reaction of cyanate esters of bisphenols (cf. Figure1).

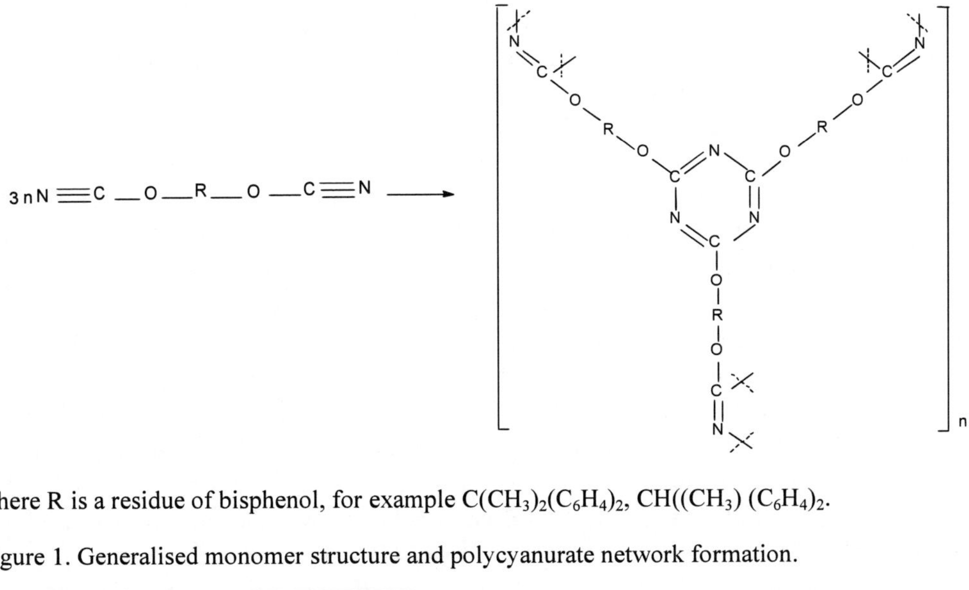

where R is a residue of bisphenol, for example $C(CH_3)_2(C_6H_4)_2$, $CH((CH_3)\ (C_6H_4)_2$.

Figure 1. Generalised monomer structure and polycyanurate network formation.

* Address for correspondence: fainleib@i.kiev.ua.

For the electronics market, attractive features of PCN are their low dielectric loss characteristics, dimensional stability at molten solder temperatures (220-270°C), high purity, inherent flame-retardancy (giving the potential to eliminate brominated flame retardants) and excellent adhesion to conductor metals at temperatures up to 250°C [1]. Since the late 1970s, cyanate ester resins have been used with glass or aramid fibers in high-speed multilayer circuit boards and this remains their primary application. Several reviews [2-6] collecting the numerous publications (papers and patents) in the field of PCN synthesis, processing, characterization, modification and application have appeared since 1990s. In addition, like conventional FR-4 diepoxides, cyanate ester laminates retain the desirable (ketone) solution processing characteristics and the ability to be drilled, making possible to employ them in printed circuit board manufacture. In the last three decades, aerospace composites have evolved into damage-tolerant primary and secondary structures utilizing both thermoset and thermoplastic resins. PCN homopolymers develop approximately twice the fracture toughness of multifunctional epoxies while qualifying for service temperatures of at least 150°C, intermediate between epoxy and bismaleimides capabilities. PCN have already flown in prototype radomes and high gain antennae, with possible applications in primary and secondary structures of the High Speed Civil Transport (HSCT) and European Fighter Aircraft. PCN are also being qualified for satellite truss and tube structures and cryogenic, radiation-resistant components in the Superconducting Supercollider [7]. This is indeed the problem, to convince a traditionally conservative industry that the superior performance of PCN (which surpass the glass transition temperature and hydrophobicity of epoxies while matching their processability and are easily toughened) makes them worthy of further investigation in spite of their price, which is currently higher than the price of the epoxies. PCN must be traditionally cured at high temperatures in order to achieve complete conversion, which increases manufacturing cost, but reactive modification of PCN [5] allows decreasing the high temperature of PCN post-curing. The primary drawback of PCN, which hinders more extensive application of the cured materials, however, is low room temperature toughness.

Cyanate esters may be modified by co-reaction with monomers or oligomers that contain active hydrogens (e.g., water, phenols, bisphenols, diamines, diepoxides, ethers, esters, etc) [2-6]. The PCN modified with reactive monomers usually have a homogeneous single phase, with properties proportional to the ratio of the two components.

PCN have been toughened, on the other hand, by physical modification through the addition of thermoplastics [2-6], with formation of so-called semi-interpenetrating polymer networks (IPNs) [3,5]. In this case, the degree of toughening depends on the final morphology of the polymer blends and composites, which, in turn, depends on the rate of phase separation during cure relative to the rate of polymerization [8]. A number of amorphous thermoplastics, including polyetherimides, polysulphones, polyethersulphones, polyarylates and polyimides, have been dissolved in aryl dicyanate formulations, and when they phase separate during cure, toughened systems are achieved [2-6]. Rubbers have also been used to toughen dicyanate esters [2-6]. Disadvantages of physical toughening are large-scale heterogeneity for interpenetrating polymer systems, difficulty in controlling domain size and poor processability. Some results have been published [9-11] where the researchers used the combination of chemical and physical methods of modification. There are two possibilities to combine those kinds of modification: simultaneous use of reactive co-monomer and inert rubber or thermoplastics to change the chemical structure and phase morphology of polycyanurates and use of a rubber or a thermoplastic containing reactive chain ends. In the combined method, the ability of additives to react with the matrix is often of great interest since the effect is to improve the adhesion between phases in the case of a biphasic material, and in any case they ensure a chemical linkage between the oligomeric modifier and the network. Polycyanurates with well-defined morphology and improved characteristics have been synthesized [9-11].

2. Hybrid Networks from Cyanate Esters and Polyethers (Polyesters)

Last years several papers describing synthesis, structure and properties of polycyanurates modified with commercial polyethers [12-15] and polyesters [16] have been published.

Several series' of polycyanurate networks (PCN), based on the dicyanate of bisphenol A monomer (DCBA), were synthesized in the presence of different contents of hydroxyl-terminated polyethers (PEth), such as poly(oxypropylene glycol) (PPG) [12,13,15] and poly(oxytetramethylene glycol) (PTMG) [14,15]. The influence of the nature of the oligomeric modifier, initially miscible with DCBA, on the chemical structure, glass transition behaviour, phase morphology and mechanical properties of modified PCN was studied. The possibility of PEth incorporation into the PCN structure through mixed cyanurate ring formation has been discussed. A maximum PEth incorporated content of 0.1 mol of PEth per 1 mol of DCBA (~26-28 wt. %) irrespective of PEth type has been detected by comparison of calculated and experimental values of gel-fraction content. The theoretical value of gel fraction ($w_{g\ theor}$) was calculated using equation 1 and with the conjecture that the PEth was completely extracted.

$$w_{g\ theor} = (1-w_{PEth}) \times (2\alpha-1)/\alpha^2 \qquad (1)$$

where α is the OCN-conversion (determined by both FTIR and DSC methods) and w_{PEth} is the weight fraction of PEth in the initial composition.

Based on the knowledge from Organic Chemistry [16,17] relating to reactions of organic cyanates with active hydrogen containing compounds and taking in account the results of investigations on modification of PCN by phenols [2], the authors of refs. [12,13] have proposed a scheme (cf. Figure 2) of chemical incorporation of hydroxyl-terminated modifiers into the PCN structure.

DMTA analysis has shown the formation of multiphase polymer compositions due to microphase separation of the components occurred during DCBA/PPG curing procedure.

On the other hand, all PCN/PTMG cured compositions exhibited [14,15] a single, broad glass transition (cf. Figure 4) that shifted to lower temperature as the modifier content was increased and the experimental T_g versus modifier content showed a slight positive deviation from the Fox equation (cf. Figure 5) for miscible polymer systems.

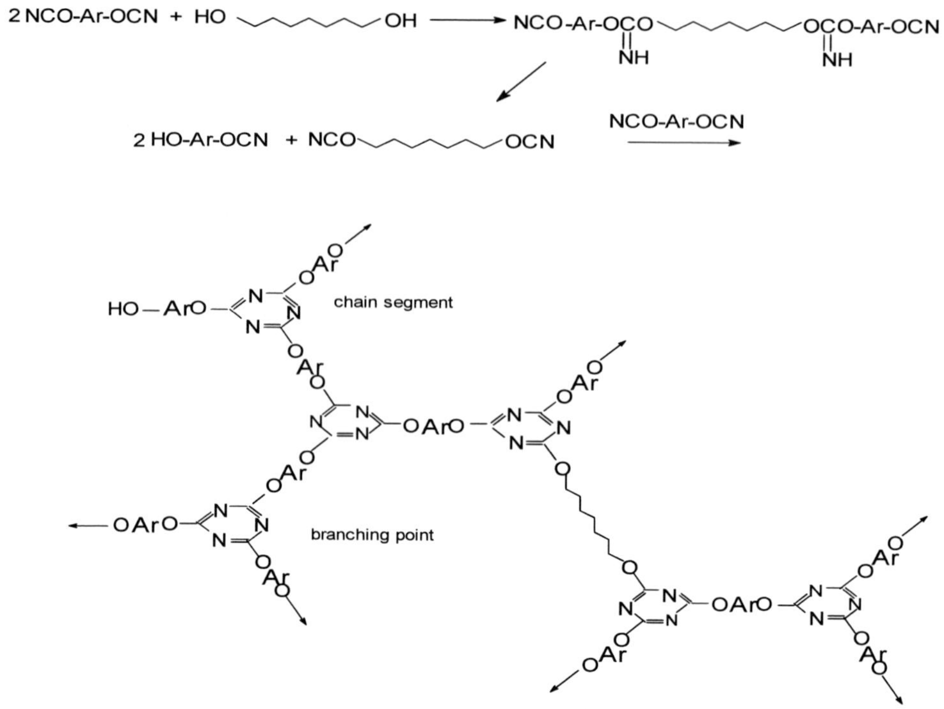

Figure 2. Scheme of chemical modification of PCN with hydroxyl-terminated polymer

The formation of very finely divided morphologies with highly interpenetrated phases, i.e. a PCN- rich phase, a mixed phase of PCN/PPG components and a PPG-rich phase can be clearly seen from the temperature dependence of loss modulus, E''(cf. Figure 3).

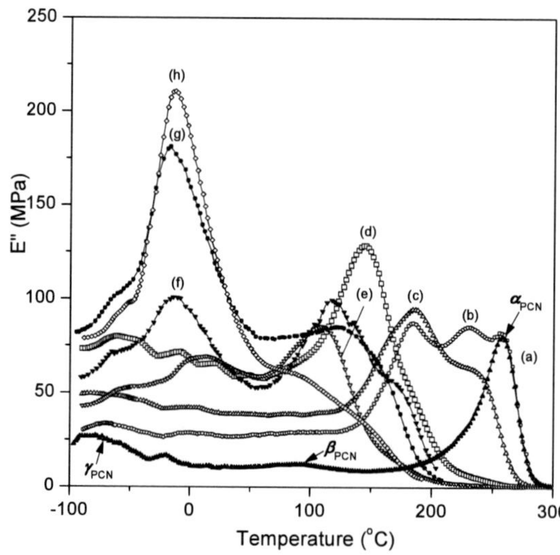

Figure 3. Loss modulus (E'') *versus* temperature for PCN/PPG composition (bending mode, 1Hz). (a) Pure PCN; (b) 98:2; (c) 95:5; (d) 90:10; (e) 85:15; (f) 74:26; (g) 69:31 and (h) 63:37 PCN/PPG (wt.%).

It was concluded [12] that the PCN and the PTMG have a higher degree of compatibility than the PCN and the PPG and, further, that the rate of the incorporation reaction in the case of PTMG is faster than in the case of PPG. Thus, the growing PCN/PTMG network has no time to phase separate. This conclusion, certainly, requires additional experimental verification.

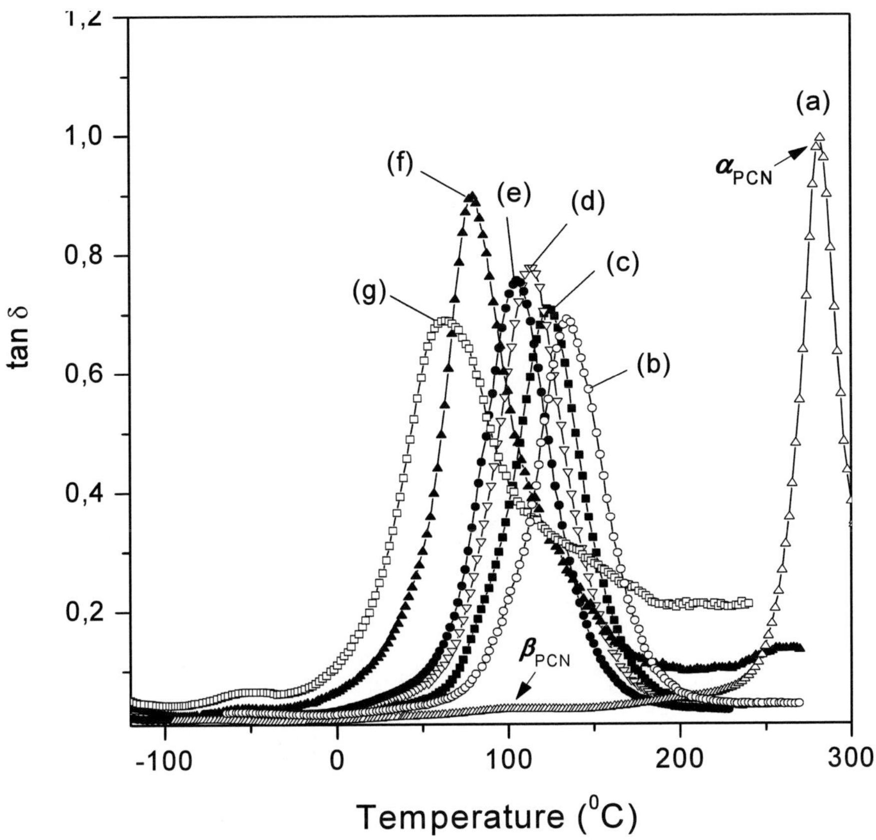

Figure 4. Loss factor (tan δ) *versus* temperature for the PCN/PTMG blends (bending mode, 10 Hz). (a) Pure PCN; (b) 74:26; (c) 71:29; (d) 69:31; (e) 66:34; (f) 62:38 and (g) 58:42 PCN/PTMG (wt.%).

The introduction of PEth into PCN improves the mechanical properties of the latter to maximum extent at the PEth content of 26-28% (cf. Figure 6), corresponding to the maximum PEth incorporation degree determined [12]. The higher values of tensile strength for PCN/PTMG can be explained by better mixing of the components.

Polycyanurate network was modified with hydroxyl–terminated polyester, poly(butylene glycol adipate), PBGA, by polycyclotrimerization of bisphenol A dicyanate in the immediate presence of PBGA [18]. The modified networks with PBGA content from 5 to 20 wt. % were characterized by a combination of FTIR, DSC, TMA, TGA, impact testing and sol-gel analysis. It has been established that almost the whole PBGA was chemically incorporated into the PCN structure. It is supposed that the chemistry of such incorporation is the same as that shown in Figure 2.

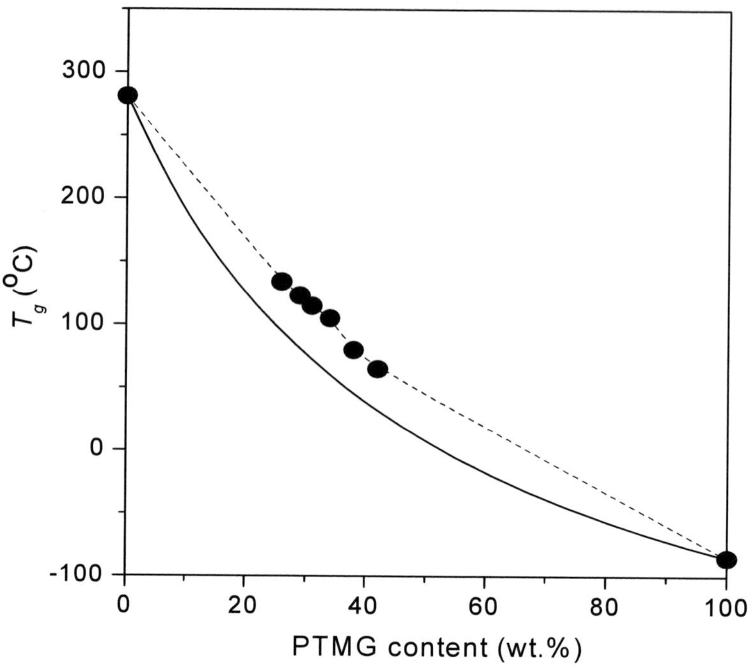

Figure 5. T_g values determined by (--●--) DMTA and calculated from the Fox's equation (—) for the PCN/PTMG compositions.

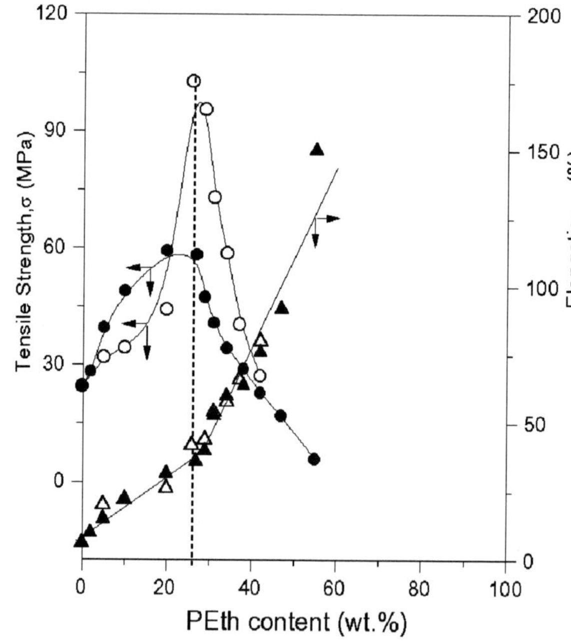

Figure 6. Tensile strength, σ (●,○) and elongation at break, ε (▲,Δ) for the PCN/PEth compositions as a function of modifier (PEth) content. (●,▲) PPG; (○,Δ) PTMG.

The T_g values for the compositions studied were determined from TMA using two static loads, 100 and 1900 mN, as well as from DSC data. The concentration dependence of T_g determined is presented in Figure 7.

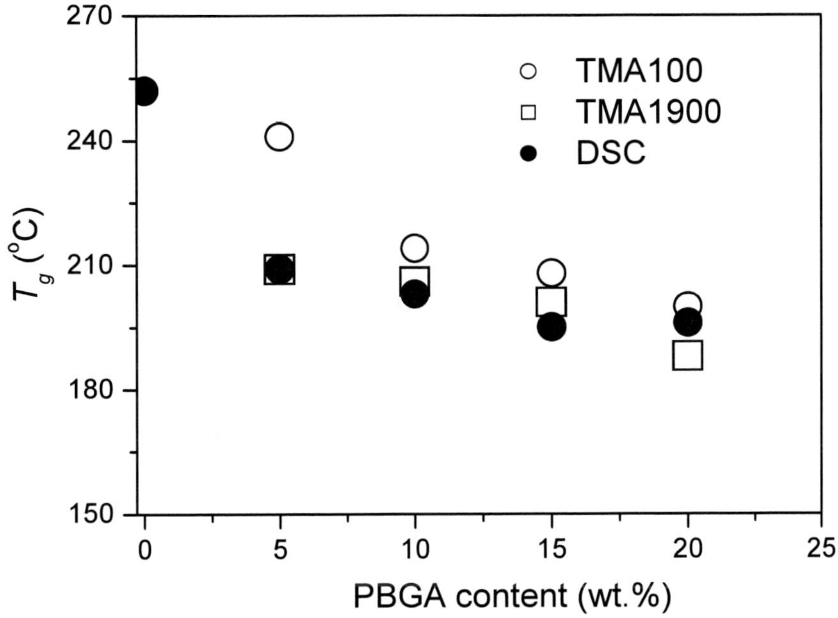

Figure 7. Composition dependence of glass transition temperature, T_g for PCN/PBGA

We can observe a quite sharp reduction of T_g for PCN modified with 5 % of PBGA (reduction of T_g by 17 %). However, further increase of PBGA content up to 20 % reduces T_g just by 5 % more. The authors of ref. [18] suppose that the sharp decrease of T_g by introducing 5 % of PBGA is connected with significant destruction of the regularity of the polycyanurate network structure. The further increase of PBGA content does not reduce the T_g value strongly because, as it was noted above, almost the whole PBGA is chemically incorporated into the network and cannot act as a usual plasticizer. It is worth to note a high convergence of the results obtained by different methods.

Figure 8 reports simultaneously the measured values of ΔC_p as a function of the concentration of PBGA and the calculated variations of ΔC_p determined from pure PCN on the basis of the law of additive assuming that PCN and PBGA do not interact. We can see that the higher the content of PBGA, the higher the difference between experimental and calculated values of ΔC_P. Thus, according to the "strong-glass" glass former liquid concept proposed by Angell [19, 20], the authors of ref. [18] conclude that the higher the content of PBGA, the stronger the interaction between the components.

It is shown [18] that the introduction of PBGA does not drastically reduce the thermal stability of the PCN network while the impact properties are improved significantly. It is seen from Figure 9 that by introducing even 20 % of PBGA into PCN the temperature of onset of degradation (T_d) of the latter decreases just by 5 %, while the impact strength decreases by 86 %. It is concluded [18] that the modified PCN developed can be successfully used simultaneously at high temperatures and high loading.

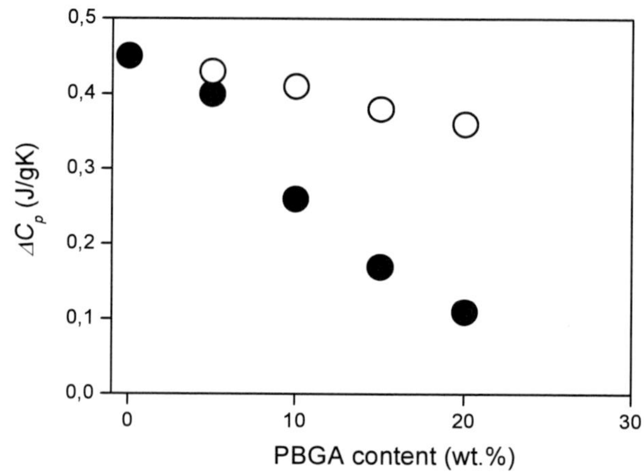

Figure 8. Composition dependence of ΔC_P at glass transition; (•) experimental (DSC) and (o) calculated for PCN/PBGA assuming both additivity and no interactions between the components.

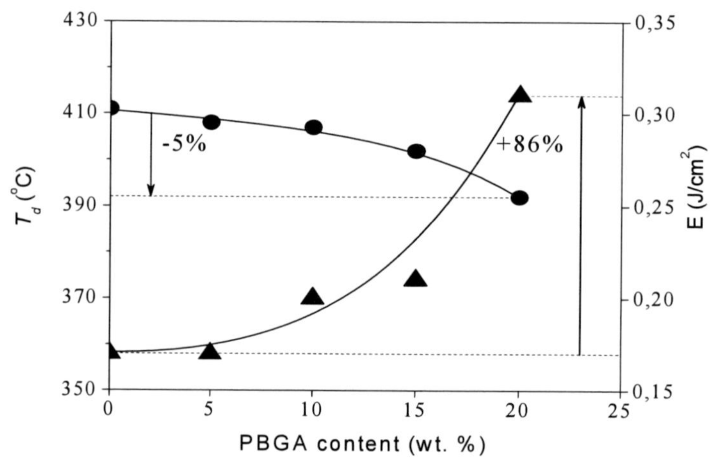

Figure 9. Composition dependence of temperature of degradation, T_d (onset) and adsorbed energy, E for PCN/PBGA.

Polycyanurate networks modified by dihydroxy-telechelic oligo(ε-caprolactone) were recently synthesized [21] *via* thermal polycyclotrimerization of dicyanate ester of bisphenol E in the presence of the reactive modifier. The TGA investigation of the networks with varying PCN/PCL compositions has permitted to define two groups of systems. The first group with a low content of oligo(ε-caprolactone) (*i.e.* up to 20 wt. %) has been described as single phase systems. It has been shown by FTIR, gel fraction content and density measurements that the major part of the oligomeric modifier was chemically incorporated into the PCN network structure, for this group. The second group with a higher content of oligo(ε-caprolactone) (*i.e.* higher than or equal to 30 wt. %) has been defined as multiphase systems, in which a significant amount of the modifier has not been covalently linked to the PCN network. Interestingly, the thermogravimetric analysis has also showed that there is no drastic reduction in thermal stability of PCN-type networks, whatever the PCL content introduced into the structures. The DSC and DMTA study of their solid-state organization has evidenced the occurrence of at least a two-phase structure with two kinds of local environments within such PCN/PCL systems: first, a PCN-rich phase, mainly constituted of PCN cross-linked chains modified by PCL embeddings, and second, a mixed phase formed by PCL-modified PCN network and unreacted free oligo(ε-caprolactone). Authors has noted that the polymer systems studied can be effectively used as precursors for the generation of porous cyanurate thermosetting materials through the quantitative hydrolysis of the oligo(ε-caprolactone) sub-chains.

Later the same authors reported [22] that porous frameworks were prepared from such PCN/PCL based hybrid networks by selective hydrolysis of a significant part of PCL sub-chains under mild conditions. The structure and properties of the precursory networks as investigated by FTIR, solid-state 13C NMR, DSC, TGA, and pycnometry were compared to those of the resulting porous materials. Thermogravimetric analyses showed that the porous hybrid systems were characterized by thermal and thermal oxidative stabilities higher than those of their non-hydrolyzed homologues. The porosity of the resulting hydrolyzed networks was examined by SEM and DSC-thermoporometry: pore sizes ranging from 10 to 150 nm- at most- were thus determined. The morphologies of such PCN based networks, before and after hydrolysis are shown in Figure 10. One can see that the nondegraded sample displays a compact structure without any pores (1) while well visible pores present in a microphotograph for the sample after hydrolysis (2).

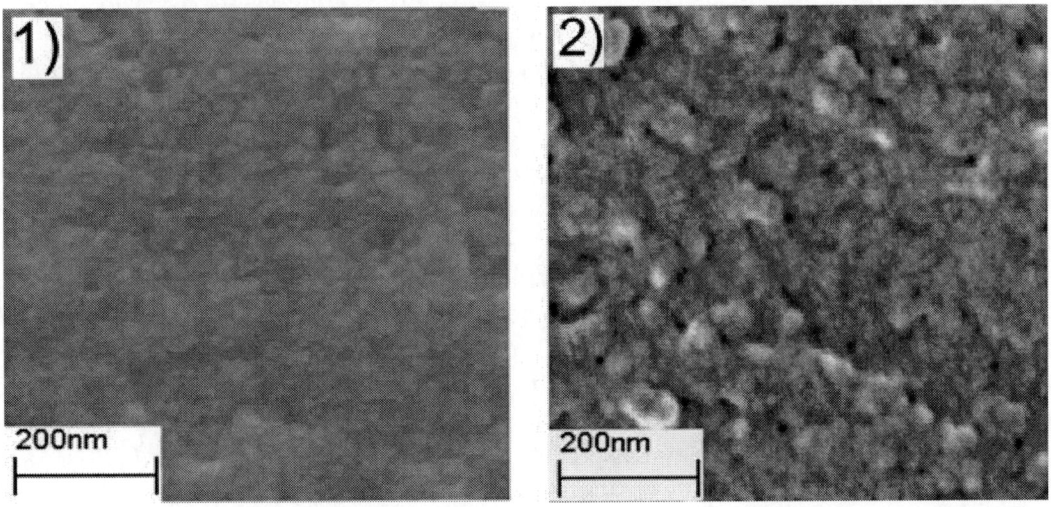

Figure 10. SEM micrographs of PCN/PCL systems (70/30 wt %) before (1) and after (2) hydrolysis.

3. Polycyanurate-Polyurethane Grafted Semi-IPNs

In 1992-1994 the first papers [21-25] on synthesis, kinetic peculiarities and characterization of structure-property relationships of polycyanurate-polyurethane semi-IPNs were published. The idea behind these semi-IPNs was to synthesize a new material from widely used linear polyurethane as a thermoplastic elastomer and intensively developing high performance cyanate ester resins (CER) as a thermoset. Linear segmented polyurethanes (LPU) exhibit rubbery characteristics and thermoplasticity that is directly connected with their structure. Linear segmented polyurethanes are characterized by internal heterogeneity specified by phase separation of soft and hard block of polymer chain. CER formulations offer a variety of excellent thermal and other properties (see section 1), which commend them for use in high performance technology (e.g. high speed electronic circuitry and aerospace composite matrices).

3.1 Synthesis, Chemical Interaction between Components, Reactive Grafting and Compatibilization

The attempt was undertaken to combine the best properties of both materials (polycyanurate and polyurethane) in their semi-IPN. First the suggestions about possible chemical interaction between components in polyurethane-polycyanurate semi-IPNs leading to formation of chemically grafted semi-IPNs were published by Fainleib and co-authors [23] and then confirmed in later works [24-27].

Polycyanurate-polyurethane semi-IPNs were prepared by polycyclotrimerization of dicyanate ester of bisphenol A (DCEBA) in the immediate presence of segmented linear polyurethane, LPU (molar mass 40000 g/mol) synthesized from 4,4'-diphenylmethane diisocyanate, oligobuthylene glycol adipate (molar mass 1000 g/mol) and 1,4-butane diol as a chain extender. Calculations made [23] by comparison of the polycyclotrimerization kinetics (analysis of infra-red spectra) of pure DCEBA and in the blend with LPU have shown that, with increase of LPU content from 10 to 80 %, the portion of the cyanate groups consumption for the polycyclotrimerization decreases from 90 to 33 %. It was concluded that during semi-IPN synthesis at least two competitive chemical processes coexist: 1) formation of the polycyanurate network by polycyclotrimerization (see Figure 1) of DCEBA in the presence of linear LPU leading to components microphase separation, which begins at a certain reaction time due to incompatibility of the resulting products that is typical for IPN systems, and 2) chemical interaction between the forming network and polyurethane that prevents the components microphase separation. By that time, no references concerning the cyanate – urethane chemical reaction were found. The chemical interaction between cyanate and urethane was first investigated with model monofunctional low molar mass organic compounds [28]. As model compounds 4-tert-butylpheny cyanate (BPC) and phenyldodecanol urethane (PDU) were used.

Chromatograms obtained by HPLC for BPC/PDU=5:1 after heating at 150°C are plotted in Figure 11 [28]. The elution times are 9,2 min and 25,4 min for BPC and PDU, respectively. One can see a decrease of the height of the mentioned peaks with increasing reaction time. Simultaneously at t_e=5,3 min, t_e=17,3 min, and t_e=26,9 min three main products are eluted, as well as two additional products at t_e=30,0 min and t_e=31,3 min are observed. The products with elution time 5,3 min and 26,9 min were identified as 4-tert-butylphenol (BP) and 4-tert-butylphenoxytriazine (BPT), respectively. As far as separate heating of BPC and PDU at the same conditions has not led to formation of the product eluted at t_e=17,3 min, it is reasonable to suppose that this one is the reaction product of an immediate interaction between BPC and PDU.

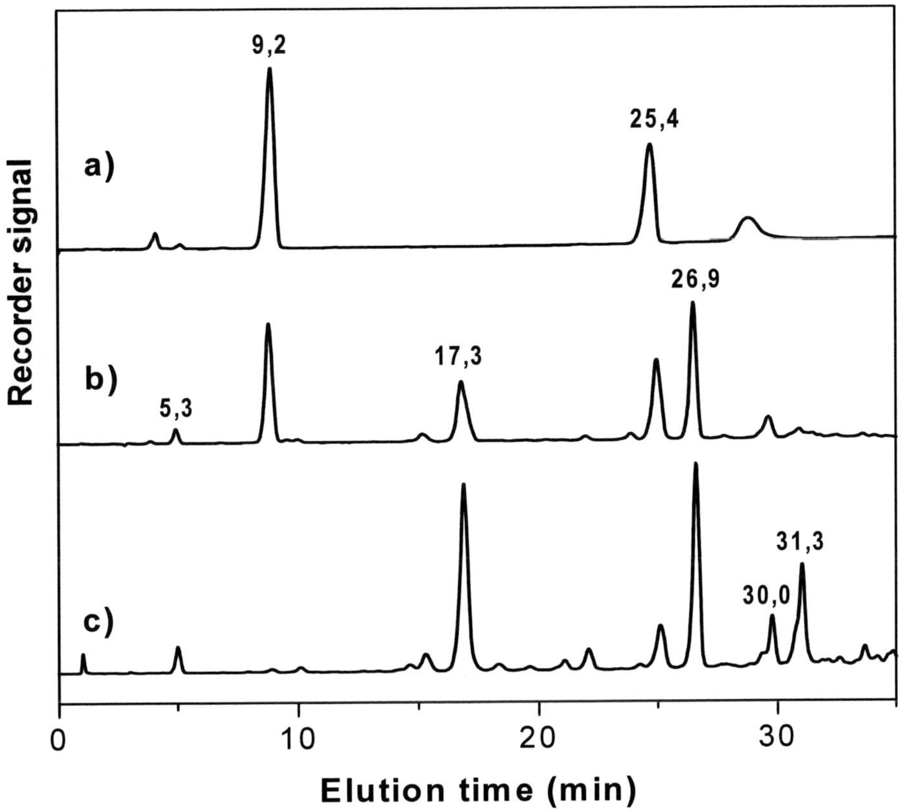

Figure 11. Chromatograms of BPC/PDU=5:1 (mol) after heating at 150°C. 0 min; b) 60 min; c) 180 min.

The normalized FTIR spectra for BPC/PDU=5:1 are presented in Figure 12. One can see the significant changes in spectra of this composition during heating at 150°C. The decrease of the normalized intensity of the 2236-2272 cm^{-1}-bands of cyanate groups confirms the participation of BPC in the chemical process. On the other hand, there are changes in position and intensity of the bands, which are attributed to the urethane group. As it can be seen, the intensity of the 1704 cm^{-1}-band ($\nu_{C=0}$ bonded) decreases with increasing intensity of the absorbance of the band at 1730 cm^{-1} ($\nu_{C=0}$ free), whereas the intensity of the 1239 cm^{-1}-band (δ_{N-H}) decreases with increasing intensity of the peak at 1215 cm^{-1}, which is attributed to aromatic OH. Besides, the appearance of peaks at 2216 and 1639 cm^{-1} characteristic for the nitrile group and at 1570 cm^{-1} of C=N for the cyanurate cycle (see Figure 1) [2-4] followed by an increase of their intensity can be noted. MALDI-TOF experiments gave the following mass peaks (M^+): 300, 336, 369, 412, 470, 505, 526, 562, 680.

A scheme of the chemistry of the process studied (cf. Figure 13) with the formation of BP, BPT, some intermediate and additional products, including possible substituted cyanurate, isocyanurate and mixed triazine cycles, is presented below.

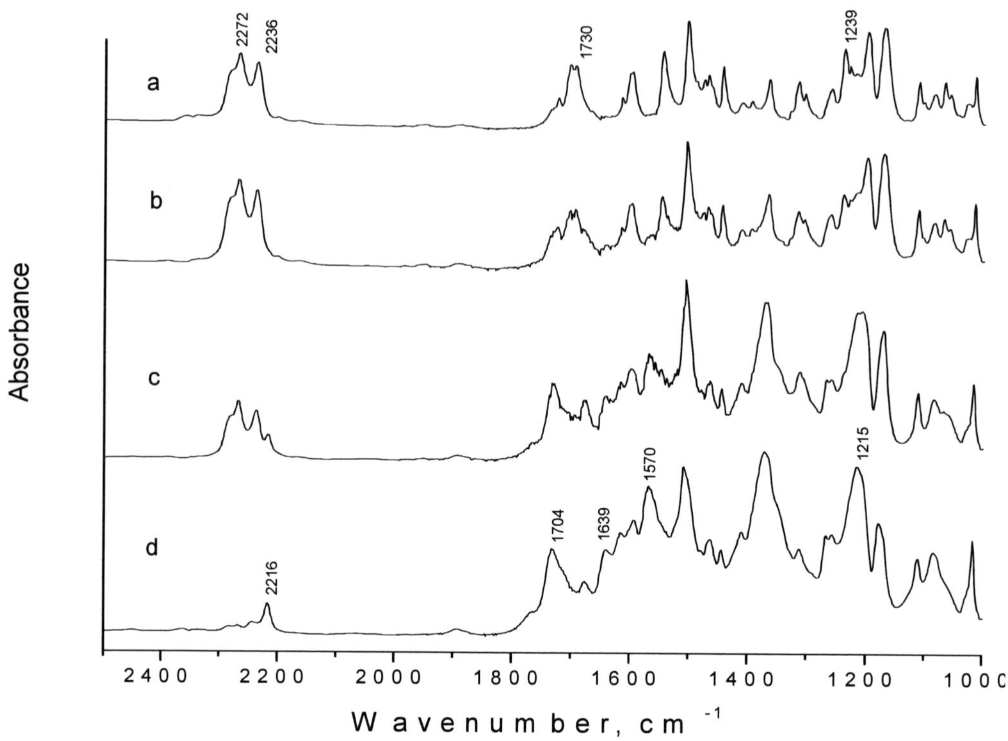

Figure 12. FTIR spectra of BPC/PDU=5:1 (mol) after heating at 150°C. a) 0 min; b) 60 min; c) 120 min; d) 180 min.

The existence of hybridisation effect in polycyanurate-polyurethane semi-IPNs via chemical interaction between cyanate and urethane groups was confirmed later by the method of multiple attenuated total reflection (ATR) infrared spectroscopy [29]. The ATR infrared spectra for some of the samples studied are given in Figure 14. The disappearance of the 2240-2270 cm^{-1} doublet in the spectra of properly prepared samples was accompanied by characteristic spectral changes depending on the polyurethane (PUR) content of the sample. These changes refer especially to appearance and enhancement of the 1560 and 1360 cm^{-1} bands of C=N and to absorption of polyurethane at 1700-1740 cm^{-1}.

One can see that pure PCN is characterized by intense 1560 and 1360 cm^{-1} absorption bands as a consequence of C≡N → C=N transformation. In PCN/PUR compositions, the intensity of these two bands decreases with PUR content but differently. Figure 15 shows these changes related to the intensity of the 1500 cm^{-1} band (benzene rings). The D_{1360}/D_{1500} ratio remained constant (~1.6) for all of compositions with 0-60% of PUR; for 80PUR/20PCN and 90PUR/10PCN networks, the 1500 cm^{-1} band could not be identified because of its overlapping with more intense PUR absorption at 1540 cm^{-1}. Contrarily, the D_{1560}/D_{1500} ratio certainly diminished with PUR content of the sample. There was no trace of the 1560 cm^{-1} band in the compositions with 80-90% PUR.

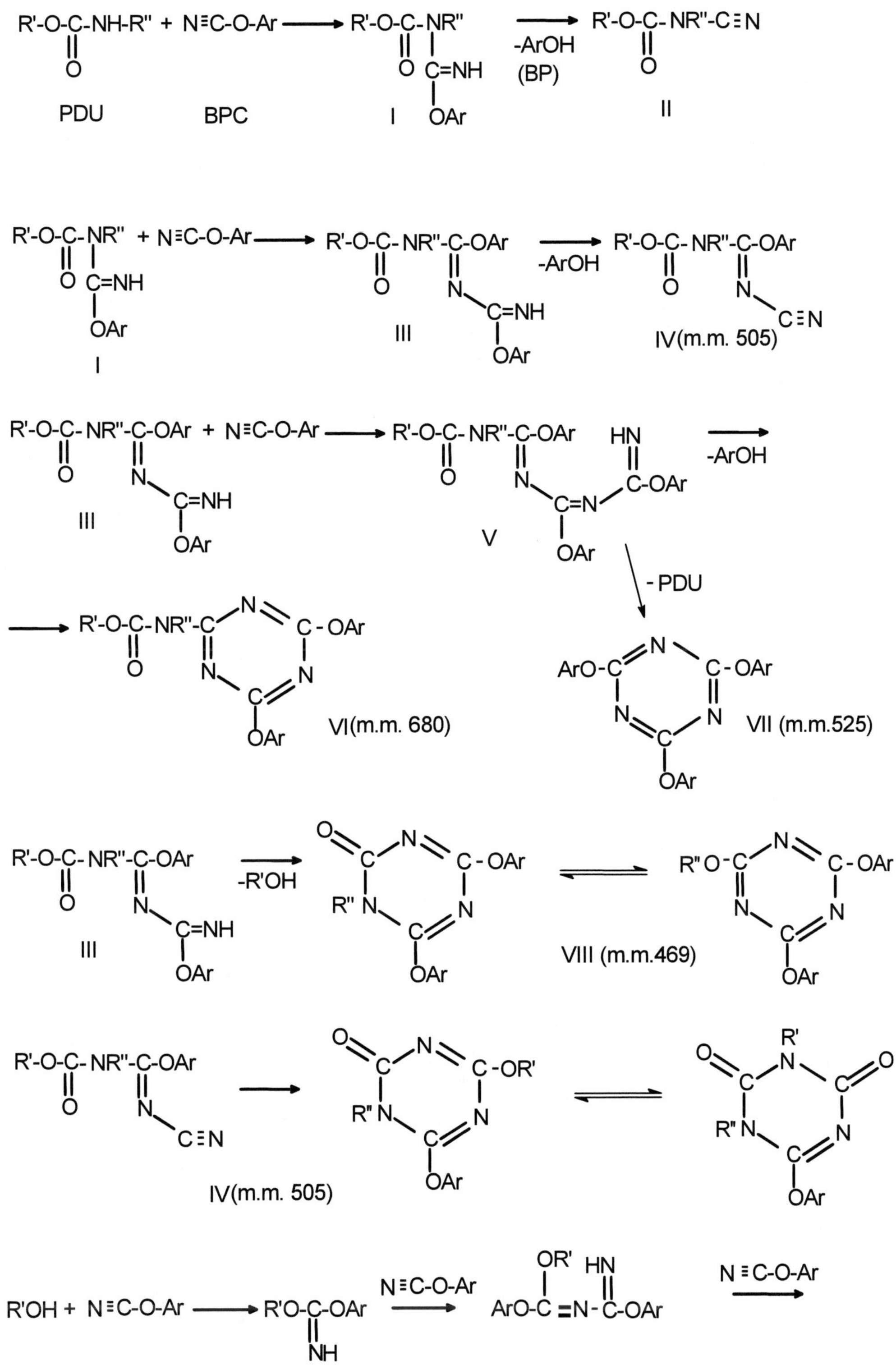

Figure 13 continued on next page.

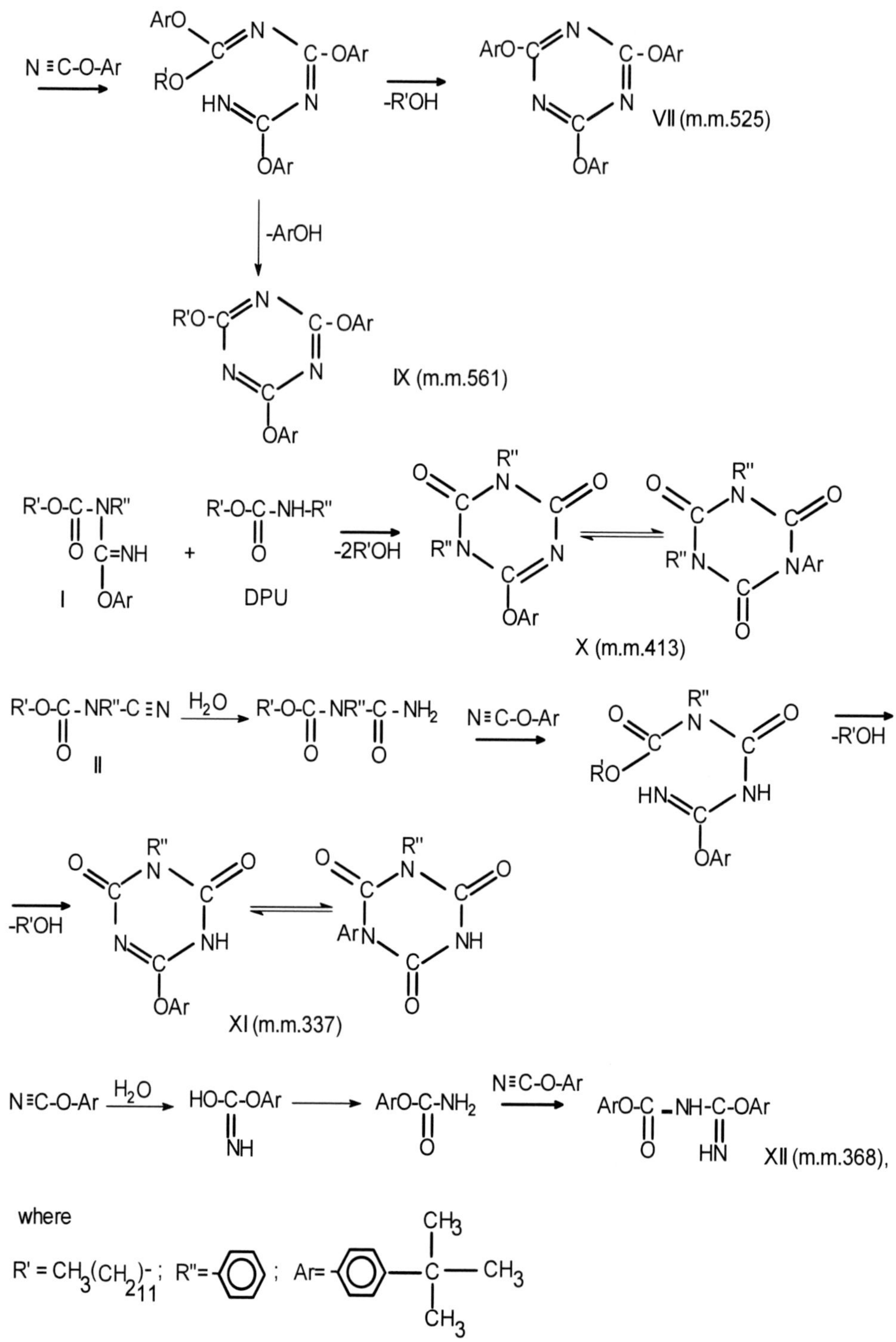

Figure 13. Chemical reactions and transformations in the mixture BPC/PDU.

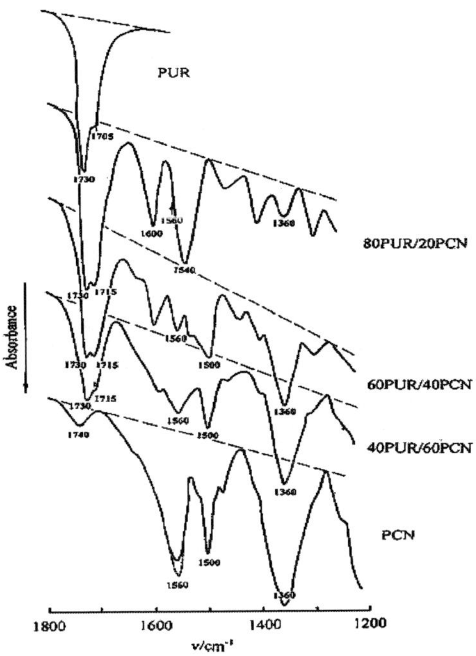

Figure 14. Multiple ATR spectra for individual PCN and PUR components, and some of the PCN/PUR IPNs. KRS crystal light guide, incidence angle $\Theta = 45^o$, $N=5$ reflections. Dashed line at 1560 cm^{-1} indicates the absorption band after PCN post-curing treatment for 20 min at 560 K.

Figure 15 compares also the plot of the D_{1560}/D_{1500} ratio versus composition with the compositional dependence of the gel fraction estimated before [26]. Practically similar dependence is observed for the curves 1 and 3, which is consistent with the generally adopted assignment of the 1560 cm^{-1} band to triazine ring vibrations, i.e. to insoluble PCN network domains. At the same time, the invariability of the D_{1360}/D_{1500} ratio (curve 2) validates the assignment of the 1360 cm^{-1} band to all of the -O-C=N- groups formed, irrespectively of their position in the network structure , triazine rings (conjugated C=N bonds) or substituted urethane (non-conjugated C=N bonds).

In fact, in case that the 1360 cm^{-1} band would be attributed to O-C=N groups in polytriazine structure only, the D_{1360}/D_{1500} ratio had to follow not to curve 2 but to the dashed curve 4 in Figure 15. Comparison of the curves 2 and 4 shows that at least 10-30% of the cyanate groups were spent for the reaction with PUR at 20-60% PUR content in a composition. Besides, the small gel fraction for the compositions with 70-80% PUR (curve 3) correlates well with the absence of any indication of the 1560 cm^{-1} band in the ATR spectrum of the 80PUR/20 PCN composition. This result suggests an obvious predominance of simple grafting of DCEBA to urethane groups over polytriazine network formation. Such small contribution of the cyclotrimerization process may be explained by insufficient cyanate concentration and low probability for the elementary act of trimerization because of the 'dilution effect'.

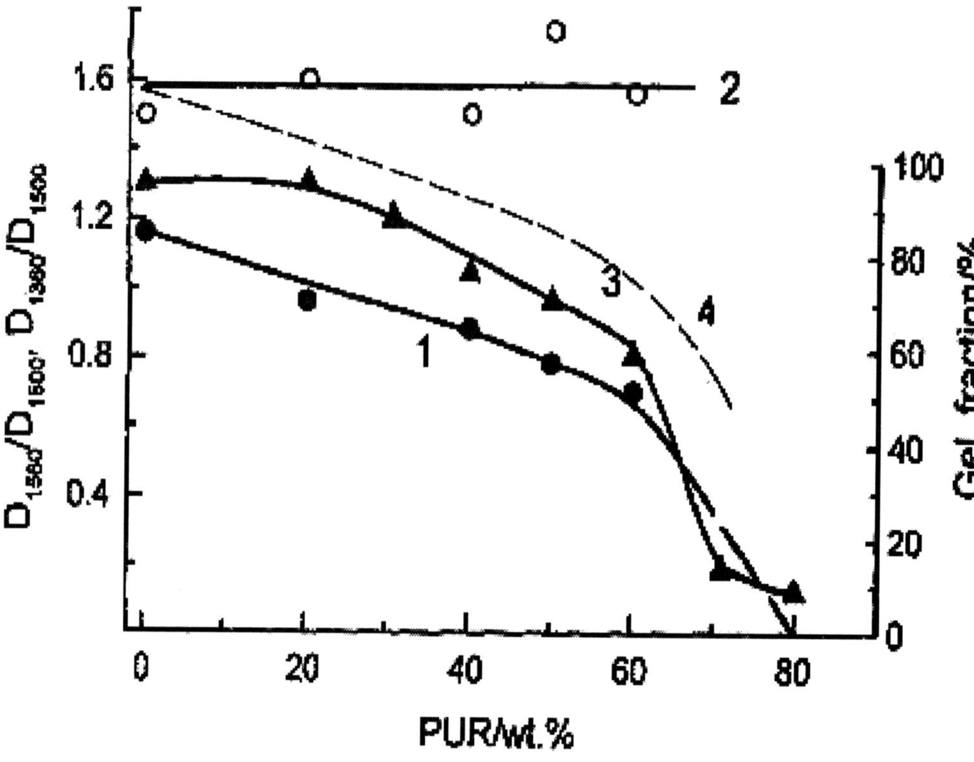

Figure 15. PCN/PUR semi-IPNs: compositional dependencies of the ratios of the band optical densities D_{1560}/D_{1500} (1) and D_{1360}/D_{1500} (2) in the ATR spectra, and the gel fraction value (3) as estimated in (4). Dashed curve 4 corresponds to the hypothetical dependence of the D_{1360}/D_{1500} ratio in the case the 1360 cm^{-1} band would be attributed to C=N groups in polycyanurate rings only.

It is noteworthy that the above estimates agree well with recent IR/HPLC experiments on low-molecular model cyanate/urethane blends described above [28]. It was found that, besides polycyclotrimerization, ca.20-30% cyanate formed co-products with urethane.

Finally, the notion about hybridization effect in PCN/PUR semi-IPNs is confirmed also by the carbonyl absorption behavior (Figure 14). Weak absorption at 1740 cm^{-1} in pure PCN is associated with some side, e.g., oxidative, reaction, which is just best registered for a surface layer analyzed by using the ATR technique. For pure PUR, as extensively hydrogen-bonded segmented elastomer, typical $\nu_{C=O}$ absorption splits into 1730 and 1705 cm^{-1} doublet corresponding to 'free' and 'H-bonded' carbonyls, respectively [30]. In PCN/PUR networks, $\nu_{C=O}$ absorption manifests itself at 1730 and 1715 cm^{-1} (Figure 14). This result is consistent with the hybridization idea. Really, cyanate/urethane reaction leads evidently to better mixing of the components, destruction of the inter-urethane hydrogen bonds, and to formation of the substituted urethanes (carbamates) as indicated above. According to [31], for substituted urethanes the $\nu_{C=O}$ bands are typically displaced by ~15±5 cm^{-1} to lower frequencies compared to those for urethane. Consequently, the 1715 cm^{-1} band may just correspond to substituted urethane groups deprived of hydrogen bonding (1730 $cm^{-1} \rightarrow$ 1715 cm^{-1}).

3.2. Kinetic Peculiarities

The kinetic curves of polyurethane-polycyanurate semi-IPNs synthesis [23, 24] are presented in Figures 16 and 17. As it can be seen from Figure 16, the process of polycyclotrimerization of DCEBA is characterized by induction period, which strongly decreases at introducing LPU to the reaction system and disappears at LPU content in the composition higher than 20 %. We can also see that the higher the content of LPU in the system, the higher the rate of increase of cyanate groups conversion.

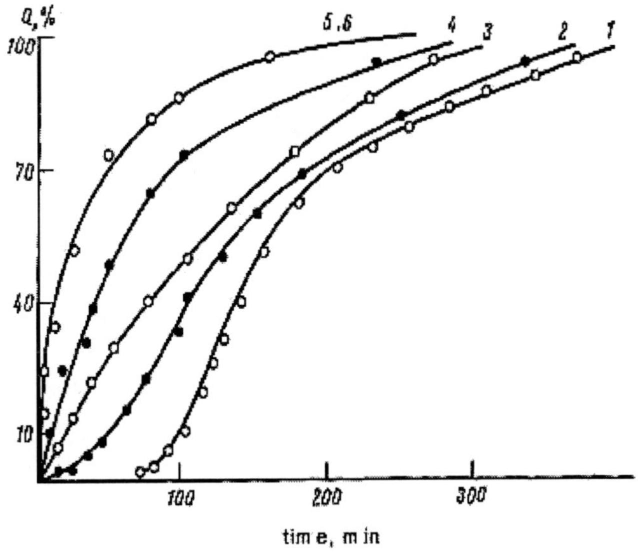

Figure 16. Conversion (Q) of cyanate groups *versus* time for PCN/LPU semi-IPNs. 100/0 (1), 80/20 (2), 60/40 (3), 50/50 (4), 40/60 (5) and 20/80 (6).

Figure 17 presents the dependence of the reaction rate on cyanate groups conversion. One can see from that a low content of LPU (up to 20 %) does not change the kinetic curve shape, and the maximal rate corresponds to 40% of cyanate groups conversion. For the compositions with higher LPU percentage the shape of kinetics curves changes – the reaction rate is observed to be maximal at the beginning (at no conversion of cyanate groups).

Vilensky and co-authors [32] have studied the kinetic peculiarities of DCEBA polycyclotrimerization in the presence of two linear polyurethanes: LPU-1 of mol. mass 40 kg/mol, based on polybutylene glycol adipate (PBGA, molar mass 1 kg/mol), 4,4'-diphenylmethane diisocyanate (DFMDI) and 1,4-butane diol (BD) with co-monomers ratio 1:2:1 and LPU-2 of mol. mass 40 kg/mol, based on polyoxytetramethylene glycol (PTMG, molar mass 1 kg/mol), DFMDI and BD with co-monomers ratio 1:2:1. Thermodynamic affinity and miscibility of LPUs used towards DCEBA were estimated by calculation of solubility parameters (δ, J/cm^3), polarities, interaction parameters and Krause's criterion.

Based on calculations, the authors concluded [32] that DCEBA had to be thermodynamically miscible with LPU-2 and immiscible with LPU-1. For both cases the chemical interactions between the components of IPNs studied have been confirmed.

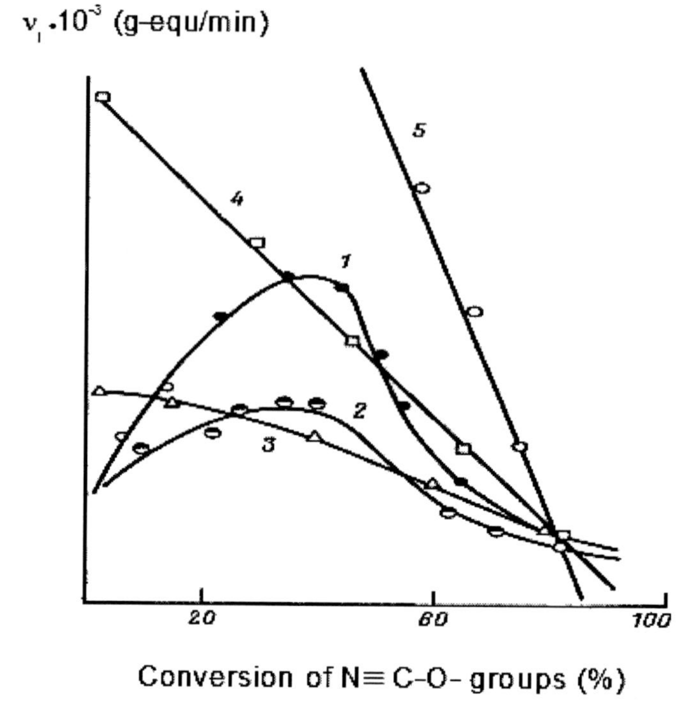

Figure 17. Reaction rate versus conversion of cyanate groups for PCN/LPU semi-IPNs. 100/0 (1), 80/20 (2), 60/40 (3), 50/50 (4), 40/60 (5)

3.3. Relaxation Behaviour and Phase Structure

The solubility parameters for PCN and LPU-1 were also calculated in [33]. In general, for polymer blends, if the values of δ for two components are similar, this implies that they will be miscible in the blends [34]. The calculated solubility parameters were δ=26.65 $(J/cm^3)^{1/2}$ for PCN component and δ=23.42 $(J/cm^3)^{1/2}$ for LPU-1 component. The authors noted that the values of solubility parameters for LPU-1 and PCN are not very close; thus one can conclude that PCN has to be thermodynamically immiscible with LPU-1 and the PCN/LPU-1 compositions should have a microheterogeneous structure.

In series of publications [25,27,29,35-40] several methods were used for characterization of the microphase structure of the semi-IPNs studied. Small-angle X-ray scattering (SAXS), differential scanning calorimetry (DSC) [27, 35-37], dynamic mechanical thermal analysis (DMTA) [27, 30-32], dielectric relaxation spectroscopy (DRS), and thermally stimulated depolarization currents (TSDC) [25, 39, 40] measurements have shown that pure PCN is characterized by a typical homogeneous structure, but for segmented LPU the microphase separation on the level of hard and soft domains due to their thermodynamic immiscibility was denoted. As for semi-IPNs, the destruction of the microphase separated morphology of LPU was observed and the microphase separation between PCN and LPU phases, expected from the difference of solubility parameters, was not found.

The miscible polyblends are characterized by a single glass transition temperature (T_g), which depends on the relative weight fractions of components and their respective T_g values.

As an example, Figure 18 shows the Arrhenius plots for the α relaxation associated with glass transition for four IPNs and LPU based on DRS, TSDC and DSC data [39]. For comparison, a plot of the β relaxation in LPU is also shown. The lines are fits of the Vogel-Tammann-Fulcher equation and of the Arrhenius equation to the data for the α and the β relaxation, respectively, with reasonable values of the fitting parameters [39]. The α relaxation in Figure 18 shifts systematically to longer relaxation times/higher temperatures with increasing PCN content. For the β relaxation, on the contrary, the data for the IPNs (not shown in the figure) practically coincide with those for LPU, indicating that local motions in LPU are not significantly affected by the presence of PCN.

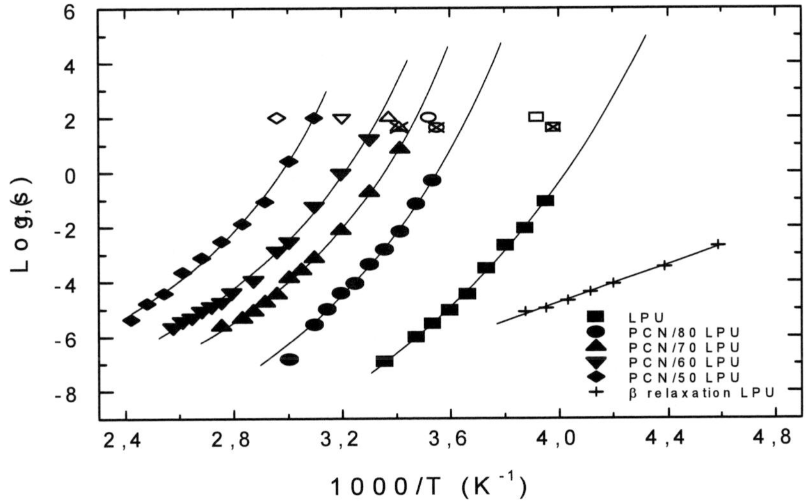

Figure 18. Arrhenius plot of the semi-IPNs indicated on the plot. Full symbols are DRS data for the α relaxation, crosses are DRS data for the β relaxation in LPU, open symbols are DSC data and centered-crossed symbols are TSDC data.

Such a "large-scale" homogeneity is confirmed by the existence of a single T_g in PCN/LPU semi-IPNs, estimated by DMTA, DSC [35-37], DRS and TSDC [39, 40] techniques, common data being presented in Figure 19. All presented T_g values show the same trend with composition. The theoretical compositional dependence of T_g of the PCN/LPU semi-IPNs can be obtained for DMTA data [27] according to the Fox [34] equation:

$$1/T_g = x_1/T_{g1} + x_2/T_{g2} \tag{2}$$

and for DSC data [29] according to the Couchman – Karasz [41] equation:

$$T_g = \frac{x_1 \Delta C_{p1} T_{g1} + x_2 \Delta C_{p2} T_{g2}}{x_1 \Delta C_{p1} + x_2 \Delta C_{p2}} \tag{3}$$

where x_1 is the weight fraction of LPU in the semi-IPNs, T_{g1} the glass transition temperature of LPU, ΔC_{p1} the specific heat increment at the glass transition of LPU and x_2, T_{g2} and ΔC_{p2}

the corresponding quantities of PCN. Figure 19 shows the concentration dependence of T_g obtained from the prediction based on the Fox and Couchman – Karasz equations.

It can be seen that a slight negative deviation from both equations is observed. Such a deviation indicates that there is some interaction between the components in the system [42]. Interestingly, at the same time, combined CRS/DSC analysis [29] indicated the pronounced nanoscale (≤ 2 nm) dynamic heterogeneity within or below the extraordinarily broad glass transition in these single-phase materials. The authors of ref. [29] noted that, despite the attainment of homogeneous structure in the hybrids, the combined CRS/DSC approach permitted to resolve a few kinds of motion that constitute the relaxation region within or close to the broad glass transition.

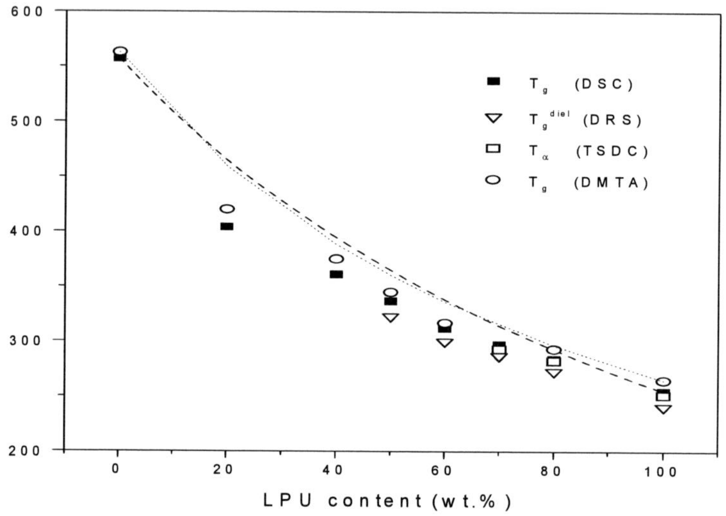

Figure 19. T_g values determined by DSC, DMTA, DRS, and TSDC versus LPU content. The dotted line is a fit of Fox's equation (eq. 2) to the DMTA data and the dashed line is a fit of the Couchman-Karasz equation (eq. 3) to the DSC data.

Their molecular assignments could be safely made, and separate contributions of the PCN network and LPU to the relaxation spectra were estimated. On this basis, first of all, owing to the analysis of creep rate spectra, besides segmental dynamics heterogeneity, also some information concerning the nanoscale heterogeneity of network cross-linking was obtained.

3.4. Influence of Carbon Fiber Filler on Formation and Phase Structure

The influence of two carbon fiber fillers: basic (CF) and with a surface modified by orthophosphoric acid residuals (PCF) on kinetics of epoxypolycyanurate-polyurethane semi-IPNs formation and their phase structure has been studied [43-45].

In Figure 20 the kinetic curves for the formation of pure epoxypolycyanurate (EPCN) and semi-IPNs with 20 and 50 % of polyurethane have been plotted. It can be seen that the reactions of cyanate and epoxy groups are accelerated with increasing LPU content in the system.

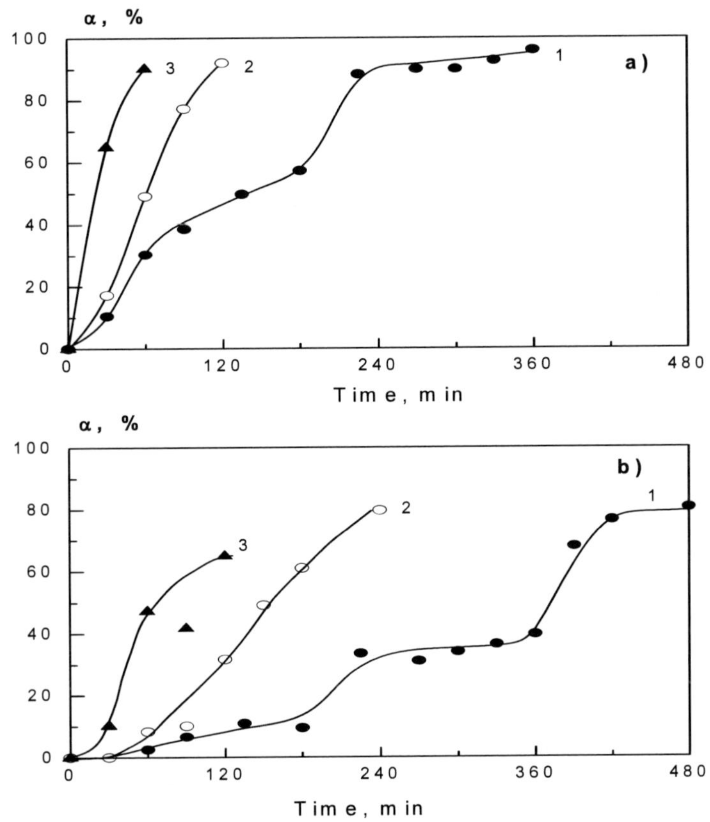

Figure 20. Kinetic curves of conversion of (a) cyanate and (b) epoxy groups for semi-IPNs with EPCN content: 100 (1), 80 (2), 50 (3) wt. % .

For example, the time to achieve the same conversion of the cyanate and epoxy groups compared to pure epoxypolycyanurate decreases for semi-IPNs with 20 % of LPU by 2 and 4 times, respectively (Figure 20a, curves 2, 3) and for semi-IPNs with 50 % of LPU by 1,5 and 3 times, respectively (Figure 20b, curves 2, 3). The additional accelerating effect for the above mentioned reactions is observed at introducing both carbon fiber fillers into the semi-IPNs studied. This can be seen from the kinetic curves given for semi-IPNs with 20 % of polyurethane component in Figure 20. The authors of refs. [44, 45] associate the acceleration of chemical processes with the additional chemical reactions between the components described in [23-25, 28, 29] and by interactions of polymers with the filler surface discussed earlier in [43].

The influence of the fillers on the phase structure of the obtained semi-IPNs was studied by DMA. For the 50/50 semi-IPN a wide single peak on the temperature dependence of E'' has been observed. The authors of [44, 45] explain this fact by forced compatibility of the components. This apparent compatibility appears as a result of a competition between the rates of the chemical processes between the components and phase separation (diffusion). As it was shown above, the rate of polycyanurate network formation increases by 4 times at introducing the LPU to the cyanate monomer. Thus, phase separation does not occur. The authors of references [44, 45] have investigated the influence of the introduction of filler, as well as of its surface modification on the degree of microphase separation in the semi-IPNs studied. In Figure 21 the mechanical spectra of unfilled 50/50 semi-IPN and that

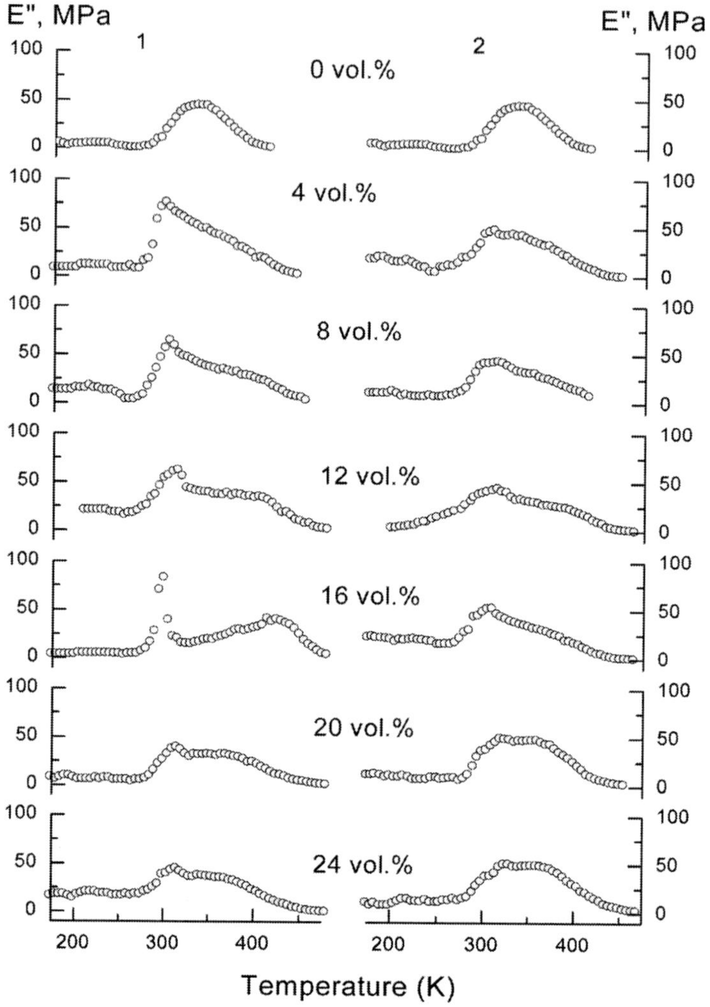

Figure 21. Temperature dependence of loss modulus for semi-IPNs 50/50 filled with CF (1) and PCF (2).

containing from 4 to 24 vol.% of CF are presented. As can be seen from Figure 21, the introduction of small additions (4 vol.%) of CF into semi-IPNs leads to the appearance of a clearly observed relaxation maximum in the mechanical spectrum related to LPU and to a downward shoulder in the temperature interval from 325 to 425 K related to relaxation processes conditioned by molecular mobility in the interface. Thus, the introduction of CF filler generates a microphase separation in the semi-IPNs studied. The further increase of filler content leads to extension of the microphase separation. As it is seen from Figure 21, the intensity of the peak in the region of 325-425 K increases and two peaks with the maxima close to T_g of LPU and EPCN are observed in the mechanical spectrum of semi-IPNs filled with 16 vol.% of CF. At concentrations of CF 20 – 24 vol.% the spectrum looks rather like a curve with one wide peak in the temperature region between T_g of LPU and EPCN. Analyzing the mechanical spectra shown in Figure 21 the authors of refs. [44, 45] supposed a preemptive adsorption of LPU onto CF surface and formation of 2 phases: a polymer phase reach in EPCN and a phase rich in LPU adsorbed onto filler particles at small concentrations of the filler. At higher content of the filler it is supposed that both the components adsorb onto the filler surface that decreases the degree of phase separation in the system. Comparative analysis of the influence of the filler surface activity on the phase structure of the semi-IPNs filled with CF and PCF has shown that the system filled by the latter is characterized by lower level of heterogeneity.

As far as the filler was introduced into the system on the stage of network formation, the physical adsorption of the LPU onto PCF surface is perhaps compensated by competitive chemisorption of EPCN due to chemical reactions of epoxy and cyanate groups of epoxycyanate oligomer with PCF surface functional groups that hampers the phase separation process.

Additionally, an essential increase of specific electrical resistance, reduction of combustibility and existence of piezoelectric effect have been found [46] for polycyanurate-polyurethane semi-IPNs filled with PCF.

3.5 Properties. Adhesion to Metals

The adhesion characteristics of polycyanurate modified with linear polyurethane based on the principle of semi-IPNs have been studied [33]. Aluminum and titanium plates were used as substrates to prepare the joints bonded by DCEBA curing in the presence of LPU. Two clear dependencies of adhesive strength, on adhesive layer thickness and on LPU content, have been observed (see Figures 22 and 23). The dependence of adhesive strength value on adhesive layer thickness has been observed for all the compositions studied. Generally, the thicker the adhesive layer, the higher the adhesive strength for the systems studied. Introduction of LPU first increases the shear strength of PCN to aluminium and titanium. The maximal values of adhesive (shear) strength are achieved at LPU content of 20-25 wt.%, corresponding to formation of the hybrid PCN/LPU network only, owing to the chemical grafting of LPU to PCN. At higher LPU contents semi-IPNs based on the hybrid network and non-grafted LPU are formed. The presence of non-grafted LPU in the adhesive layer leads to reduction of the adhesive strength. Authors note [47] that polycyanurates modified with polyurethanes could be used potentially as high-temperature heating-melt adhesives, coatings, and sealants, as well as matrices for high-performance composites.

4. Polycyanurate-Polyurethane Linked Full-IPNs

The first publications related to synthesis and characterization of polycyanurate-polyurethane full IPNs have appeared in 2000 [48]. Full sequential interpenetrating polymer networks (seq-IPNs) of cross-linked polyurethane (CPU) and polycyanurate (PCN) based on thermally cured dicyanic ester of Bisphenol A (DCEBA) were synthesized by swelling of CPU film by DCEBA followed by polymerization of the latter inside CPU. The CPU, PCN and CPU/PCN seq-IPNs were characterized by small-angle X-ray diffraction, dynamic mechanical analysis, stretching calorimetry and microhardness measurements. Neat CPU was shown to be a microphase-separated system characterized by a regular, three-dimensional macrolattice of network junctions, embedded in uniform-size microdomains of stiff chain fragments, which spanned the continuous matrix of soft chain fragments. In contrast, no large-scale structural heterogeneities were detected in the PCN. The X-ray long spacing *(L)*, the degree of microphase segregation (DMS), the α-relaxation temperature and the mechanical properties (elastic modulus, *E* and microhardness, *H*) were studied as functions of PCN content. Results were explained [48] in the light of a model that discussed the maximum degree of CPU swelling by molten DCEBA as a function of composition. It has been suggested that predominantly chemical interactions between the molten DCEBA and the stiff chain fragment microdomains, reinforcing primary physical interactions, are responsible for the observed transition at 40% PCN content to more homogeneous phase morphology of seq-IPNs.

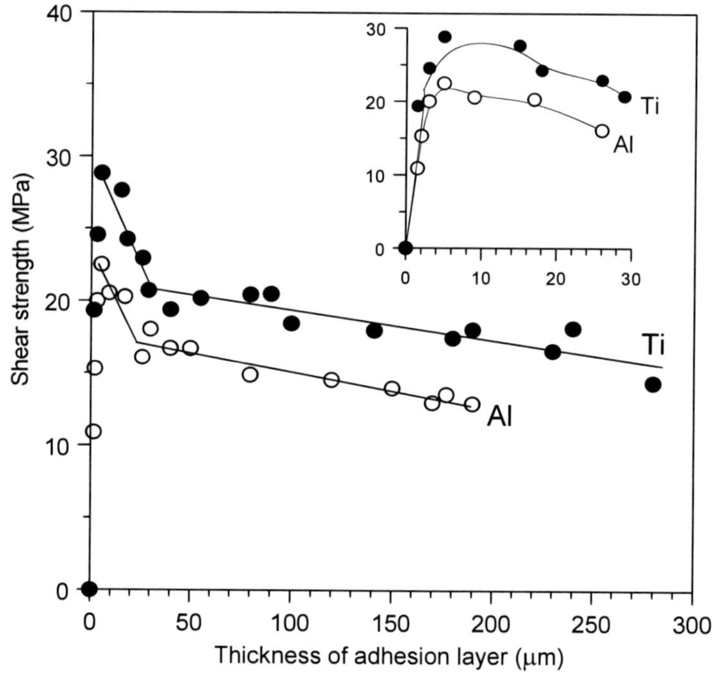

Figure 22. Adhesive strength versus adhesive layer thickness for pure PCN network (metal substrates – Al and Ti).

The characteristic dependence of *L*, DMS, T_g, *E* and *H* of a series of full, seq-IPNs upon composition has been explained according to the following model [48, 49]. The maximum degree of the CPU swelling by molten DCEBA is reached already in the first composition interval φ<0.15.

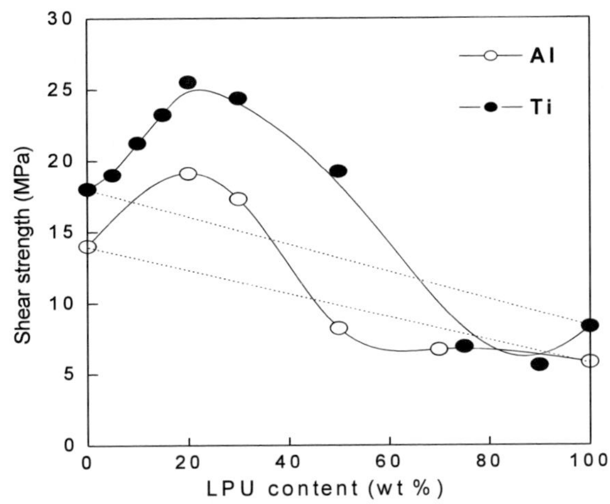

Figure 23. Composition dependence of adhesive strength for aluminium (○) and titanium (●) (layer thickness h ≈150 μm).

However, the volume increment of the swollen CPU exceeds, by far, the actual content of the PCN within corresponding seq-IPNs. Hence, in the second composition interval $0.15 \leq \varphi \leq 0.5$ only a relatively small fraction (i.e. of the order of $\varphi \approx 15$ of the available molten DCEBA) was in a direct interaction with the expanded soft chain fragments of the CPU, whereas the remaining DCEBA simply filled the empty space created by swelling. This, according to the authors of ref. [48], facilitated the accessibility of unreacted functional groups at the periphery of stiff domains of the CPU to strong interactions with the excess of the molten DCEBA. As a result, with increasing PCN content the sharp interface between components is gradually smeared out until a completely homogeneous structure is developed in the composition interval $\varphi>0.5$. Predominantly, chemical interactions between the molten DCEBA and the stiff chain fragments microdomains of the CPU complementing primary physical interactions (dilution) are suggested [48] to be responsible for the observed cross-over to a more homogeneous phase morphology of seq-IPNs at $\varphi>0.4$. The main physical characteristics determined for these seq-IPNs are presented in Table 1.

Table 1. Physical properties of polycyanurate-polyurerthane full seq-IPNs.

CPU/PCN	T_g (K)	E (GPa)	10^5 x α_L (K^{-1})	H (MPa)
100/0	365	0.20	18.4	43.6
88/12	360	0.26	10.0	31.6
78/22	375	0.52	6.1	63.6
62/38	405	1.10	5.3	178.1
51/49	435	2.60	4.0	247.1
37/63	455/540	3.50	3.2	236.2
0/100	485/570	4.05	2.2	241.7

The temperature dependences of heat capacity of the same full seq-IPNs in the 6-340 K range were studied [50] using adiabatic vacuum calorimetry. Using a calorimeter equipped with a static bomb and an isothermal shell, the energies of combustion of the CPU and of the three samples seq-IPNs containing 10, 30 and 50 wt. % of the PCN were measured. On the basis of the experimental data the thermodynamic functions of the research subjects $H^o(T)$-$H^o(0)$, $S^o(T)$- $S^o(0)$ and $G^o(T)$– $H^o(0)$ for the range from $T \to 0$ to $T = 340$ K were calculated and the standard enthalpies of combustion and the thermodynamic parameters of formation ΔH^o_f, ΔS^o_f and ΔG^o_f at 298.15 K were determined

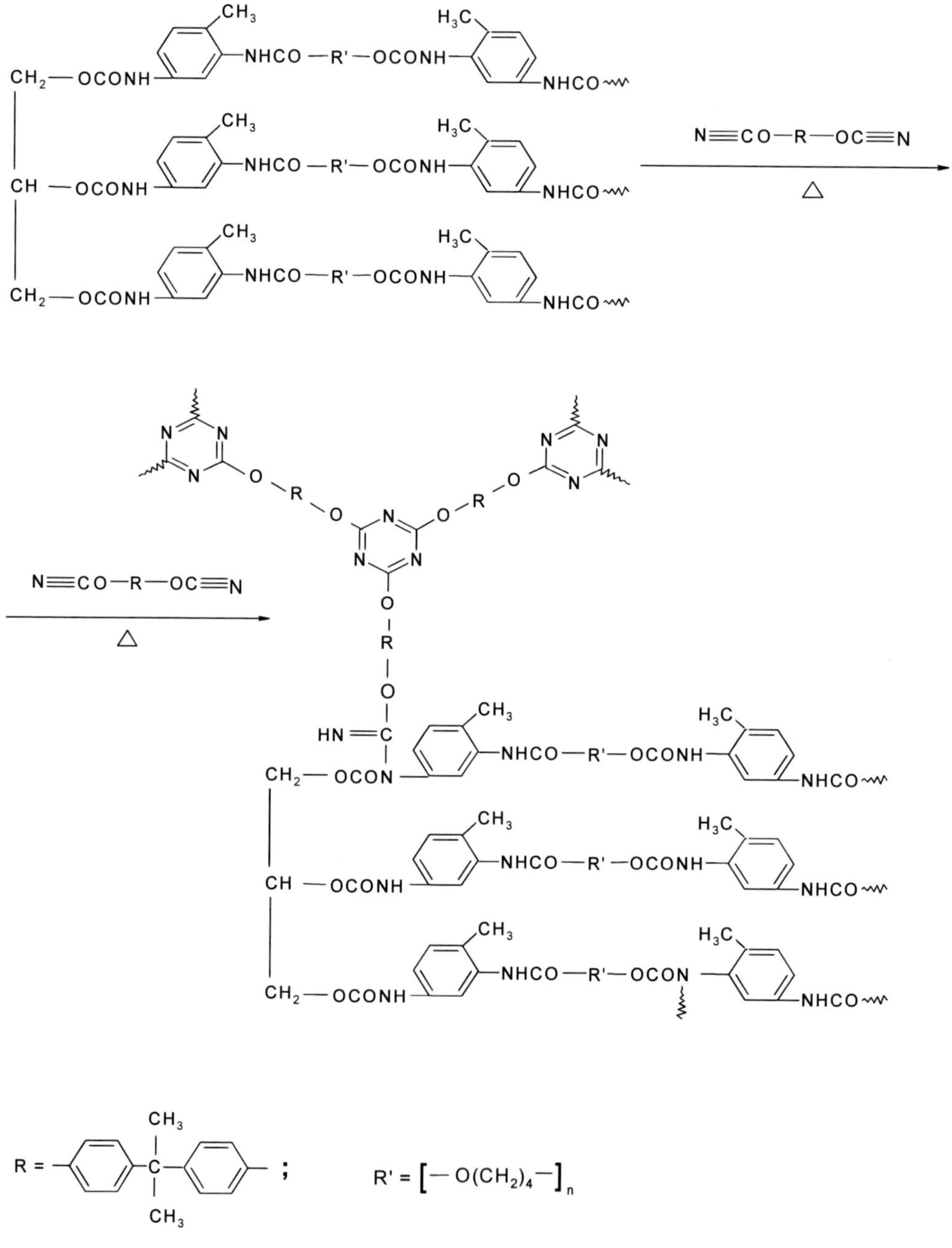

Figure 24. Scheme of chemical linking of PCN and CPU in full seq-IPNs.

The data obtained were used to calculate enthalpy, entropy and the Gibbs functions for the seq-IPNs synthesis. It was shown that the isotherms of diverse thermodynamic properties of interpenetrating polymer networks plotted versus their composition, in particular the molar fraction of the CPU per conditional mole, can be described by straight lines. This made it possible to estimate the thermodynamic behavior of the seq-IPNs of any compositions at standard pressure within a wide temperature range. It was determined [50] that at molar content > 0.50 of PCN in seq-IPNs studied ΔG^{o}_{p} (ΔG^{o} of process) < 0 and this has allowed authors to conclude about thermodynamical miscibility of the components for seq-IPNs of these composition

The authors of ref. [51] have confirmed the high homogeneity of these full seq-IPNs with PCN content > 40 wt. % of PCN. They explained this fact by strong chemical interactions (linking) of the components in these seq-IPNs, especially in ones with high PCN content and confirmed their conclusions by the FTIR data. They have proposed a scheme of covalent links formation during polycyanurate-polyurethane full seq-IPNs synthesis (cf. Figure 24).

Using the method of thermogravimetry (TGA), the authors of ref. [51] have studied the stages of thermal oxidative degradation (cf. Figure 25) for polycyanurate-polyurethane full seq-IPNs of different composition. Based on FTIR and TGA results they have established the formation of a new structure, a so-called hybrid network. They have shown that the coexistence of CPU, PCN and hybrid structures in these linked full seq-IPNs depends on the ratio of the components. According to the number of degradation stages three structures were found for the 90/10 CPU/PCN; PCN and hybrid structures were observed in the 70/30 seq-IPNs and the hybrid network is supposed [51] to be the main structure detected for the 50/50 seq-IPNs.

Broadband dielectric relaxation spectroscopy (DRS) and thermally stimulated depolarization currents (TSDC) techniques were employed [40] to investigate molecular mobility in relation to morphology in PCN/CPU full seq-IPNs in comparison to semi-IPNs from PCN and linear LPU. As it was mentioned above [33] the semi-IPNs were found to be homogeneous at length scales larger than about 2 nm, whereas heterogeneity was suggested [29] at shorter length scales. The full IPNs are characterized [40] by microphase separation. Non-additivity of several physical properties with composition in both systems was explained in terms of increased free volume due to loosened segmental packing of chains confined to nano volumes and of chemical bonds between the components. For example, Figure 26 shows in a log-log plot the real part of dielectric permittivity ε' against frequency f of the full PCN/CPU seq-IPNs at room temperature (298 K). The decrease of ε' with increasing f, observed for all the samples except for pure PCN, corresponds to the segmental (α) relaxation, associated with the glass transition of the CPU phase, as indicated by measurements at various temperatures not shown here. The main result in Figure 26 is the observed non-additivity, i.e. ε' does not change with composition as predicted by mean field theories for non-interacting PCN and CPU phases.

The authors of references [52, 53] have synthesized new sequential and *in-situ* sequential polycyanurate-polyurethane full IPNs and studied their structure-properties relationship. PCN based on 1,1-bis-4-cyanatophenyl-ethane (CPE) and CPU synthesized from an adduct of 1,1,1-trimethylolpropane with 2,4-toluene diisocyanate (1:3, mol) and poly(tetramethylene) glycol were used as the components for IPNs.

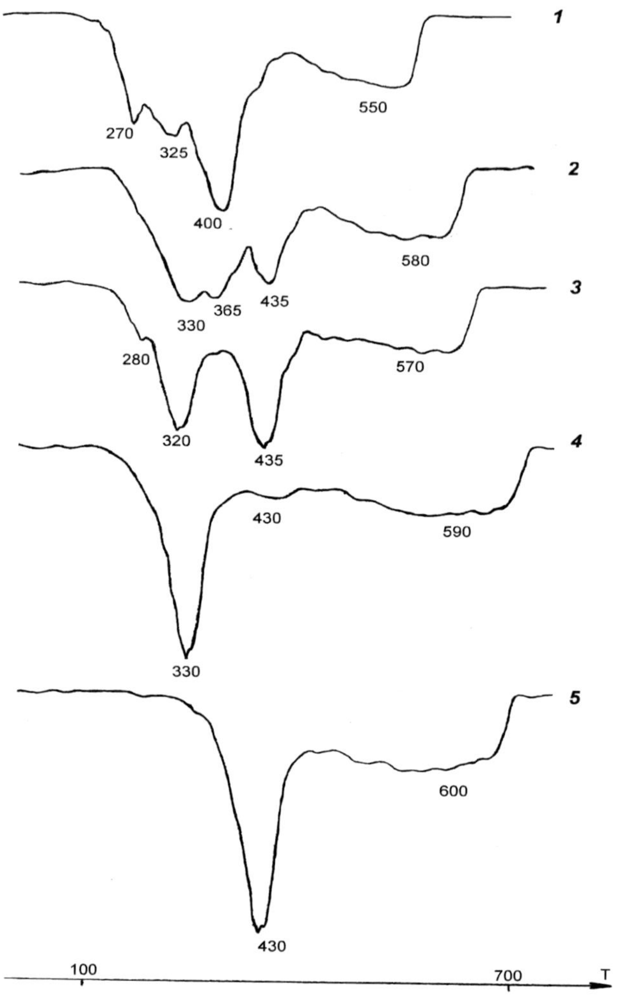

Figure 25. DTG curves. CPU (1), CPU/PCN = 90/10 (2), CPU/PCN = 70/30 (3), CPU/PCN = 50/50 wt. % (4), PCN (5).

Sequential and *in-situ* sequential full IPNs with different weight composition (PCN/CPU = 100/0, 75/25, 50/50, 25/75) were prepared. Sequential IPNs were prepared by swelling of a film of preliminarily synthesized CPU by CPE followed by thermal polycyclotrimerization of the latter inside CPU. On the contrary, the *in-situ* method (some kind of simultaneous one) consisted of mixing all the monomers together, then synthesis of CPU in the presence of CPE monomer at low temperature and next synthesis of PCN by polycyclotrimerization of CPE inside CPU at higher temperature.

The investigation of the thermal and mechanical characteristics of the full-IPNs synthesized was performed in the temperature regions above and below the glass transition. It has been observed that *in-situ* full IPNs exhibit higher values of density (cf. Figure 27) than seq-IPNs, as a consequence of enhanced intermolecular packing.

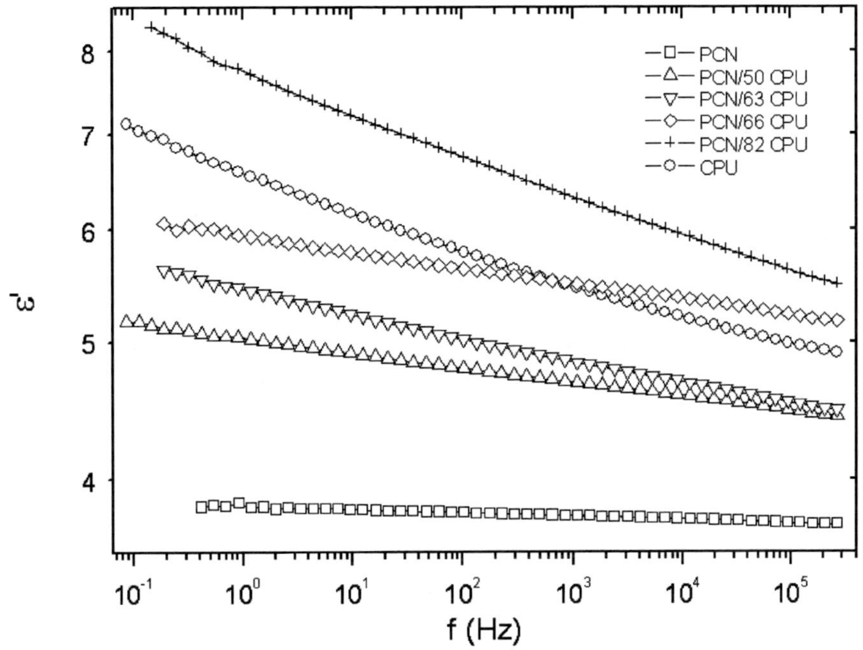

Figure 26. Log-log frequency plot of real part of dielectric permittivity, ε′ (f), at 298 K for the PCN/CPU IPNs indicated on the plot.

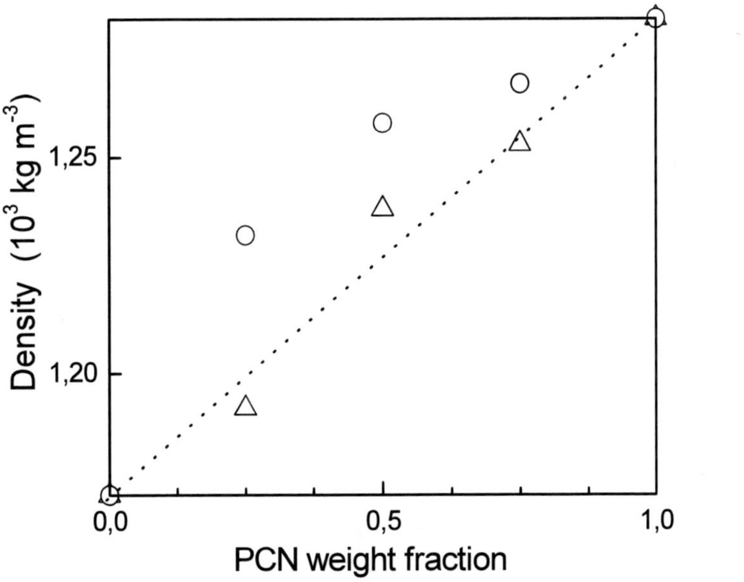

Figure 27. Dependence of the density ρ on the PCN weight ratio in *in-situ* (○) and sequential (Δ) full-IPNs. The diagonal dashed line represents the behavior predicted by the law of additive volumes.

In blends characterized by adhesion between interfaces and no molecular mixing at the boundary, the density of the blend would be expected to follow the rule of mixtures (dotted line in Figure 27). Generally, an abrupt variation of density values implies significant morphology changes. *In-situ* IPNs possess densities higher than those predicted by additive volume law and also systematically larger than those of the corresponding *se*q-IPNs. This result indicates unambiguously that *in-situ* synthesis improves the degree of interpenetration and chain packing leading to a marked reduction of free volume in the network.

According to the authors of ref. [53], it is believed that, in the interpenetration process, the junctions of PCN network between CPU soft segments can lock CPU chains by making disentanglement very difficult. Moreover, the hard-segment domains of PCN act as cross-links and filler particles to the soft segment matrix of CPU. This should lead to a decreasing mobility of CPU molecular groups and to a corresponding increase of the energy required for cooperative motions. On the contrary, the CPU phase, increasingly soft with increasing temperature, acts as a sort of internal diluent for the PCN matrix and lowers its *Tg*.

Table 2. Calorimetric glass transition temperatures Tg and activation energies Ei of the mechanical γ_1 and γ_2-relaxation in *in-situ* and *seq*-IPNs

PCN/CPU weight ratio	T_g (K)		E_i (kJ/mol^{-1})	
	CPU	PCN	E_1	E_2
0/100	314	-	39	-
seq-25/75	320	377	40	61
seq-50/50	326	387	35	66
seq-75/25	333	395	40	60
in-situ-25/75	333	385	41	61
in-situ-50/50	340	392	42	67
in-situ-75/25	345	400	43	60
100/0	-	445	-	55

Two well-defined Tg's are observed in all the IPNs samples (cf. Figures 28) revealing that microsegregation of CPU and PCN phases takes place. A significant convergence of $T_g s$ of CPU and PCN components in IPNs can bee seen from Table 2. It is very easy to calculate that the difference between the T_g values of the individual networks is 131 K while it is around 52-62 K between the components in IPNs. This can be explained by interpenetration process, as well as by the above mentioned chemical linking between the individual networks in IPNs. Thus, it emerges that PCN/CPU blends have a structure where the PCN phase is plasticized (α_a-peaks at increasingly lower temperatures than in pure PCN) while CPU soft segments are partially locked (α_a-peaks at increasingly higher temperatures than in pure CPU).

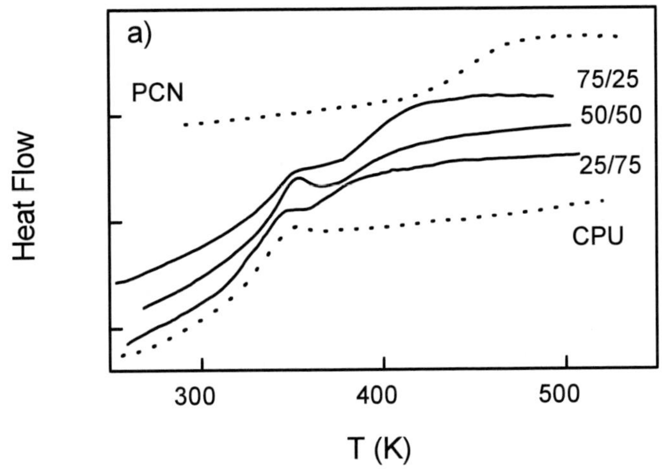

Figure 28. DSC traces through the glass transition region for the neat CPU and PCN and for the three samples of *in-situ* IPNs. The curves have been duly shifted for easier com

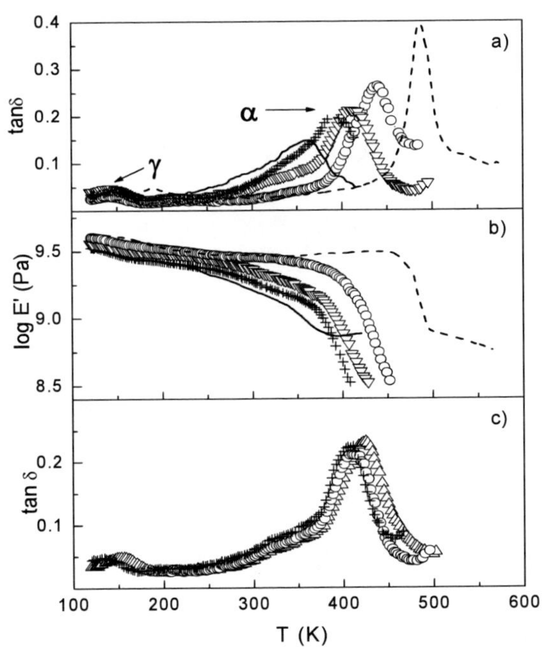

Figure 29. Temperature dependencies of (a) the internal friction tan δ and (b) the dynamic modulus E in pure CPU and PCN and *in-situ* IPNs at selected frequency of 3 Hz: continuous line, pure CPU; (+), PCN/CPU =25/75; (∇), PCN/CPU =50/50; (o), PCN/CPU =75/25; (---) pure PCN. (c) The effect of the driving frequency on the temperature dependence of tan δ in the 50/50 *in-situ IPN*:(+), 0.3 Hz; (∇), 3 Hz; (o), 30 Hz.

Figures 29 and 30 show the comparison between the mechanical spectra of two samples containing the same PCN/CPU ratio (25/75), but prepared by sequential or by *in-situ* sequential method, demonstrating significant differences between the width and the temperature locations of the primary relaxation curves.

In-situ IPNs show higher calorimetric (cf. Table 2) and mechanical (cf. Figure 30) glass transition temperatures, T_g than *se*q-IPNs, having the same composition. In the glassy region, the relaxation dynamics of *in-situ* IPNs probed by dynamical mechanical analysis reveals the existence of distinct local segmental motions associated to the individual components. It has been proved that the interpenetration process associated to the *in-situ* synthesis reduces markedly the free-volume in the system, giving rise to significant differences between the local and cooperative molecular mobilities of *in-situ* and *se*q-IPNs. In particular it has been observed a larger "γ-suppression" effect for the γ_2-relaxation of PCN phenylene units, as a consequence of enhanced local intermolecular packing, which prevents the group rearrangements.

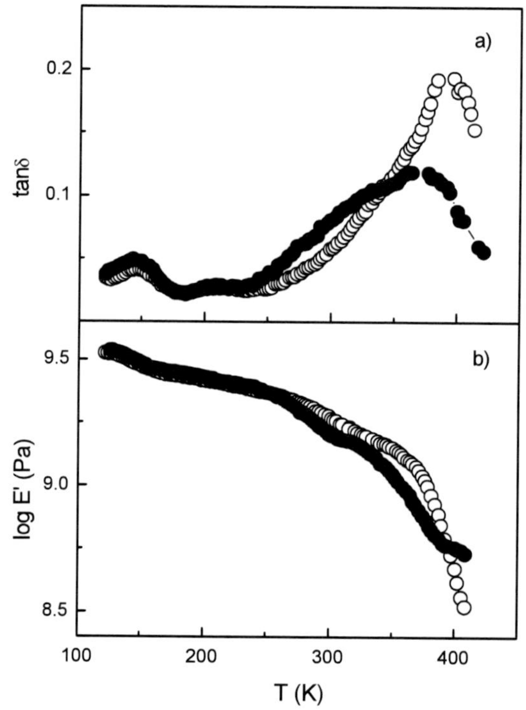

Figure 30. Comparison between the temperature dependencies of (a) the dynamic modulus E' and (b) *tan* δ data in *in-situ* (○) and *seq*-(●) *IPNs* samples having the same weight ratio (25/75).

Recently [54] a series of IPNs has been synthesized from oligourethane methacrylate (OUMA) and DCEBA by a combined *in-situ* sequential photo/thermo curing. First, oligourethane diisocyanate, OUDI was synthesized by reaction of poly(butylene glycol adipate with toluene diisocyanate at molar ration 1:2. Then, OUMA was synthesized by reaction of OUDI with 2-Hydroxyethyl methacrylate (HEMA) at molar ration 1:2. Next,

PUMA was synthesized by photo-initiated curing reaction of OUMA in the presence of DCEBA. Finally, the PCN was synthesized by thermal curing reaction of DCEBA. In addition to these reactions (chemical equations can be found in [54]), we have to notice that some chemical grafting of the two networks can also occur by reaction of cyanate groups of growing PCN with urethane groups of PUMA as described above in this Chapter.

The chemical structure of IPNs has been studied by FTIR spectroscopy, gel fraction content and density measurements. The phase structure has been studied by means of X-Ray diffraction, DSC and TMA, while the thermal stability has been estimated using Thermogravimetry technique. The interpenetration of the two network chains and possible chemical linking between the networks is discussed. The X-Ray, DSC and TMA data have shown that at low polycyanurate network (PCN) content, the structure of polyurethanemethacrylate (PUMA) network is semi-crystalline, while the compositions with PCN content greater than 25% and up to 50% are fully amorphous with a phase inversion process occurred. For IPNs with PCN content varying from 50% up to 95% a mixed phase is formed. The thermal stability of the IPN samples possessing mixed phase with greater content of PCN component is found to be higher.

5. CONCLUSIONS

The development of high-performance, high-temperature composites is an application area where cyanate esters are pitted directly against the epoxy resins, a class of materials, which has earned a reputation for combining both high load-bearing characteristics together with ease of processing. Cyanate esters exhibit all the characteristics inherent to epoxy systems: excellent adhesion to most metallic alloys and to various other substrate types, wide range of processing, cure and property characteristics, absence of volatile products during curing, easy processing without the necessity for applying high pressures during bonding operations, good wetting properties and relatively low shrinkage during cure. Since cyanate esters also have a much better hot/wet performance, much lower moisture and solvent uptake, much lower dielectric loss properties and no shrinkage during cure, this makes them outstanding candidates to challenge the dominance of epoxy resin systems. However, the polycyanurate networks (PCN) often display poor toughness due to their high crosslink density that along with a high price limits their industrial application.

Modification of PCN by reactive oligomers and polymers with formation of hybrid or interpenetrating (semi- and full grafted or linked IPNs) polymer networks with controlled degree of phase separation (often nano-scale) of the components provides development of composites with an optimal combination of mechanical and thermal characteristics for transportation and electronics industries. Curing the cyanate ester resins in the presence of monomers, oligomers or polymers of certain reactivity (the known content of the functional groups of the known reactivity) towards them provides the needful limited concentration of chemical bonding between the components (reactive compatibilization) leading to formation of fine morphology. The chemical grafting of polymer chains of modifiers during network formation reduces the heterogeneity of the system to the degree depending on the concentration of graftings (interbondings). Selection of modifiers with the required content of the functional groups of the correct reactivity allows achieving the lowest level of heterogeneity, nano-scale heterogeneity, producing a nanostructure.

Certainly, using of the additional components and additional reactions at PCN composites processing requires control of selective physical and chemical adsorption of the polymer components into the filler surface, which will lead to improved adhesion at the polymer matrix / filler interface and correspondingly to improved operating characteristics.

Finally, the fine morphology described provides improved properties; in particular, a toughened material of cost effective formulation with good thermal stability for electronics and structural composite applications as well as for membrane technology.

REFERENCES

[1] Shimp DA, Christenson JR, Ising SJ. *Proc 34th SAMPE,* 1989, p.222.

[2] Fainleib AM, Shantalii TA, Pankratov VA. Copolymers of cyanate esters and plastics based on them. *Compos. Polym. Mat.* (Kiev, in Russian), 49, 39-53 (1991).

[3] Fainleib AM, Sergeeva LM, Shantalii TA. Triazinecontaining interpenetrating polymer networks. *Compos. Polym. Mat.* (Kiev, in Russian), 50, 63-72 (1991).

[4] Hamerton I, editor. Chemistry and Technology of Cyanate Ester Resins. Glasgow: Chapman and Hall, 1994.

[5] Grigoryeva O, Fainleib A, Sergeeva LM. Thermoplastic polyurethane elastomers in interpenetrating polymer networks, in: "Handbook of Condensation Thermoplastic Elastomers", editor S Fakirov, WILEY-VCH, Germany, 2005, Chapter 12, p. 325-354.

[6] Nair CPR, Mathew D, Ninan KN. Advantages in cyanate ester resins. *Adv. Polym. Sci.*, 155, 1-99 (2000).

[7] McConnel V.P. *Advanced Composites* 1992, May/June issue, 28.

[8] Pascault, J.P. *Macromol.Chem., Macromol Symp.* 1995, 93, 43-51.

[9] Cao ZQ, Mechin F, Pascault JP. *Polym Int* 1994;34:41-48.

[10] Uhlig C, Bauer J, Bauer M. *Macromol Chem Macromol Symp* 1995;93:69-79.

[11] Srinivasan SA, Joardar SS, Kranbeuhl D, Ward TC, McGrath JE. *J Appl Polym Sci* 1997, 64, 179-190.

[12] Fainleib A, Grigoryeva O, Hourston D. Synthesis of inhomogeneous modified polycyanurates by reactive blending of bisphenol A dicyanate ester and polyoxypropylene glycol. *Macromol. Symp.,* 164, 429-442 (2001).

[13] Fainleib A, Hourston D, Grigoryeva O, Shantalii T, Sergeeva L. Structure development in aromatic polycyanurate networks modified with hydroxyl-terminated polyethers. *Polymer,* 42, 8361-8372 (2001).

[14] Fainleib A.M., Grigoryeva O.P., Hourston D.J. Structure-Properties Relationships for Bisphenol A Polycyanurate Network Modified with Polyoxytetramethylene Glycol. *Int. J. Polym. Mat.,* 2001, 51(1-2), 57-75.

[15] Fainleib AM, Shantalii TA, Klochok OO, Galatenko NA. Synthesis and some properties of poly(bisphenol A)cyanurate modified with oligoethers. *Compos. Polym. Mat.*, Kyiv, 23(1), 14-19 (2001).

[16] Martin D, Bacaloglu R. Organishe Synthesen mit Cyansäureestern. Berlin: Akademie-Verlag, 1980.

[17] Martin D, Schwarz KH, Rackow S, Reich P, Gründemann E. Chem Ber 1966;99:2302-2308.

[18] Fainleib A, Grenet J., Garda MR., Saiter JM , Grigoryeva O, Grytsenko V, Popescu N, Enescu MC. Poly(bisphenol A)cyanurate network modified with poly(butylene glycol adipate). Thermal and mechanical properties. *Polym. Degr. Stab*, 81(3), 423-430 (2003).

[19] Angell Ca. *J Non-Cryst Solids* 1999;131-133:13-31.

[20] Angell Ca. In: Ngai Kl, Wright Gb, Editors. *Relaxation In Complex Systems.* Washington DC: Naval Research Labaratory, 1984. P.3.

[21] Fainleib A, Grigoryeva O, Garda MR, Saiter JM, Lauprêtre F, Lorthioir C, Grande D. Synthesis and characterization of polycyanurate networks modified by oligo(ε-caprolactone) as precursors of porous thermosets. *J Appl Polym Sci* 2007;106:3929-3938.

[22] Grande D, Grigoryeva O, Fainleib A, Gusakova K, Lorthioir C. Porous Thermosets via Hydrolytic Degradation of Poly(ε-caprolactone) Fragments in Cyanurate-Based Hybrid Networks. *Eur. Polym. J.* – in press.

[23] Fainleib A M, Novikova T I, Shantalii T A and Sergeeva L M (1992) Kinetic of formation semi-interpenetrating polymer networks based on crosslinked polycyanurate and linear polyurethane, *Polym. Sci., Ser B* 33:60-67.

[24] Fainleib A M, Novikova T I, Shantalii T A and Sergeeva L M (1992) Synthesis, structure and some properties of the polycyanurate-polyurethane semi-IPNs, *Polym Mater Sci Eng* 66:131-132.

[25] Lipatov S Yu, Fainleib A M, Shantalii T A and Sergeeva L M (1992) Semi-interpenetrating networks based on the oligomers of dicyanates and linear polyurethanes, *Polym Sci* 34:407-410.

[26] Fainleib A M, Shantalii T A and Sergeeva L M (1993) Polycyanurate-polyurethane semi-IPNs, *Compos Polym Mat* 54:14-16.

[27] Brovko A A, Fainleib A M, Shantalii T A, Sergeeva L M and Davidenko V V (1994) Structure and viscoelastic properties of polycyanurate-polyurethane semiinterpenetrating polymer networks, *Polym Sci* 36:934-938.

[28] Fainleib A M, Kay M, Buffel K, Bauer J and Bauer M (2000) Chemical transformations in blends of monofunctional organic cyanate and urethane, *Reports of the National Academy of Sciences of Ukraine* N12:170-174.

[29] Bershtein VA, Egorova LM, Ryzhov VP, Yakushev PN, Fainleib AM, Shantalii TA and Pissis P (2001) Structure and segmental dynamics heterogeneity in hybrid polycyanurate-polyurethane networks, *J Macromol Sci Phys, B* 40:105-131.

[30] Srichatrapimuk V.W.and Cooper S.L., *J. Macromol. Sci.-Phys., B15,* 267 (1978).

[31] Bellamy LJ. *The Infra-Red Spectra of Complex Molecules*. Wiley, New York, 1954.

[32] Vilensky V A, Fainleib A M, Goncharenko L A and Danilenko I Yu (2002) Influence of "inclusion" polymer and components thermodynamic affinity on bisphenol A dicyanate ester polycyclotrimerization, *Reports of the National Academy of Sciences of Ukraine* N1:142-148.

[33] Grigoryeva O., Fainleib A., Pissis P, Boiteux G. Effect of hybrid network formation on adhesion properties of polycyanurate/polyurethane semi-interpenetrating polymer networks. *Polym. Eng. Sci.*, 42 (12), 2440-2448 (2002).

[34] Van Krevelen D.W. *Properties of polymers*. Their correlation with chemical structure; their numerical estimation and prediction from additive group contributions , Elsevier, Amsterdam, 189 (1990).

[35] Bartolotta A, Di Marco G, Lanza M, Carini G, D'Angelo G, Tripodo G, Fainleib A M, Slinchenko E A and Privalko V P (1997) Molecular mobility in semi-IPNs of linear polyurethane and heterocyclic polymer networks, *J Adhesion* 64:269-286.

[36] Bartolotta A, Di Marco G, Carini G, D'Angelo G, Tripodo G, Fainleib A, and Privalko V (1998) Relaxation in semi-interpenetrating polymer networks of linear polyurethane and heterocyclic polymer networks, *J Non-Cryst Solids* 235-237:600-604.

[37] Bartolotta A, Di Marco G, Lanza M, Carini G, D'Angelo G, Tripodo G, Fainleib A M, Slinchenko E A, Shtompel V I and Privalko V P. (1999) Synthesis and physical characterization of semi-IPNs of linear polyurethane and heterocyclic polymer network, *Polym Eng Sci* 39:549-558.

[38] Balta Calleja F J, Privalko E G, Fainleib A M, Shantalii T A, and Privalko V P (2000) Structure-microhardness relationships for semi-interpenetrating polymer networks, *J Macromol Sci Phys, B* 39:131-141.

[39] Georgoussis G, Kyritsis A, Bershtein V A, Fainleib A M and Pissis P (2000) Dielectric studies of chain dynamics in homogeneous semi-interpenetrating polymer networks, *J Polym Sci, B Polym Phys* 38:3070-3087.

[40] Pissis P, Georgoussis G, Bershtein V A, Neagu E and Fainleib A M (2002) Dielectric studies in homogeneous and heterogeneous polyurethane / polycyanurate interpenetrating polymer networks, *J Non-Cryst Solids* 305:150-158.

[41] Couchman R. P. *Macromolecules*, 11, 1156 (1978).

[42] Zhang Y. and Hourston D.J. *J.Appl.Polym.Sci.,* 69, 271 (1998).

[43] Seminovych G M, Fainleib A M, Slinchenko E A, Brovko A A, Sergeeva L M and Dubkova V I (1999) Influence of carbon fibre on formation kinetics of cross-linked copolymer from bisphenol A dicyanate and epoxy oligomer, *React Funct Polym* 40:281-288

[44] Brovko O O, Fainleib A M, Slinchenko E A, Dubkova V I and Sergeeva L M (2001) Filled semi-interpenetrating polymer networks: formation kinetics and properties, *Compos Polym Mat* 23(2):85-91.

[45] Fainleib A.M, Brovko O.O., Slinchenko E.A., Sergeeva L.M. Compatibilization of components in interpenetrating polymer networks. Influence of carbon fiber filler on formation kinetiks and phase structure. *Nonlinear Optics. Quantum Optics*, 32, 149-160 (2004).

[46] Sergeeva L M, Dubkova V I, Fainleib A M, Alekseenko V I, Brovko O O and Maevskaya O I (2000) The role of active carbon-fibrous filler in decrease of combustibility of semi-interpenetrating polymer networks, *Intern J Polym Mat* 47:31-41.

[47] *Pat.USSR* N1807066, 1992. *Composition for coating*. Fainleib A.M., Shantalii T.A., Sergeeva L.M.

[48] Balta Calleja F.J.,.Privalko E.G., Sukhorukov D.I., Fainleib A.M., Sergeeva L.M., Shantalii T.A., Shtompel V.I., Monleon Pradas M., Gallego Ferrer G., Privalko V.P. Structure-properties relationships for cyanurate-containing, full interpenetrating polymer networks. *Polymer*, 41(12), 4699-4707 (2000).

[49] Fainleib A.M, Shantalii T.A., Shtompel V.I., Monleon Pradas M., Sergeeva L.M., Privalko Y.P. Polycyanurate-polyurethane interpenetrating polymer networks. Synthesis and phase structure. *Reports of the NAS of Ukraine*, 2002, N7, 155-160.

[50] Lebedev B.V., Kulagina T.G., Bykova T.A., Fainleib A.M., Grytsenko V.V., Sergeeva L.M. Thermodynamics of interpenetrating polymer networks based on crosslinked polycyanurate and polyurethane in region from T →0 to 340K. *Polym. Sci.* A, 45(4), 649-659 (2003).

[51] Fainleib A., Kozak N., Grigoryeva O., Nizelskii Yu., Gritsenko V., Pissis P., Boiteux G. Structure-thermal property relationships for polycyanurate-polyurethane linked interpenetrating polymer networks. *Polym. Degr. Stab.*, 76(3), 393-399 (2002).

[52] Bartolotta A., Di Marco G., Lanza M., Carini G., D'Angelo G., Tripodo G., Fainleib A., Danilenko I., Sergeeva L. Mechanical behavior of polycyanurate-polyurethane sequential full-Interpenetrating polymer networks. *J. Non-Crystal. Solid,* 307-310, 698-704 (2002).

[53] Bartolotta A., Di Marco G., Lanza M., Carini G., D'Angelo G., Tripodo G., Fainleib A., Danilenko I., Grytsenko V., Sergeeva L. Thermal and mechanical properties of simultaneous and sequential full-interpenetrating polymer networks. *Mat. Sci. and Eng. A.*, 370, 288-292 (2004).

[54] Fainleib A., Saiter J.M., Youssef B., Garda M.R., Danilenko I., Grenet J. Novel in situ sequential IPNs from oligourethanemethacrylate and dicyanate ester of bisphenol A synthesized by combined photo/thermo curing. *Scientific Israel-Technological Advantages,* 10(1), 15-25 (2008).

In: Polyester: Properties, Preparation and Applications
Editors: Hina Yamashita, Yui Nakano
ISBN: 978-1-60456-753-3

Chapter 4

BIODEGRADABLE ALIPHATIC POLYESTERS DERIVED FROM 1,3-PROPANEDIOL: CURRENT STATUS AND PROMISES

***George Z. Papageorgiou and Dimitrios N. Bikiaris*[*]**

Laboratory of Organic Chemical Technology, Department of Chemistry, Aristotle University of Thessaloniki, Macedonia, Greece.

ABSTRACT

Among biodegradable polymers, polyesters derived from aliphatic dicarboxylic acids and diols are of special importance. Polyesters of 1,3-propanediol were overlooked till recently, since the specific monomer was not available in the quantities and price that might enable production of polymers. However, in recent years more attractive processes have been developed for the production of 1,3-propanediol from renewable resources. Nowadays, research on biodegradable poly(1,3-propylene alkanedioate)s, such as poly(propylene succinate) (PPSu), poly(propylene adipate) (PPAd) and poly(propylene sebacate) (PPSe), has gained increasing interest, due to their fast biodegradation rates and their potential uses in biomedical or pharmaceutical applications, such as drug delivery systems. The odd number of methylene units in the diol segment is responsible for the lower melting points, lower degree of crystallinity and higher biodegradation rates of the specific polymers compared with their homologues based on ethylene-glycol or 1,4-butanediol. In this chapter synthesis and properties of the 1,3-propanediol based aliphatic polyesters and especially their biodegradation characteristics are reviewed. Specific attention has been paid to preparation of related copolymers and blends with other important polymers, since these techniques may offer routes for optimizing properties and producing tailor-made materials. Copolymerization of 1,3-propanediol with mixtures of aliphatic or even aromatic acids, leads to linear polyesters with improved or balanced

[*]Send correspondence to Dimitrios N. Bikiaris, Laboratory of Organic Chemical Technology, Department of Chemistry, Aristotle University of Thessaloniki, 541 24 Thessaloniki, Macedonia, Greece; Tel.: +30 2310 997812; Fax: +30 2310 997769; E-mail address: dbic@chem.auth.gr

biodegradation and mechanical properties. Blends with other biodegradable polymers have been studied recently. Finally, potential pharmaceutical applications of poly(1,3-propylene alkanedioate)s as solubilizing and stabilizing carriers for drugs are exemplified.

1. Introduction

Polyesters of 1,3-propanediol only recently were studied, since the specific diol monomer was not available in the quantities and price that might enable production of polymers. Since in recent years more attractive processes have been developed for the production of 1,3-propanediol (1,3-PD) from renewable resources, research on related polymers has attracted interest from both an industrial and academic point of view [1].

Poly(propylene terephthalate) (PPT), the first and most studied polyester of 1,3-propanediol, is available in the market [2]. The polymer is suitable for industrial fiber production. PPT fibers are characterized by much better resilience and stress/recovery properties than PET and PBT. These properties are due to the crystal structure of PPT. PPT chains are much more angular structured than PET and PBT chains due to the odd number of methylene groups of the diol segment. Therefore these chains can be stretched up to 15% with a reversible recovery [3]. PPT is anticipated to gain a significant share in the thermoplastic polyesters market in the next years. However, like the other terephthalate polyesters, PPT is not susceptible to degradation in the environment.

Recently, biodegradable aliphatic polyesters, and especially poly(propylene succinate) (PPSu) and poly(propylene adipate) (PPAd) have gained increasing interest [4-6]. Their synthesis, thermal properties and biodegradation were studied in comparison with the familiar polyesters poly(butylene succinate) (PBSu), poly(ethylene succinate) (PESu), poly(butylene adipate) (PBAd) and poly(ethylene adipate) (PEAd), which are also important biodegradable polymers. Poly(butylene terephthalate-co-adipate) copolymers are industrially produced in an attempt to arrive at polymers having both the advantages of high performance and being susceptible to degradation in environmental conditions [4]. Furthermore, high molecular weight PBSu is already available in the market.

The odd number of methylene units in the diol segment is responsible for the lower melting points, lower degree of crystallinity and higher biodegradation rates of the polymers prepared form 1,3-propanediol, compared with their homologues based on ethylene-glycol or 1,4-butanediol [5]. In the following chapter an effort is made to summarize results of such works dealing with biodegradable polyesters of 1,3-propanediol.

2. Discussion

2.1. 1,3-Propanediol as a Monomer for Polymer Production

1,3-Propanediol became available in the market in sufficient quantity and purity, only a few years ago. Nowadays, more attractive processes have been developed for its production such as selective hydration of acrolein, followed by catalytic hydrogenation of the intermediate 3-hydroxypropionaldehyde or hydroformylation of ethyleneoxide (Figure 1). Recently, Du Pont and Shell Chemical companies have succeeded in producing 1,3-propanediol commercial products with low cost and high quality by using different methods. [1]. 1,3-PD is a valuate chemical intermediate that has recently found extended applications as monomer for the production of polyesters [2, 7]. It is considered to be one of the bulk chemicals that will be produced in large scales in the future.

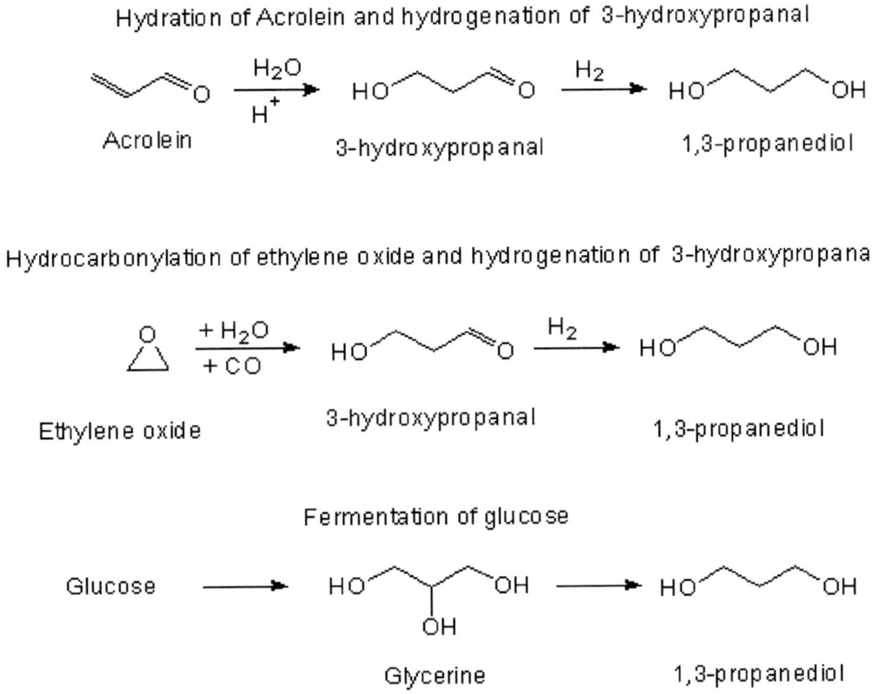

Figure 1. Industrial processes for synthesis of 1,3-propanediol.

Certain polymers like PPSu, PPAd and poly(propylene sebacate) (PPSe) beyond biodegradability, present the additional advantage that they can be produced from oligomers arriving from renewable sources. The production of chemicals and fuels by using alternative sources instead of hydrocarbons (oil), has received great attention as a green feed stock manufacture. 1,3-Propanediol, succinic, adipic and sebacic acid that are used as monomers for the preparation of biodegradable polyesters, are chemicals produced by using green techniques [4]. Until now 1,3-PD has been manufactured mainly by chemical synthesis, which requires high temperature, high pressure and expensive catalysts. Thus, in the last decade much effort has been paid to its production by bioprocesses on large scales. Many microorganisms like *Klebsiella*, *Citrobacters*, *Enterobacter*, *Lactobacillus* and *Clostridia* are able to convert glycerol or glucose to 1,3-PD via fermentation processes [8-11]. Experimental investigations showed that the fermentation is a complex bioprocess, taking place mostly in two stages while microbial growth is subjected to multiple inhibitions of substrate and by-products. On the other hand, succinic acid is a dicarboxylic acid used as a monomer for the preparation of aliphatic and fully biodegradable polyesters. It can be produced pertochemically from butane and also by anaerobic fermentation, requiring glucose and CO_2, decreasing the pollution caused from the petrochemical process. Glycerol and wood hydrolysates can be also used as carbon sources for succinic acid production, during fermentation by *Mannheimia succiniciproducens* bacterium [12,13].

2.2. Synthesis and Characterization of the Polyesters of 1,3-PD

The synthesis of aliphatic polyesters with high molecular weight, in order to achieve satisfactory mechanical properties, is considered as being one of the most difficult problems to be solved. Till today this can be achieved only by either using techniques such as ring-opening polymerization of cyclic monomers (lactones) or with the use of chlorides of acids, which are very expensive and inappropriate for industrial scale use [14,15]. The production of high molecular polyesters using diacids and diols can proceed only by the addition of chain extenders or branched comonomers as is the case of Bionolle® [16].

Synthesis of the aliphatic polyester samples from 1,3-PD has been performed in laboratory scale following the two-stage melt polycondensation method (esterification and polycondensation) in a glass batch reactor [17]. The synthesis is almost identical to that used for the preparation of PET with the only difference the lower temperature used for the polycondensation step. In brief, the proper amount of diacid (e.g. succinic acid or adipic acid) (0.55 mol) and 1,3-propanediol in a typical molar ratio 1/1.2 and the catalyst tetrabutoxy titanium TBT (typically 3×10^{-4} mol TBT/mol of diacid) were charged into the reaction tube (250 ml) of the polycondensation apparatus. The reaction mixture was heated at 190°C in an argon atmosphere until the collection of almost all the theoretical amount of H_2O. In the second step of polycondensation, polyphosphoric acid (PPA) was added (5×10^{-4} mol PPA/mol of diacid) and a vacuum (5.0 Pa) was applied slowly over a time period of about 30 min. For each polyester, the polycondensation temperature was kept stable at 230°C while stirring speed was slowly increased to 720 rpm. The polycondensation reaction was finished after 2h of heating. From the studies it was found that monomer ratio, initial produced oligomers and polycondensation temperature are crucial parameters, which should be taken into consideration in order to obtain high molecular weight polyesters [18]. As reported above, the polymerization process involves two different steps according to the well-known process used for polyesters synthesis. In the first stage (esterification), aliphatic diacid reacts with 1,3-propanediol and water is removed as by-product. At the temperature that reaction takes place (190°C), water can be easily removed by distillation from the reactor and oligomers are formed. At this stage it is very important to remove as much of the formed water as possible and oligomers with the highest possible molecular weight to be produced. In order to increase the molecular weight at the appropriate level in the second stage (polycondensation), the prepared oligomers condensate at high temperature (230°C) with the application of high vacuum. This temperature is probably the maximum possible for the preparation of high molecular weight aliphatic polyesters, as was concluded from our study [19,20]. Above this temperature decomposition reactions can take place, which despite their very slow rate can reduce the molecular weight of the final polyester and color the sample. The reactions that occur during these stages (esterification and polycondensation) and the procedures used for the synthesis of the 1,3-PD polyesters are presented in Figure 2.

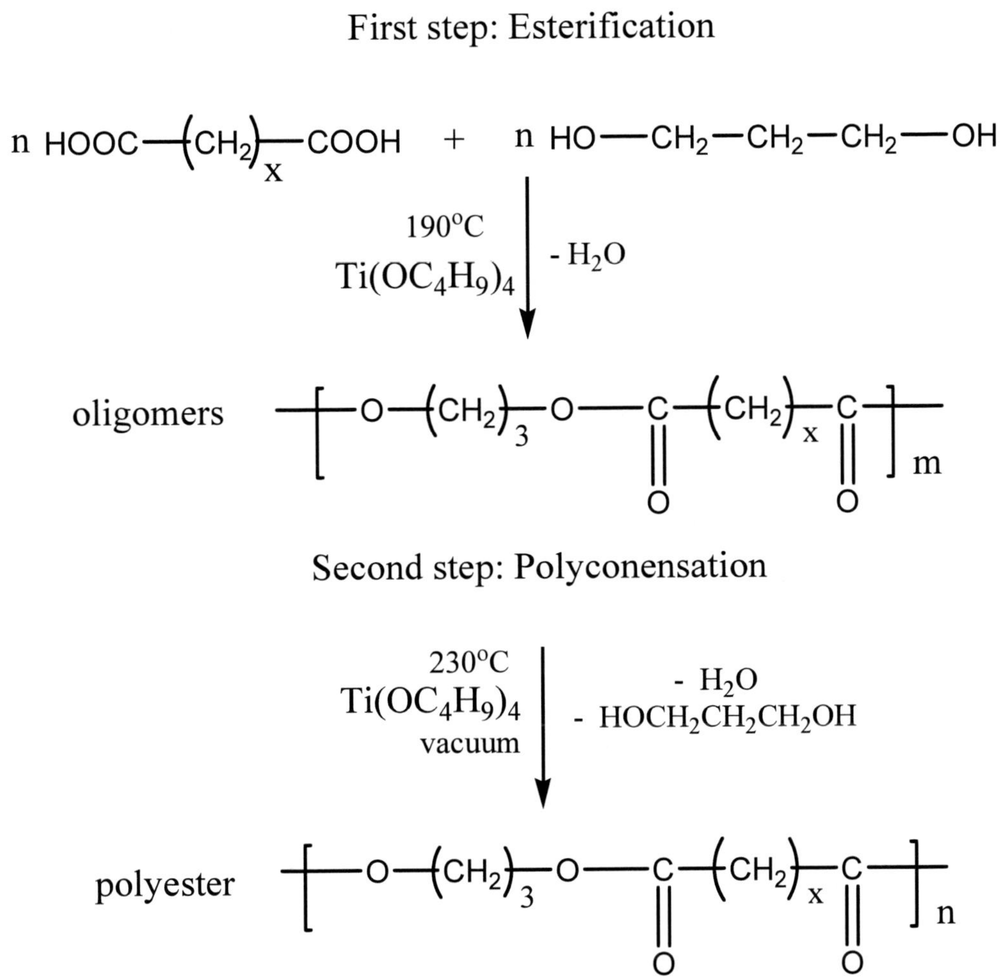

Figure 2. Synthesis of poly(propylene alkanedioate)s following the polycondensation method.

Using different aliphatic diacids with increasing number of methylene units (X), from succinic with two methylene units to sebacic acid with eight methylene units, a series of polyesters of 1,3-propanediol was prepared. Thus, poly(propylene succinate) (X=2), poly(propylene glutarate) (X=3), poly(propylene adipate) (X=4), poly(propylene pimelate) (X=5), poly(propylene suberate) (X=6), poly(propylene azelate) (X=7) and poly(propylene sebacate) (X=8) samples have been synthesized.

Bikiaris and Achilias, following the polycondensation method and with varying the catalyst amount and/or the temperature of the second step of the process, prepared a series of PPSu samples with different molecular weight values [18]. In general it was found that with increasing the catalyst/diacid molar ratio in the reaction mixture, higher average molecular weight could be achieved, for given polycondensation temperature. Also, polymers with higher number average molecular weight (M_n) values were prepared by increasing the temperature of polycondensation up to 230°C. Finally, from theoretical simulation results it was found that although higher initial ratios of glycol to succinic acid are useful to increase the esterification rate, they lower the number average degree of polymerization of the oligomers at a fixed conversion of acid end groups.

The ^{1}H-NMR spectra of PPSu and PPAd were reported [5,6]. In the ^{1}H-NMR spectra for PPSu (Figure 3) a single peak appeared at 2.63 ppm which was attributed to methylene protons of succinic acid, *c*, and a triple peak 4.09-4.21 ppm attributed to *b* protons and a multiple peak between 1.9-2.02 ppm corresponding to *a* protons were also observed [5]. In the spectra of PPAd, also shown in Figure 3, peaks for the *a* and *b* protons of 1,3-propanediol are recorded as for PPSu, while peaks for *d* and *e* protons are also observed at 1.67ppm and 2.36ppm respectively [6].

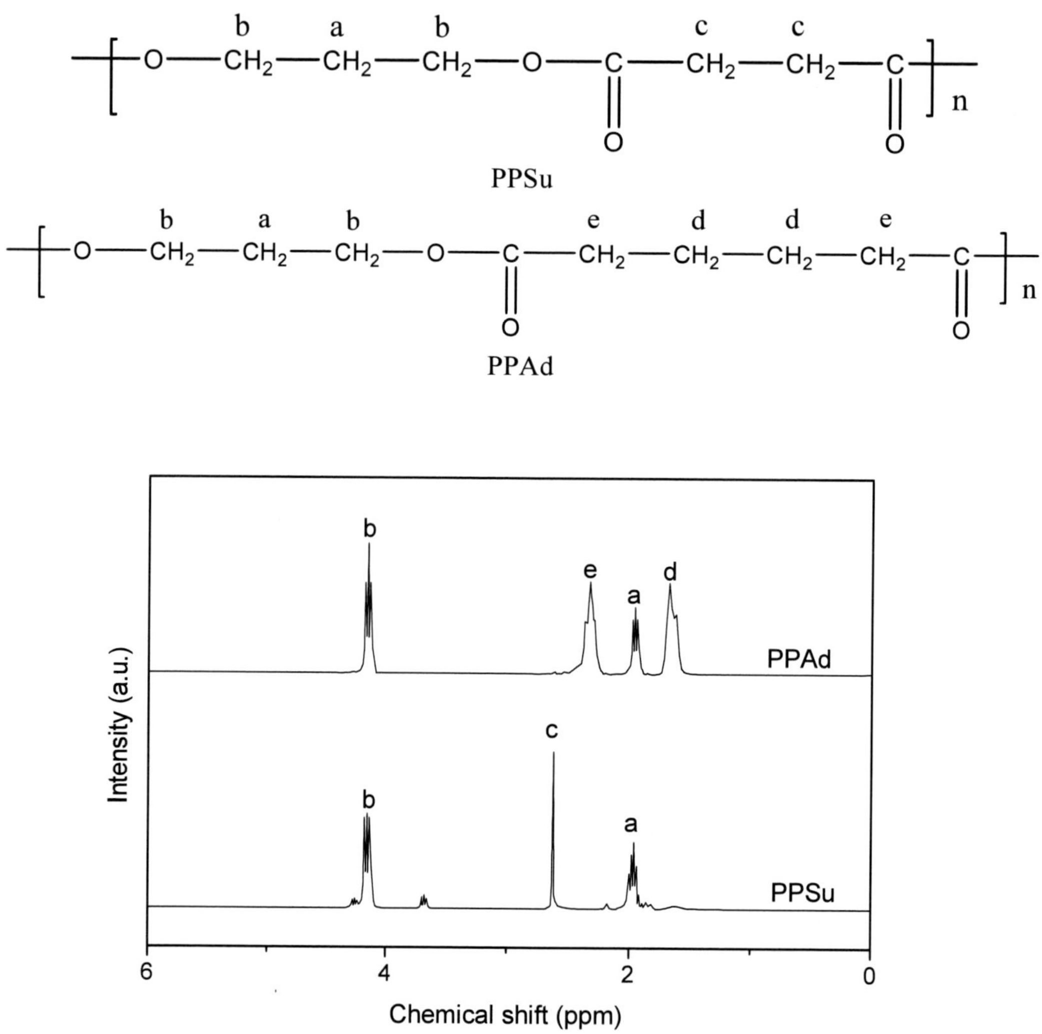

Figure 3. Chemical structures and ^{1}H-NMR spectra of PPSu and PPAd.

The WAXD patterns of the prepared series of polymers, from poly(propylene succinate) (X=2) to poly(propylene sebacate) (X=8), are quite different as one can see in Figure 4. In general the parameters of the unit cell characterizing the crystals of the polymers have not been reported. However, the crystal structure of poly(propylene sebacate) (PPSe) has been studied [21]. PPSe has an orthorhombic unit cell of dimensions a=0.532, b=0.7532 and c=3.133nm (fiber axis) and belongs to the $P2_12_12_1$ space group. There are two chemical units per fiber repeat and the two chains within the unit cell are positioned on 2_1 screw axis, parallel to the c direction. The fiber repeat is 0.25nm shorter than that for an all-trans conformation. The non-trans torsion angles are located in the glycolic moiety of the chemical repeat. It is anticipated in general the unit cells for polyesters of 1,3-PD to be orthorhombic in contrast to the butylene homologues for which monoclinic unit cells have been reported [21].

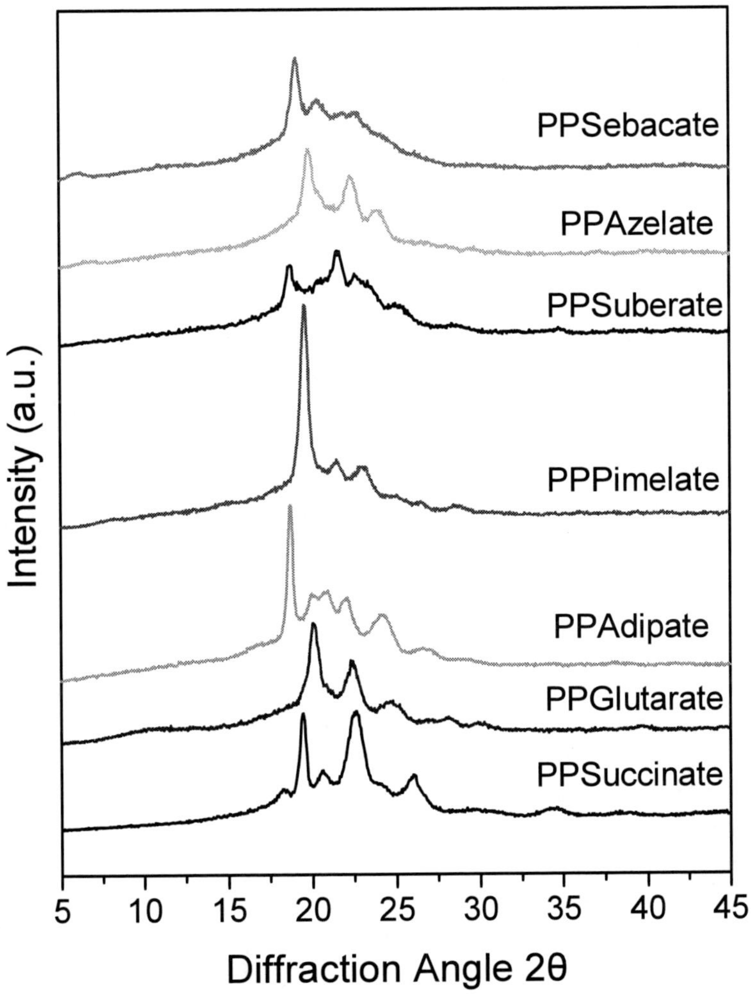

Figure 4. WAXD patterns of poly(propylene alkanedioate)s.

The melting point variation for these polyesters with increasing methylene units in the diacid moiety, in comparison to the corresponding for 1,2-ethanediol and 1,4-butanediol can be seen in Figure 5. To construct the plots, data from the paper of Müller et al were also used [4]. It is obvious that for the 1,3-propanediol polyesters the variation of the melting points is small compared to those polyesters of the other two diols.

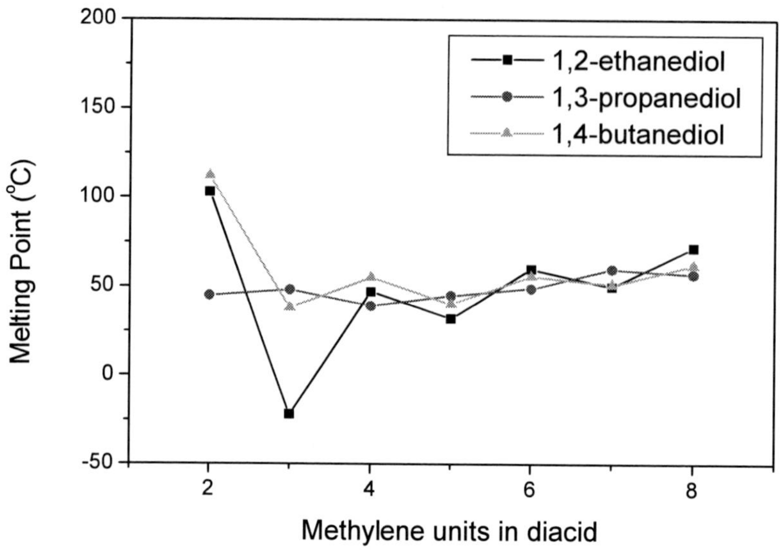

Figure 5. Melting points of aliphatic polyesters of 1,2-ethanediol, 1,3-propanediol and 1,4-butanediol.

It was observed that the crystallization rates of the polyesters of 1,3-propanediol in general increased with increasing the number of methylene groups in the respective diacid. Poly(propylene succinate) is a slowly crystallizing polyester which shows a glass transition temperature T_g at about -35°C and a melting temperature T_m=46°C. The latter is much lower than the corresponding values for PESu (T_m=104°C) or PBSu (T_m=112°C), as a result of the well-known odd-even effect for the melting temperatures of polyesters. On the other hand the T_g value is intermediate to those for PESu (-10°C) and PBSu (-43°C). The polymer does not crystallize during DSC cooling scans from the melt or on heating from the glassy state, unless very slow rates such as 2.5°C/min or even slower, are applied [20]. Figure 6a shows the respective heating scans for PPSu after melt-quenching. Poly(propylene adipate) crystallizes faster than PPSu due to the increased number of methylene groups in its repeating unit [6]. This is obvious in the DSC traces of Figure 6b, where the behavior of both quenched PPSu and PPAd on heating from the amorphous state is presented. In contrast to PPSu, PPAd shows cold-crystallization on heating by 10°C/min. Also, it has a T_g at –57°C, much lower than that of PPSu and a melting point of about 40°C for well-crystallized samples or lower for cold-crystallization ones. The values for the degree of crystallinity that the two polymers usually achieve are comparable, ranging close to 35% for both.

For PPSu similar to the other two important succinates i.e. poly(ethylene succinate) (PESu) and poly(butylene succinate) (PBSu), multiple melting behavior was observed [20]. As can be seen in Figure 6a, double melting peaks are observed after cold crystallization. This behavior however, was more clearly shown for samples after isothermal crystallization from the melt. For the interpretation of these observations the partial melting-recrystallization-final melting scheme was adopted. Multiple peaks were more clearly revealed in DSC traces after crystallization in the temperature range from 0 to 20°C. The equilibrium melting temperature (T_m^o) of PPSu and the heat of fusion for the pure crystalline polymer were estimated to be 58°C and 22kJ/mol respectively. This T_m^o value compared to 114°C and 133.5°C for the T_m^o of PESu and PBSu respectively, is quite lower.

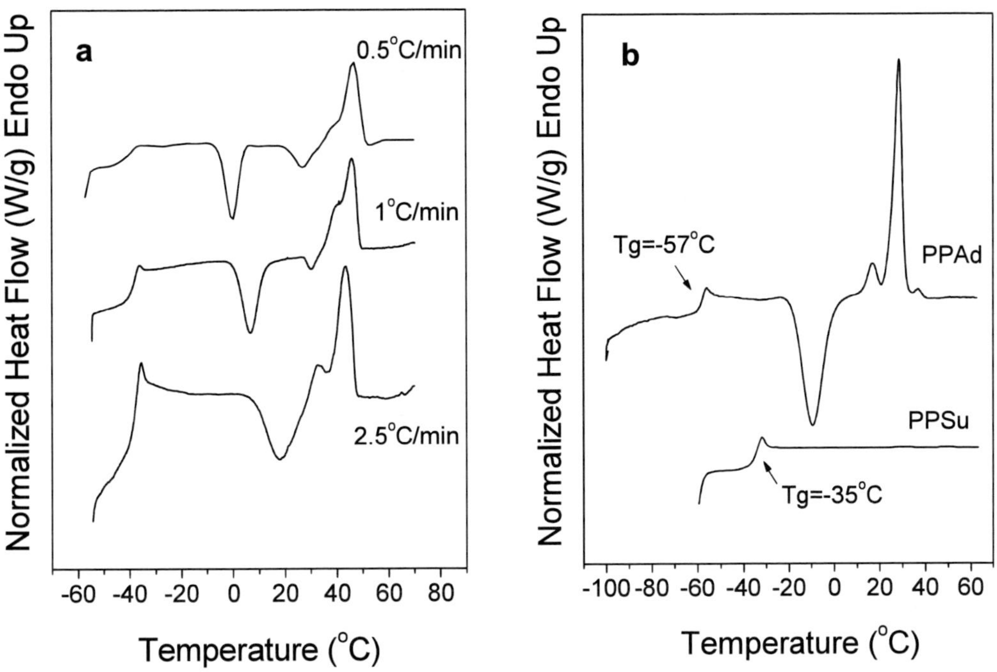

Figure 6. a) DSC heating traces for quenched PPSu at different slow heating rates and b) DSC traces of amorphous PPSu and PPAd recirded at 10°C/min.

Dielectric spectra measurements, upon crystallization of PPSu, have been performed by Soccio et al [22]. Crystallization was performed at 25°C. After a controlled period of time the sample was rapidly cooled to –25°C at 5°C/min to perform a frequency swept. PPSu at –25°C that is a temperature above the glass transition temperature exhibits two main relaxations processes characteristic of most polymers, α and β, in order of increasing frequency. It was found that for the polyester the a and β relaxations appear simultaneously and are well resolved in the experimental frequency window. The β relaxation was used to characterize the crystalline structural development while the a relaxation was used to collect information about the evolution of the amorphous phase dynamics. In this way structural development could be characterized by a single experiment during the crystallization process. The analysis of the dielectric loss supported the existence of precursors of crystallization in the induction period.

The thermal degradation behavior of the polyesters has been studied [6,23,24]. From the thermogravimetric curve of Figure 7 it can be seen that PPSu presents a relatively good thermostability since no significant weight loss (only1.2 %) occurred until 300°C. As it can be seen in the curve of derivative weight loss curve (DTG), in the early stages of the decomposition, there is a small shoulder peak. Until 366 °C, where the maximum rate of this decomposition step appears, the volatile matter corresponds to about 7 wt% of initial weight. Such a pre-major weight loss stage was also mentioned for poly(propylene terephthalate) (PPT) with low number average molecular weights ranging between 13000 and 23000 g/mol, and PCL [25-27].

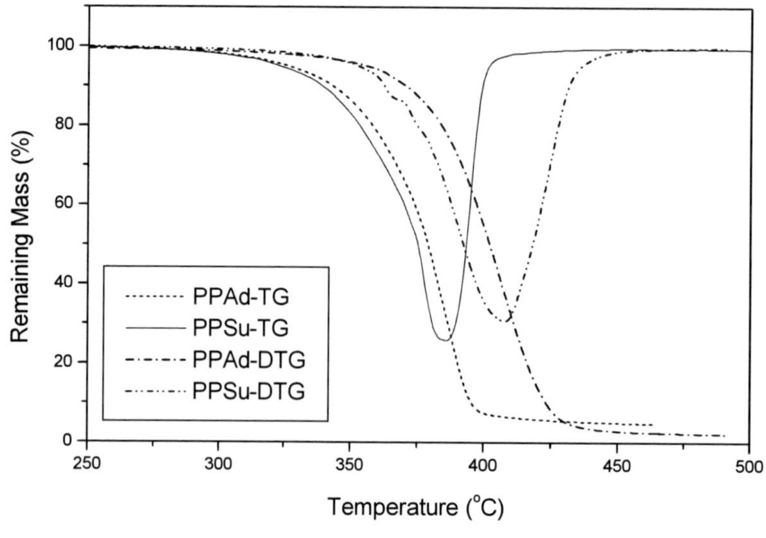

Figure 7. TGA and derivative-TGA curves for PPSu and PPAd recorded at 10°C/min.

The temperature at the maximum weight-loss rate of this stage increases significantly with molecular weight while the weight loss decreases steadily. The first decomposition step (shoulder) is attributed to degradation and volatilisation of the oligomers detected with ^{1}H-NMR spectroscopy. From preliminary studies of the monomers it was found that succinic acid degrades at temperatures close to 200°C while 1,3-propanediol at slightly higher temperatures. However, both are fully decomposed at temperatures lower than 300°C. These data verified the hypothesis that the first decomposition step is due to oligomer degradation. The temperature at which PPSu decomposition gains the highest rate is at 404°C for heating rate 10 °C/min (Figure 7). This temperature is very high taking into account that PPSu is an aliphatic polyester and the temperature value is comparable to decomposition temperatures reported for aromatic polyesters like terephthalates (PET, PBT, PPT) and naphthalates (PEN) [17]. Furthermore, this temperature is much higher compared to other aliphatic polyesters like poly(L-lactide) for which major degradation step appears at temperatures lower than 370°C, as well as poly[(*R*)-3-hydroxybutyrate] and poly(ε-caprolactone) [28, 29]. From the above it can be concluded that PPSu although it has low melting point presents a very high stability against thermal degradation.

Thermal degradation of PPAd was also studied by determining its mass loss during heating [6]. In Figure 7 the remaining mass (TG %) and the derivative (DTG) curves at heating rate 10°C/min are presented for the polyester. In both TG and DTG thermograms one stage of mass loss is followed and as can be seen the weight loss takes place gradually and smoothly. Additionally, there is no differentiation between using low or high heating rates. PPAd presents a relatively good thermostability since no significant weight loss, ~0.25%, occurred until 250°C. After that temperature the polyester decomposes quickly and looses about 95.8 wt% until 500°C. The temperature at which PPAd decomposition gains the highest rate is 385°C, for a heating rate of 10°C/min. However, the behaviour of the polyester is not similar to that of poly(butylene adipate) (PBAd) and poly(propylene succinate) (PPSu) and has a lower thermal stability. Finally, the decomposition temperature of PPAd is 20-25°C lower than the corresponding temperature of PPSu.

The mechanical properties of PPAd were also reported [6]. It has a tensile strength of 11.6 MPa, which is similar with that of low density polyethylene 13.2 MPa (LDPE). However its homologues PEAd and PBAd have much higher tensile strength, reaching 17.3 and 16.2 MPa, respectively. Furthermore, for high molecular weight PPAd samples satisfactory elongation at break was observed, higher than 400%. PPSu on the other hand has a very low tensile strength 1.6 up to 3.8 MPa depending on its molecular weight and breaks before yielding [5].

2.3. Biodegradation

Biodegradability of a certain polymer in the form of enzymatic hydrolysis is controlled by several factors. The most important one is the nature of the polymer itself, meaning its chemical structure and the occurrence of specific bonds, which might be susceptible to hydrolysis, along its chains. Such groups are those of esters, ethers, amides, etc. [30,31]. Enzymatic hydrolysis studies of PPSu have been carried out [5]. For comparison, the behavior of PESu and PBSu samples having the same molecular weight values, (number average molecular weight of about 6800g/mol), were also studied, at the same conditions.

The polyesters in the form of films of 5x5 cm in size and approximately 2 mm thickness, prepared in a hydraulic press, were placed in petries containing phosphate buffer solution (pH 7.2) with 1 mg/mL *Rhizopus delemar* lipase. Usually enzymatic hydrolysis studies are performed at 37°C. However this temperature is very close to the melting point of PPSu. So, in this case the tests were performed at 30±1°C. The degree of biodegradation was estimated from the mass loss and molecular weight reduction as measured by GPC.

The results showed that PPSu exhibits the highest degradation rates (Figure 8a), while PBSu the lowest ones as one can see in [5]. For PESu weight loss rate seemed to be similar to that of PBSu, but final mass loss was slightly larger. Since the used polyester samples had almost identical molecular weights, crystallinity seems to be the predominant factor that controls the biodegradation rates. It is therefore the higher degree of crystallinity of the PBSu samples, as was found by DSC and WAXD, which resulted in lower degrees of biodegradation comparing to that of PESu or PPSu. PPSu, which shows the lower crystallinity and melting point, hydrolyzes faster comparing to the other polyesters.

In recent studies the biodegradation characteristics of PPSu, PPAd and PPSe, as well as of some relative copolymers were reported [32, 33]. The biodegradability of the polyesters was detemined by monitoring the normalized weight loss of polyester films with time in phosphate buffer (pH=7.2) without and with Rhizopus delemar lipase at 37°C. It was found that the biodegradation rates of the homopolyesters follow the path PPSu>PPAd>PPSe. Poly(propylene sebacate) did not show significant weight loss in presence of enzyme which may be due to its highest degree of crystallinity and melting point compared to PPSu, PPAd and copolyesters. The PPSu and PPAd copolyesters with PPSe showed slower degradation rates than the PPSu and PPAd homopolyester respectively. However, from recent work we have concluded that PPAd degrades faster than the other adipate polyesters, PEAd and PBAd, as can be seen from Figure 8b. At very sort time, only 6 days, the whole amount of PPAd has been enzymatically hydrolyzed, while at the same time only the 40 wt% of PBAd has been hydrolyzed.

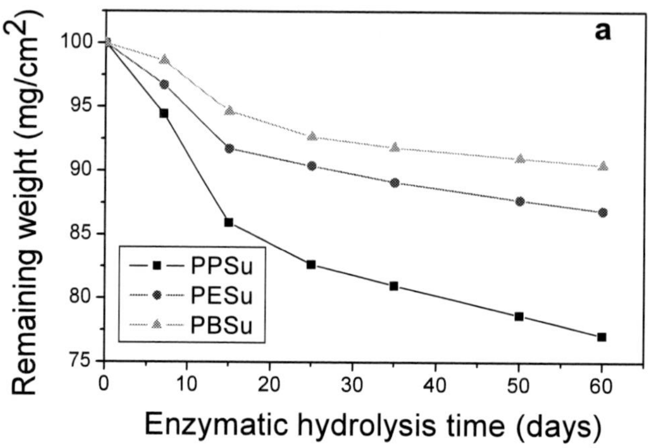

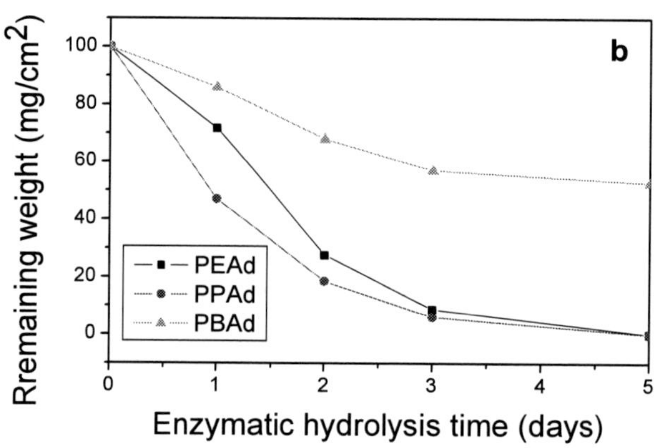

Figure 8. Variation of the weight of specimens of a) PPSu, PESu and PBSu during enzymatic hydrolysis using *Rhizopus delemar* lipase and b) PPAd, PEAd and PBSu during enzymatic hydrolysis using 0.09 mg/mL *Rhizopus delemar* lipase and 0.01 mg/mL of *Pseudomonas Cepacia* lipase.

SEM microphotographs revealed specific features indicative for surface erosion in the case of PPSu as can be seen in Figure 9. In general the observations lead to conclusions consistent with those derived from the weight loss and molecular weight measurements during the experiments. As it is well known enzymatic hydrolysis is a heterogeneous process and enzymes are attached on the surface of an insoluble substrate. Hydrolysis takes place via surface erosion. During this procedure small fragments and water soluble monomers or oligomers are generated, during hydrolysis of ester bonds. In general, the interior of polyesters specimens is not attacked until extended holes are created on the surface allowing the enzymes to enter and attack the main body. After consumption of the amorphous material of the surface a layer of crystalline domains remains, where only slow degradation may occur. As can be seen in Figure 9 the crystallites of PPSu are well distinguished after 35 days of enzymatic hydrolysis and remain almost unaffected. In principle, hydrolysis affected the amorphous material surrounding the spherulites and numerous holes were then created.

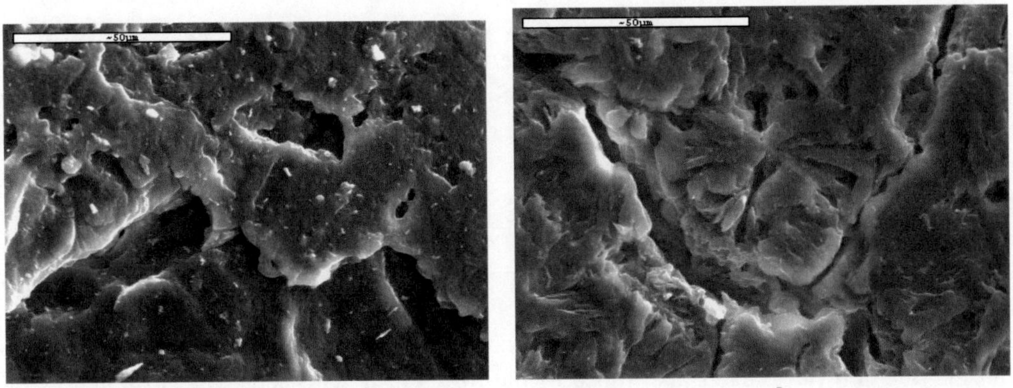

a b

Figure 9. SEM micrographs of PPSu polyester during hydrolysis at a) 15 days and b) 35 days.

2.4. Copolymers

Recently, except from homopolymers, copolymers based on 1,3-propanediol, succinic acid and other diacids or diols were synthesized [34]. A full series of random poly(butylene-co-propylene succinate) (PBPSu) copolymers with high molecular weight were prepared, by the polycondensation method. The intrinsic viscosity of the copolymers ranged from 0.6 up to 0.75 dL/g, but without showing any clear trend for the dependence on composition. The copolyesters showed a melting point depression with increasing comonomer content, as can be seen in Figure 10.

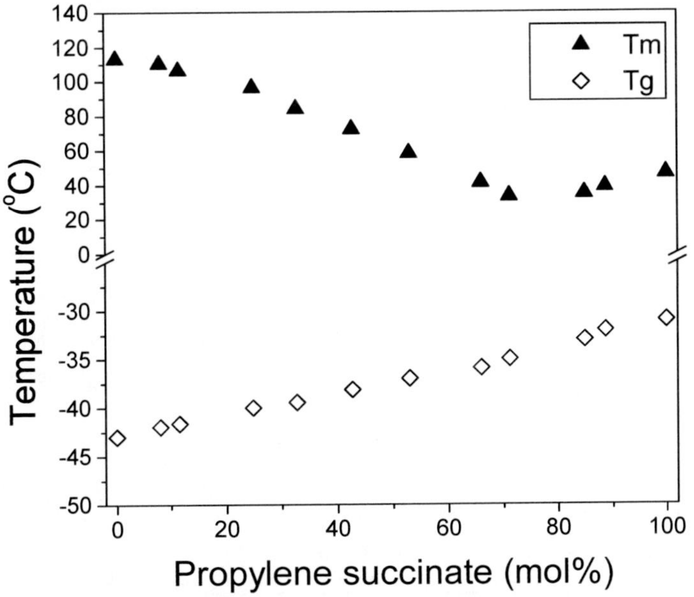

Figure 10. Melting points and glass transition temperatures for the poly(butylene-co-propylene succinate) copolymers as a function of propylene succinate units content.

Also, on the basis of WAXD observations and the decrease in melting points, isodimorphic cocrystallization was concluded. The probability, comonomer propylene succinate units to be incorporated in the PBSu crystals, was higher for high comonomer content. As the comonomer content increases, the mean length of sequences of the same repeating unit along the macromolecular chains, are expected to reduce. This in turn results in formation of non-crystallizable chain segments, and finally the crystallinity decreases. In the case that comonomer units are inserted in the crystals, they act as crystal defects, lowering the melting point of the copolymers. The most important feature of these PBPSu copolymers however, was that of their fast enzymatic hydrolysis rates.

Figure 11 shows that the copolymers with high propylene succinate content experienced faster degradation rates, even than those of the neat PPSu. This fact was attributed to their lower degree of crystallinity, lower melting points and increased mobility in the amorphous phase, which is known that is general the phase that is mainly affected by enzymatic hydrolysis.

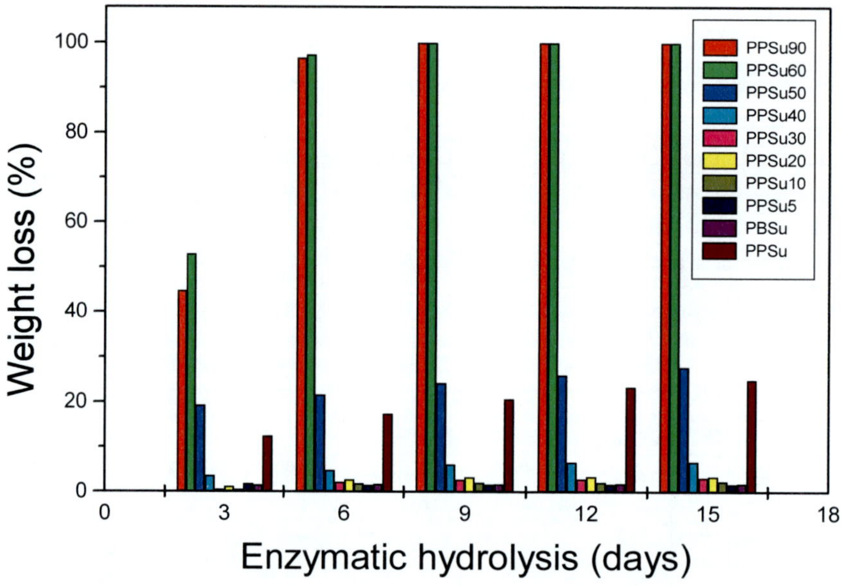

Figure 11. Weight loss as a function of time of enzymatic degradation of PBPSu copolyesters compared to the neat PPSu and PBPSu.

Figures 12 and 13 show SEM photographs of the surfaces of copolymer specimens taken after different times of enzymatic hydrolysis. Surface erosion is more pronounced for higher content of propylene succinate units in the copolymers.

Also, two PBPSu copolyesters were studied with respect to their isothermal crystallization kinetics [35]. The Avrami exponent was found to be between 2.2 and 2.8, showing that the crystallization mechanism was a heterogeneous nucleation with spherical growth geometry in the crystallization process of the polyesters. Multiple melting peaks were observed. PBPSu copolymers were identified to have the same crystal structure with that of PBSu by using WAXD. It was assumed that only butylene succinate units could crystallize, while the propylene succinate segments remain in the amorphous state. Other studies showed that this hypothesis might hold for low propylene succinate content [34].

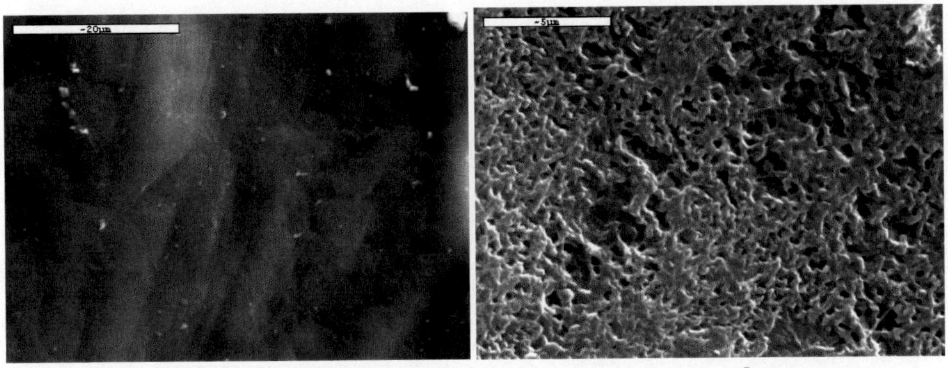

a b

Figure 12. SEM micrographs of PBPSu 70/30 w/w copolymer during enzymatic hydrolysis at several times a) 0 days and b) 15 days.

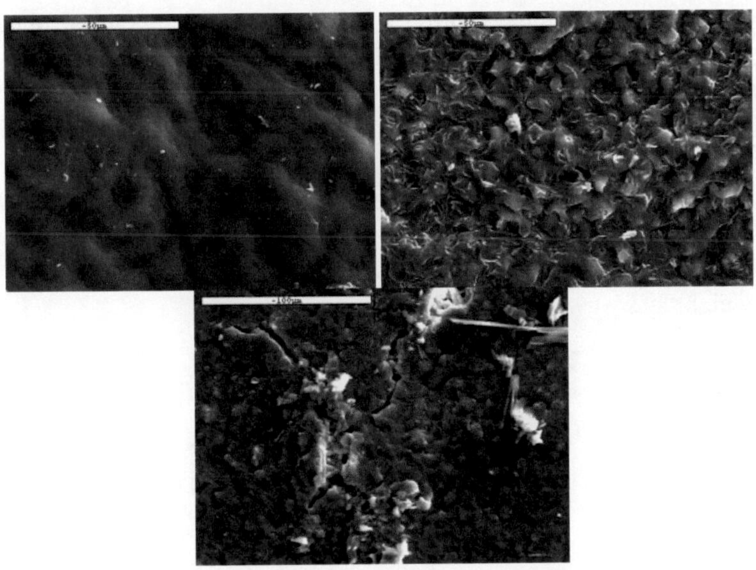

Figure 13. SEM micrographs of PBPSu 10/90 w/w copolymer during enzymatic hydrolysis at several times. From left to the right: 0 days, 3 days and 15 days.

Poly(propylene succinate-co-ε-caprolactone) PPSu/PCL copolymers with compositions 95/5, 90/10, 25/75, 50/50 and 75/25, w/w were also prepared according to the procedures described by Seretoudi et al., using a combination of polycondensation and ring opening polymerisation [36]. The degree of randomness of the copolymers was characterized by using ^{1}H-NMR and ^{13}C NMR and was verified that at concentrations up to 25 wt% of PPSu, block copolymers were prepared while at compositions 50 and 75 wt% random copolymers instead of block were prepared [37]. This is due to the extensive transesterification reactions that are taking place simultaneously with the ring-opening polymerization of ε-CL. The molecular weight distribution of the prepared polyesters was determined by GPC and it was found that as the amount of PCL increases in the copolymer the molecular weight, also, increases. This is attributed to the fact that as the amount of initial PPSu increases, higher amounts of hydroxyl end groups contained in PPSu are available to act as initiators for ring opening polymerization of ε-CL monomer. However, despite the high molecular (Mw) of the copolymers, which ranged between 55000 and 87000 g/mol, their tensile strength is very low (1.3 up to 3.4 MPa) and only their elongation at break is higher than 560% for all copolymers. This behavior could be also attributed to their crystallinity. Random copolymers were completely amorphous in contrary to the block, which were semicrystalline.

In DSC thermograms recorded during heating of the copolymers containing 50 and 75 wt% PPSu only single glass transitions could be observed, which are located at –53.8 and at –43.9°C respectively. These temperatures are lower than the glass transition temperature of PPSu (–35°C) and higher than the T_g of PCL (-61.6°C). So, the T_gs of the prepared copolymers were between the T_gs of the neat polymers. For the other copolymers two separate melting peaks were observed close to the melting temperatures of the neat polyesters. This is an indication that the different blocks can crystallize separately.

Thermal degradation of PPSu/PCL 25/75, 50/50, and 75/25 w/w copolymers was studied by determining their mass loss during heating. The prepared copolymers showed satisfactory stability against thermal degradation since no significant weight loss, (~0.5%), occurred until 250°C while their maximum decomposition rate occurred at temperatures about 410°C. The corresponding temperatures for PPSu and PCL are 403.9 and 415.3 °C, respectively. Concerning their thermal decomposition mechanism it is proved that all copolymers degraded at two stages. The first stage corresponded to small mass loss taking place at 280-365°C.

The most characteristic behavior of these copolymers is their high enzymatic hydrolysis at phosphate buffer solution (pH 7.2, 30°C) using 0.09 mg/mL *Rhizopus delemar* lipase and 0.01 mg/mL of *Pseudomonas Cepacia* lipase. PCL it is well known that due to its high hydrophobicity and crystallinity has a slow degradation profile and after 9 days of enzymatic hydrolysis its weight loss was less than 5 wt%.

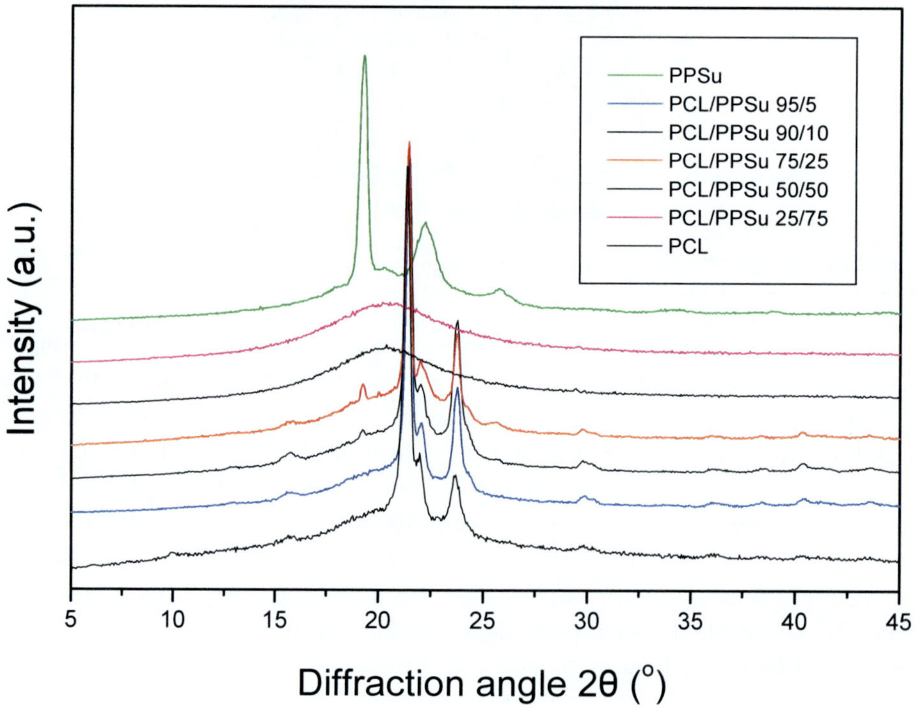

Figure 14. XRD patterns of PPSu, PCL and respective copolymers.

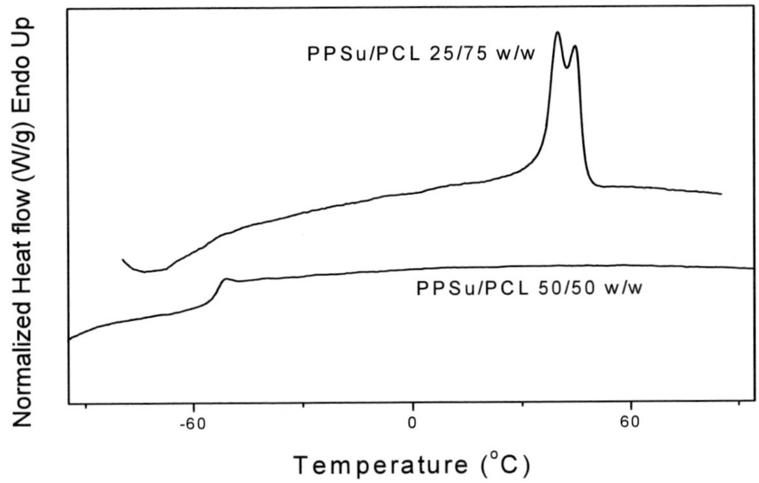

Figure 15. DSC thermograms of PPSu/PCL copolymers.

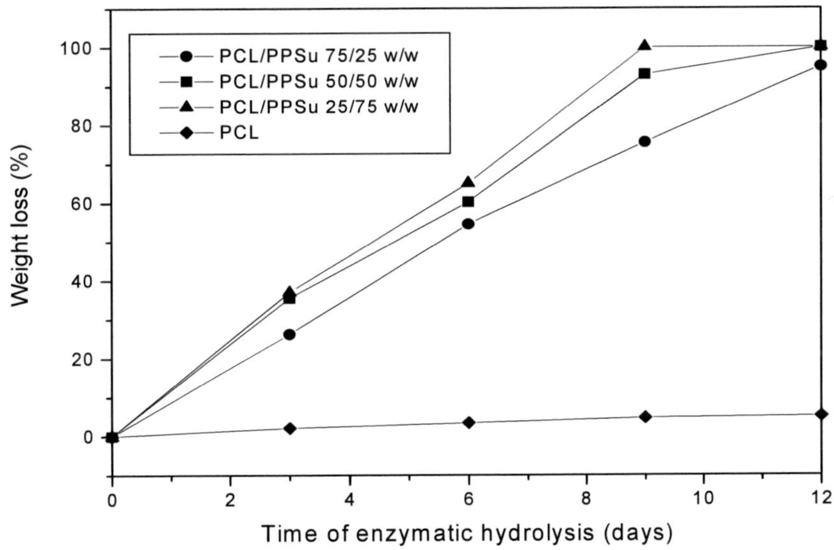

Figure 16. Weight loss of PCL/PPSu copolymers during enzymatic hydrolysis for several days.

On the contrary almost all the copolymers at the same time were completely degraded as can be seen from Figure 16. The rate of enzymatic hydrolysis was slightly dependent on the PPSu content and increased by increasing its amount in the copolymer. This behaviour should be attributed to the higher biodegradation rate that PPSu has compared with PCL and to the fact that the copolymers were amorphous. It is well known that first the amorphous phase is degraded during enzymatic hydrolysis, while the crystalline phase due to the low water absorption rate is much less affected.

Soccio et al reported synthesis of random poly(propylene isophthalate-co-succinate) (PPISu) copolyesters [38]. The copolymers also showed depression in the melting temperature with comonomer content. It was found that they could crystallize at room temperature, except those with 70 or 80 mol% propylene succinate (PSu) units. For up to 75 mol% propylene succinate units in the copolymers, crystals of poly(propylene isophthalate) were formed, while the copolyester with 90 mol% PSu gave crystals of PPSu. TGA measurements showed high thermal stability for the copolymers. Initiation of decomposition for PPSu was found to occur at 393°C, whereas the maximum rate appeared at 427°C, during scans by 10°C/min in air. The respective temperatures increased for the copolymers with increasing isophthalate content.

Similar observations were reported for random poly(propylene isophthalate-co-adipate) (PPIAd) copolyesters [39]. For up to 60 mol% propylene adipate (PAd) units in the copolyesters they formed PPI crystals. The copolymers with 70 or 80 mol% propylene adipate could not crystallize at room temperature, while that with 90 mol% PAd units formed PPAd crystals. The thermal stability of the samples was checked by TGA measurements in air by heating scans by 10°C/min. The polyesters were found to be stable. For PPAd the temperatures for initialization of decomposition and that of maximum decomposition rate were reported to be 359 and 392°C respectively. The respective temperatures for the copolymers increased with increasing propylene isophthalate content.

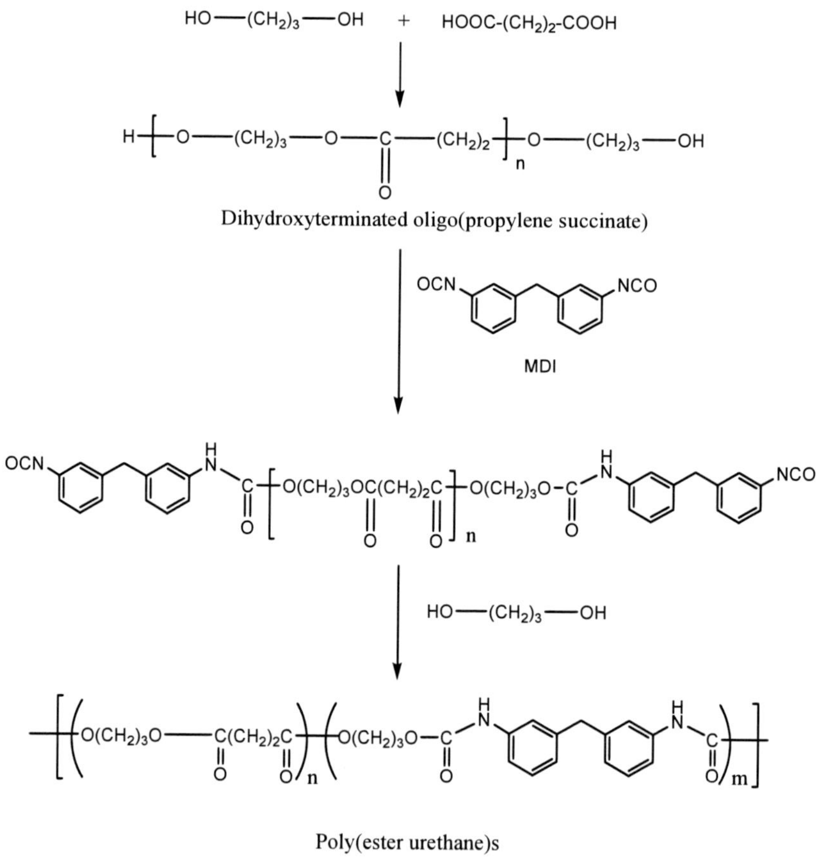

Figure 17. Synthesis of oligo(propylene succinate), by thermal polycondensation of 1,3-propanediol and succinic acid, and chain extension reaction by diisocyanate syntheis [42].

Albertsson et al have reported the synthesis of poly(ester urethane)s (PEU), and poly(ester carbonate)s prepared by chain extension [40-43]. High molecular weight poly(ester urethane)s were prepared via the diisocyanate synthesis (Figure 17). Oligo(propylene succinate)s with molecular weights of about 2300-2400 g/mol were chain extended after reaction with 4,4′-diisophenylmethane diisocyante (MDI). In this way PEUs with 46 up to 63wt% aliphatic polyester were prepared. The T_g values ranged from –7.8 to 4.8°C, while the melting points were found to be between 176 and 208°C. The sample with the highest polyester content in the reactor feed showed the lowest melting point. The Young's modulus decreased when the content of the aliphatic polyester increased.

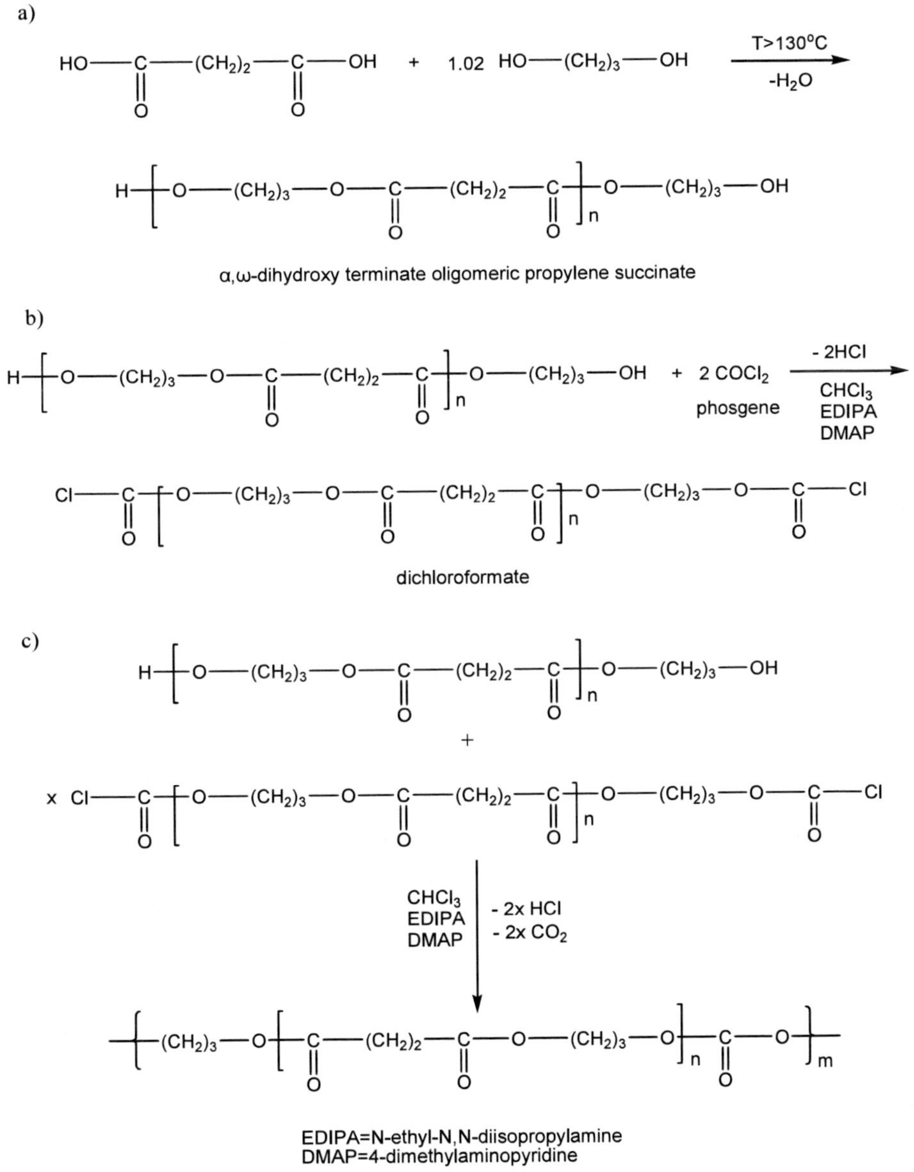

Figure 18. Chain extension reaction by dichloroformate synthesis of oligo (propylene succinate) [40].

The oligomers were first dissolved in dry chloroform in the presence of acid acceptors like ethyldiisopropylamine. Reaction with excess phosgene resulted in α,ω-dichloroformate. The latter was then polycondensated with an equivalent amount of α,ω-dihydroxy-terminated oligo(propylene succinate)s to receive high molecular weight poly(ester carbonate)s. However the melting points and mechanical properties remained low even for high molecular weight poly(ester carbonate)s. For this reason chain extension of oligomers obtained by the esterification of succinic acid in the presence of 1,3-propanediol and 1,4-cyclohexanedimethanol was also performed [43]. The resulting polymers were characterized by glass transition temperatures and melting points, which increased with increasing the 1,4-cyclohexanedimethylene succinate units content.

Finally, segmented copolymers based on oligo(propylene succinate)s and poly(ethylene glycol) were synthesized [43]. These showed a higher affinity towards polar solvents than oligo(propylene succinate)s and their water absorption capability was increased to 10-15wt%.

Another strategy to prepare polyesters is from lactones (cyclic esters) [43]. There has been established a method for the preparation of 1,4-dioxepan-2-one from 1,3-propnadiol after reaction with chloroacetic acid sodium salt.

Then, ring-opening polymerization in the presence of catalysts leads to polymers with high molecular weight (Figure 19).

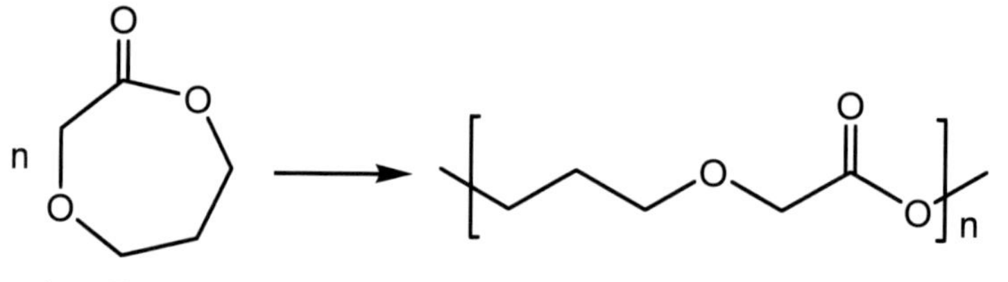

Figure 19. Ring-opening polymerization of 1,4-dioxepan-2-one [43].

As a matter of fact aliphatic polyesters of 1,3-propanediol have rather low thermal and mechanical properties, compared to aromatic polyesters. Copolymerization of 1,3-propanediol with aliphatic and aromatic acids might enable preparation of materials having both biodegradability and sufficient properties. Terephthalate polyesters such as poly(ethylene terephthalate) (PET), poly(propylene terephthalate) (PPT) and poly(butylene terephthalate) (PBT) are very important for industry, due to their high thermal and mechanical properties and finally their many applications. Although poly(propylene terephthalate) is a new polyester, it has been used in many applications already, especially in the form of fibers. It has a meting point of about 230°C, a glass transition temperature 44°C and crystallizes faster than PET, but slower than PBT. There is an increasing interest about this polymer and its applications are expanding. Terephthalates and in general aromatic polyesters cannot be degraded in the environment [44]. However, Witt et al. in relatively recent works, prepared oligo esters of terephthalic acid with 1,2-ethanediol, 1,3-propanediol, and 1,4-butanediol and these oligomers were investigated with regard to their biodegradability in different biological environments [4,45]. Well-characterized oligomers with weight-average molar masses of from 600 to 2600 g/mol exhibit biodegradation in aqueous systems, soil, and compost at 60°C. Size exclusion chromatography (SEC) investigations showed a fast biological degradation of the oligomer fraction consisting of 1 or 2 repeating units, independent of the diol component used for polycondensation, while polyester oligomers with degrees of polymerization higher than 2 were stable against microbial attack at room temperature in a time frame of 2 months. At 60°C in a compost environment chemical hydrolysis also degrades chains longer than two repeating units, resulting in enhanced degradability of the oligomers. Metabolization of the monomers and the dimers as well by the microorganisms could be confirmed by comparing size exclusion chromatography (SEC) measurements and carbon balances from specific experiments. Finally it was concluded that based on these results degradation characteristics of potential oligomer intermediates resulting from a primary chain scission from copolyesters consisting of aromatic and aliphatic dicarbonic acids can be predicted depending on their composition. These results will have an evident influence on the evaluation of the biodegradability of commercially interesting copolyesters and lead to new ways of tailor-made designing of new biodegradable materials as well.

2.5. Blends

Poly (propylene succinate) and poly(ethylene succinate) (PESu) are both biodegradable polymers. PPSu shows much faster enzymatic hydrolysis rates than PESu. On the other hand PESu shows better thermal and mechanical properties. Blending is an alternative to prepare tailor-made materials. Thus, blends of poly(propylene succinate) and poly(ethylene succinate) were prepared in order to explore their miscibility [46]. It was thought that, due to their similar chemical structures the polymers might show miscibility or compatibility. Indeed, they were found to be miscible. For the preparation of the PESu/PPSu blends polymers with the same intrinsic viscosity, about 0.28 dL/g were used. The number average molecular weight as determined by GPC, was close to 7000g/mol. These binary blends were prepared in chloroform as solvent. The solution of both polymers (0.0125g/ml) was cast on a petri dish at room temperature. The solvent was allowed to evaporate in air for 3 days. The compositions of the blends were PESu/PPSu 80/20, 60/40, 50/50, 40/60, 80/20 in weight ratio. As it was observed in the respective DSC traces, the blends showed single composition-dependent glass transition temperature (T_g) between the T_g values of the neat PESu and that of the neat PPSu. Usually, the glass transition temperature criterion can be safely applied if the T_gs of the components of the polymer blends differ by 15-20°C. The glass transition temperatures of PPSu and PESu differ by 24°C. Furthermore, for better resolution slow DSC scans for quenched samples were also preformed, by slow heating at 2.5 or 1°C/min. These traces also showed single glass transition. The evidence of single composition-dependent glass transition between those of the neat components for the PESu/PPSu blends showed that they are probably miscible over the entire range of composition. The variation of the T_g values of the blends was found that could be described by the Gordon-Taylor equation. The crystallization of the blends was studied using DSC. Though the neat PPSu, if amorphous, cannot show cold-crystallization the blends with less than 40wt% PPSu content showed cold-crystallization during scans by 20°C/min. Furthermore, for the specific blends after crystallization at low temperature, melting peaks for both components were observed. This showed that they could crystallize in the blends. In fact it was found that PESu crystallizes first, while PPSu crystallizes at a much lower temperature, in the confined environment, due to the presence of PESu crystals. Since PPSu is still liquid when PESu crystallizes, there is a melting point depression of the higher melting temperature component, PESu. Its melting point in the blends decreased steadily with increasing PPSu content. However, similar conclusions were derived for PPSu, when the blends crystallized at low temperature. The equilibrium melting points were determined for the neat polymers and the blends and the melting point depression for the high melting temperature component, PESu, was analyzed applying the Nishi-Wang theory. After that analysis the calculated values for the interaction parameter were found to be negative. This was also evidence that the specific blends were miscible in the melt phase. WAXD showed that the polymers could form their own crystals in the blends, since both diffractions associated with the PESu and the PPSu crystals were revealed in the blend patterns. However, the relative intensity of the peaks varied with blend composition. The fact that both components crystallized separately was attributed to the significant difference in the crystallization rates of the two polymers.

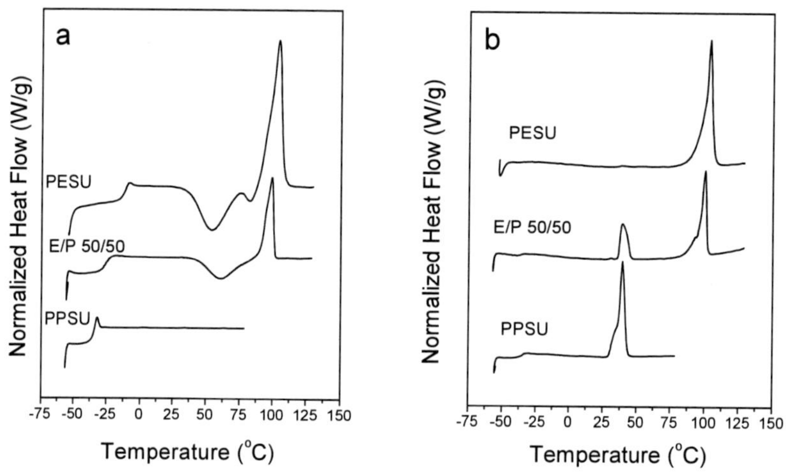

Figure 20. DSC traces for amorphous and crystallized PESu/PPSu blends.

The specific blends were classified as semicrystalline/semicrystalline. In contrast to blends between semicrystalline and amorphous polymers, which have been extensively studied, blends in which both components are semicrystalline and especially in which both are biodegradable are not often reported in the literature. In general it is believed, that such blends could be used alternatively to copolymers for pharmaceutical applications.

In contrast to PPSu which is a new polyester, PCL is well-known and extensively studied polymer. It is also biocompatible and has found many applications in pharmaceutical technology and medicine. Thus, it was also an interesting idea to explore miscibility and biodegradation behavior of blends made of these two very important polyesters. PCL/PPSu blends with concentrations 90/10, 80/20, 70/30 and 60/40 w/w were prepared by solution-casting [47]. Proper amounts of both polymers were dissolved in chloroform as common solvent, at room temperature. Sonication was applied in order to achieve complete dissolution and fine mixing of the components. The blends in the form of thin films (200-250 μm) were set up after solvent evaporation at room temperature, under a gentle air stream. They were characterized by DSC, WAXD, ^{1}HNMR, SEM, and Tensile testing. Finally, their enzymatic hydrolysis was studied. The PCL/PPSu blend system however proved to be only partially miscible. DSC studies on the melting point depression showed that there was a slight decrease in the melting point of the high temperature component, PCL, in the blends. Results after the analysis of the melting point depression using the Nishi-Wang model were marginal most probably showing a limited miscibility. Tensile properties testing however proved that the blends are at least compatible if not miscible. This was concluded because despite the decrease in the tensile properties of the blends with increasing PPSu content, there was no indication for any minimum. Crystallization rates of PCL were retarded with increasing second component content in the blends. Biodegradation of the blends was also explored. Enzymatic hydrolysis for several days of the prepared blends was performed using *Rhizopus delemar* lipase at pH 7.2 and 30°C. SEM micrographs of thin film surfaces revealed that hydrolysis affected mainly the PPSu polymer as well as the amorphous phase of PCL. For all polymer blends an increase in the values of the melting temperature and the heat of fusion was recorded after the hydrolysis. This was associated with the fact that enzymatic hydrolysis affected mainly the amorphous phase of the blends.

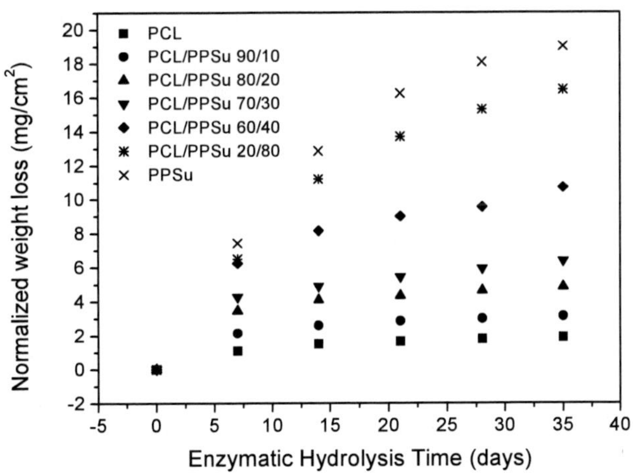

Figure 21. Weight loss of PCL/PPSu blends during enzymatic hydrolysis for several days.

The consumption of the amorphous portion of the specimens resulted in an increase of the estimated crystallinity after hydrolysis. The biodegradation rates as expressed in terms of weight loss were faster for the blends with higher PPSu content (Figure 21). PPSu was proved to have much higher biodegradation rates compared to PCL, due to the lower crystallinity, the decreased melting point (T_m for PPSu is about 45°C) comparing to PCL (T_m about 60-65°C), and the number of the ester bonds along the macromolecular chains for given molecular weight. SEM observations lead to conclusions aligned with those from the weight loss measurements. That is, only for blends with high PPSu content, extended surface erosion occurred even from the first days of enzymatic hydrolysis.

Poly(propylene adipate) (PPAd) was reported to be miscible with poly(ethylene oxide) PEO. In fact Lin and Woo studied the miscibility of PEO with a series of polyesters [48]. They found that PEO is miscible with polyesters such as poly(ethylene adipate), poly(propylene adipate), poly(butylene adipate), and poly(ethylene azelate). On the other hand PEO was found to be immiscible with poly(1,6-hexamethylene adipate) and poly(1,6-hexamethylene sebacate). Optical observations showed that the melts of the blends of PEO with the last two polyesters were cloudy and heterogeneous in contrast to all the others. Also, the Nishi-Wang model was used to analyze the melting point depression observed for PEO in the blends. The conclusion was that the ratio CH_2/CO in the polyesters is crucial for the miscibility with PEO. PEO was miscible with polyesters with a CH_2/CO ratio from 3 up to 4.5, while miscibility was favored especially in the case of a value 3.5, which is the case of PPSu.

2.6. Application of PPSu in Drug Delivery Systems

Aliphatic polyesters can be used in pharmaceutical applications, like drug delivery systems. As was reported above PPSu has a low melting temperature (40 to 46°C depending on the molecular weight), slightly above the temperature of the human body. Consequently, PPSu might be an alternative for the preparation of solid dispersions of drugs. Solid dispersions are dosage forms whereby the drug is dispersed in a biologically inert matrix. They can be used for the increase of the dissolution rates and thus the bioavailability of sparingly water-soluble drugs. As a matter of fact an increasing number of poorly water-soluble drugs are synthesized, and thus there is a demand for finding routes to increase their solubility. Alternatively solid dispersions can be used for controlled release of drugs. In a recent research, four PPSu samples with intrinsic viscosities 0.1, 0.18, 0.28 and 0.50 dL/g measured in chloroform solutions, were used in the preparation of solid dispersions of the drug Fluvastatin, with a drug load 30 wt%. A second series of solid dispersions were prepared, where the PPSu 0.28 was used, while the drug load varied from 10 to 50 wt%. The solid dispersions were prepared by the melt method, which involves dissolving the drug in the molten polymer, and then cooling to the room temperature. WAXD patterns of the solid dispersions proved that the drug was amorphous in all cases, since only the peaks corresponding to the crystalline polymer were revealed (Figure 22).

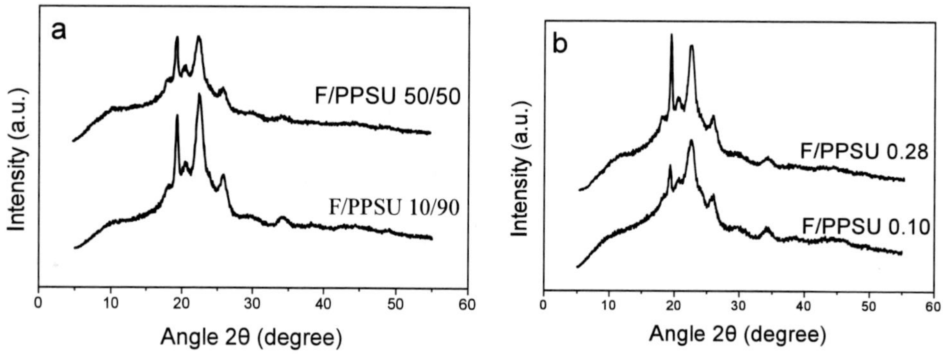

Figure 22. WAXD patterns of a) Fluvastatin solid dispersions in PPSu 0.28, with different drug loadings and b) Fluvastatin solid dispersions in PPSu samples of different molecular weight and containg 30wt% drug.

The release rates of the drug from the solid dispersions increased with decreasing the molecular weight of PPSu, probably because low molecular weight polymers are brittle and increased porosity allows drug to come in contact to the aquatic dissolution medium (Figure

23). Also, for given molecular weight of PPSu the rates increased with increasing the drug content in the solid dispersions.

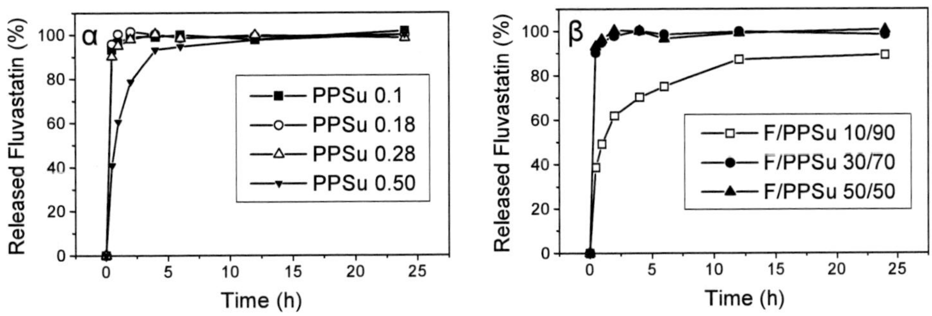

Figure 23. Release of Fluvastatin from solid dispersions in PPSu during in-vitro tests: a) solid dispersions prepared using PPSu matrices with different molecular weights and constant drug load (30wt%), and b) solid dispersions prepared using the PPSu 0.28 sample and different drug loads.

3. Conclusion

New industrial processes allowed production of 1,3-propanediol, a very important monomer for the synthesis of polyesters. Biodegradable polyesters based on 1,3-propanediol have gained an increasing interest. Especially polymers like poly(propylene succinate) are fully biodegradable and can also be produced from monomers from renewable resources. The polyesters can be synthesized following the polycondensation method. Recent experimental work on enzymatic hydrolysis has shown that polyesters of 1,3-propanediol have faster degradation rates compared to the corresponding from ethylene glycol or 1,4-butanediol. Poly(propylene succinate) seems to be faster degraded than poly(propylene adipate) or poly(propylene sebacate). Copolymers prepared from polycondensation of succinic acid with 1,3-propanediol and 1,4-buanediol showed improved biodegradation characteristics compared even to the neat PPSu. Also, copolymers synthesized from oligo(propylene succinate) and ε-caprolactone, combined characteristics of the homopolymers, showing improved hydrolysis behavior. Combination of succinic or adipic acid with isophthalic acid as starting materials for polycondensation with 1,3-propanediol gave examples of copolymers containing both segments which can hydrolyze and aromatic ones in the macromolecular chains. Presence of aromatic moieties is desirable in order to increase strength of the materials. Alternatively, starting from oligo(propylene succinate)s, methods have also been developed to produce polymers like poly(ester urethane)s or poly(ester carbonate)s with high molecular weight, which finally are also biodegradable.

Fully biodegradable blends of PPSu with PESu have been found to be miscible. Blends of PPSu with poly(ε-caprolactone) showed only very limited miscibility, but satisfactory tensile properties.

In general, results from the research on 1,3-propanediol based polymers are very encouraging. Polyesters of this class are expected to find applications, especially in pharmaceutical technology, and their uses in such applications have already been tested.

However, generally speaking and taking into account the parameter cost of biodegradable polymers, much work should be carried out to arrive at optimal solutions and wide applications.

References

[1] 1. Haas, T.; Jaeger, R.; Weber, R.; Mitchell, S.F.; King, C.F., *Appl. Catalys. A: Gener.* 2005, *280*, 83-88

[2] 2. Wang, B.; Li, C.Y.; Hanzlicek, J.; Cheng, S.Z.D.; Geil, P.H.; Grebowicz, J.; Ho, R.M. *Polymer* 2001, *42*, 7171-7180.

[3] 3. Ward, I.M.; Wilding, M.A.; Brody, H. *J. Polym. Sci. Polym. Phys. Edit.* 1976, *14*, 263-274.

[4] 4. Müller, R.J.; Witt, U.; Rantze, E.; Deckwer, W.D. *Polym. Degrad. Stab.* 1998, *59,* 203-208.

[5] 5. Bikiaris, D.N.; Papageorgiou, G.Z.; Achilias, D.S. *Polym. Degrad. Stab.* 2006, *91*, 31-43.

[6] 6. Zorba, T.; Chrissafis, K.; Paraskevopoulos, K.M.; Bikiaris, D.N. *Polym. Degrad. Stab*. 2007, 92, 222-230.

[7] 7. Karayannidis, G.P.; Roupakias, C.; Bikiaris, D.; Achilias, D.S. *Polymer* 2003, *44*: 931-942.

[8] 8. Xiu, Z.L.; Song, B.H.; Wang, Z.T.; Sun, L.H.; Feng, E.M.; Zeng, A.P. *Biochem. Eng. J.* 2004, *19,* 189-197.

[9] 9. Chen, X.; Zhang, D.J.; Qi, W.T.; Gao, S.J.; Xiu, Z.L., Xu, P. *Appl. Microbiol. Biotechnol.* 2003, *63,* 143-146.

[10] 10. Hartlep, M.; Hussmann, W.; Prayitno, N.; Meynial-Salles I.; Zeng, A.P. *Appl. Microbiol. Biotechnol.* 2003, *60*, 60-66.

[11] 11. Nakamura, C.E.; Whited, D.M. *Cur. Opin. Biotechnol.* 2003; *14*, 454-459.

[12] 12. Kim, D.Y.; Yim, S.C.; Lee, P.C.; Lee, W.G.; Lee, S.Y.; Chang, H.N.; *Enz. Microbial. Techn.* 2004, *35,* 648-653.

[13] 13. Lee, P.C.; Lee, W.G.; Lee, S.Y.; Chang, H.N. *Biotechn. Bioeng.* 2001; *72*, 41-47.

[14] 14. Chatti, S.; Behnken, G.; Langanke, D.; Kricheldolf, H.R. *Macrom. Chem. Phys.* 2006, *207*, 1474-1484.

[15] 15. Corden, T.J.; Jones, I.A.; Rudd, C.D.; Christian, P. Downes, S.; McDougall, K.E. *Biomaterials* 2000, *21*, 713-724.

[16] 16. Edlund, U.; Albertsson, A.C. *Adv. Drug Deliv. Rev.* 2003, *55*, 585-609.

[17] 17. Roupakias, C.P.; Bikiaris, D.N.; Karayannidis, G.P. *J. Polym. Sci. Part A, Polym. Chem.* 2005, *43*, 3988–4011.

[18] 18. Bikiaris, D.N.; Achilias, D.S. *Polymer,* 2006, *47*, 4851-4860.

[19] 19. Chrissafis, K.; Paraskevopoulos, K.M.; Bikiaris, D.N.. *Thermochim. Acta* 2006, *440*, 166-175.

[20] 20. Papageorgiou, G.Z.; Bikiaris, D.N. *Polymer* 2005, *46*, 12081-12092.

[21] 21. Jourdan, N.; Deguire, S.; Brisse, F. *Macromolecules* 1995, *28*, 80-86-8091

[22] 22. Chrissafis, K.; Paraskevopoulos, K.M.; Bikiaris, D.N. *Polym. Degrad. Stab.* 2006, *91*, 60-68.

[23] 23. Bikiaris, D.N.; Chrissafis, K.; Paraskevopoulos, K.M.; Triantafyllidis, K.S.; Antonakou, E.V. *Polym. Degrad. Stab*. 2007, 92, 525-536.

[24] 24. Wang, X.S.; Li, X.G., Yan, D. *Polym. Test.* 2001, *20*, 491-502.
[25] 25. Wang, X.S.; Li, X.G.; Yan, D. *Polym, Degrad, Stab,* 2000, *69*, 361-372.
[26] 26. Persenaire, O.; Alexandre, M.; Degée, P.; Dubois, P. *Biomacromolecules* 2001, *2*, 288-294.
[27] 27. Aoyagi, Y.; Yamashita, K.; Doi, Y. *Polym. Degrad. Stab.* 2002, *76*, 53-59.
[28] 28. Fan, Y.; Nishida, H.; Hoshihara, S.; Shirai, Y.; Tokiwa, Y.; Endo, T. *Polym. Degrad. Stab.* 2003, *79*, 547-562.
[29] 29. Soccio, M.; Nogales, A.; Lotti, N.; Munari, A.; Ezquerra, T.A. *Phys. Rev. Lett.* 2007, *98*, 37801-37804.
[30] 30. Bikiaris, D.; Aburto, J.; Alric, I.; Borredon, E.; Botev, M.; Betchev, C.; Panayiotou, C. *J. Appl. Polym. Sci.* 1999, *71*, 1089-1100.
[31] 31. Aburto, J.; Alric, I.; Thiebaud, S.; Borredon, E.; Bikiaris, D.; Prinos, J.; Panayiotou, C. *J. Appl. Polym. Sci.* 1999, *74*, 1440-1451.
[32] 32. Umare, S.S.; Chandure, A.S.; Pandey, R.A. *Polym. Degrad. Stab.* 2007, *92*, 464-479.
[33] 33. Chandure, A.S.; Umare, S.S. *Int. J. Polym. Mater.* 2007, *56*, 339-353.
[34] 34. Papageorgiou, G.Z.; Bikiaris, D.N. *Biomacromolecules* 2007, 8, 2437-2449.
[35] 35. Xu, Y.; Xu, J.; Guo, B.; Xie, X. *J. Polym. Sci. Part B: Polym. Phys.* 2007, *45*, 420-428.
[36] 36. Seretoudi, G. ; Bikiaris, D. ; Panayiotou, C. *Polymer* 2002, *43*, 5405-5415.
[37] 37. Papadimitriou, S.; Bikiaris, D.N.; Chrissafis, K.; Paraskevopoulos, K.M.; Mourtas, S. *J. Polym. Sci. Part A. Polym. Chem. 2007,* 45, 5076-5090.
[38] 38. Soccio, M.; Finelli, L.; Lotti, N.; Gazzano, M.; Munari, A. *J. Polym. Sci. Part B: Polym. Phys*. 2007, *45,* 310-321.
[39] 39. Soccio, M.; Finelli, L.; Lotti, N.; Gazzano, M.; Munari, A. *Eur. Polym. Jour.* 2006, *42*, 2949-2958.
[40] 40. Ranucci, E.; Söderqvist Lindblad, M:; Albertsson, A.C. *Macromol. Rapid Commun.* 2000, *21,* 680-684.
[41] 41. Liu, Y.; Ranucci, E.; Söderqvist Lindblad, M.; Albertsson, A.C. *J. Polym. Sci. Part A: Polym. Chem.* 2001, *39,* 2508-2519.
[42] 42. Liu, Y.; Söderqvist Lindblad, M.; Ranucci, E.; Albertsson, A.C. *J. Polym. Sci. Part A: Polym. Chem.* 2001, *39,* 630-639.
[43] 43. Söderqvist Lindblad, M.; Liu, Y.; Albertsson, A.C.; Ranucci, E.; Karlsson, E. *Adv. Polym. Sci.* 2002, *157*, 141-161.
[44] 44. Witt, U.; Müller, R.J.; August, J.; Widdecke, H.; Deckwer, W.D. *Macrom. Chem. Phys.* 1994, *195*, 793-802.
[45] 45. Marten, E.; Müller, R.J.; Deckwer, W.D. *Polym. Degrad. Stab.* 2005, *88,* 371-381.
[46] 46. Papageorgiou, G.Z.; Bikiaris, D.N. *J. Polym. Sci. Part B: Polym.* 2006, *44,* 584-597.
[47] 47. Bikiaris, D.N.; Papageorgiou, G.Z.; Achilias, D.S.; Pavlidou, E.; Stergiou, A. *Eur. Polym. J.* 2007, 43, 2491-2503.
[48] 48. Lin, J.H.; Woo, E.M. *Polymer* 2006, *47*, 6826-6835.

In: Polyester: Properties, Preparation and Applications
Editors: Hina Yamashita, Yui Nakano
ISBN: 978-1-60456-753-3

Chapter 5

Compatibility of Cotton/Nylon and Cotton/Polyester Warp-Knit Terry Towelling with Industrial Laundering Procedures

Adine Gericke, L Viljoen and R. de Bruin
Department of Chemistry and Polymer Science, University of Stellenbosch, Private Bag X1, Matieland Stellenbosch, South Africa, 7602; agericke@sun.ac.za

Abstract

Large institutions, such as hotels and hospitals, often use specialized industrial laundries for laundering sheets, towels or uniforms. The main purpose of this study was to determine the effect of industrial laundering procedures on the durability of cotton warp knitted towels with a synthetic ground structure of either nylon or polyester. The durability of cotton/nylon and cotton/polyester terry towelling fabric samples that were subjected to repeated industrial laundering procedures, was compared by measuring the tensile strength of fabric samples after 50 washing cycles and 50 washing/tumble-drying cycles. The difference between the tensile strengths of the cotton/polyester and cotton/nylon terry towelling samples after washing alone was not significant. The tensile strength of the cotton/nylon samples, however, was significantly less than that of the cotton/polyester samples after tumble-drying. It was concluded that industrial laundering procedures, especially tumble-drying, have a more detrimental effect on the durability of the nylon ground structure than on the polyester ground structure of warp-knitted terry towelling fabrics.

Key words: towelling, warp knit terry, laundering, tensile strength, durability

INTRODUCTION

Large institutions, such as hotels and hospitals often use specialized industrial laundries for laundering sheets, towels or uniforms. Continuous washing machines, used by industrial launderers faced with large amounts of relatively uniform laundry, are designed for efficient water and energy use. The laundering process involves interaction of numerous physical and chemical effects on the laundry.

A growing demand and the quest for improved efficiency have led South African towel manufacturers to faster and more economical production methods. Cotton warp-knitted terry towels with a synthetic ground structure are becoming more popular because of the high production speed [1] and added advantage of enhanced durability because of the synthetic component that can be incorporated into the structure [14].

Warp-knitted terrycloth is manufactured on a Raschel warp-knitting machine, using three sets of warp yarns [6][14][22]. The ground structure of pillar stitches, similar to chain stitching, is formed by the first set of warp yarns. Pillar stitches of adjacent rows are not connected. A second set of warp yarns isare looped into this base structure in the weft direction – each set connecting four adjacent rows of pillar stitches – adding stability and strength in the weft direction to the structure. The third set of yarns, referred to as the pile yarns, forms the pile loops. When warp knitting terrycloth, one needlebar is replaced with a point or pinbar, around which the pile yarns are overlapped when the pile loops are formed [22][25]. An illustration of the structure is shown in figure 1 (the length of the pile yarns is not depicted according to scale in the illustration).

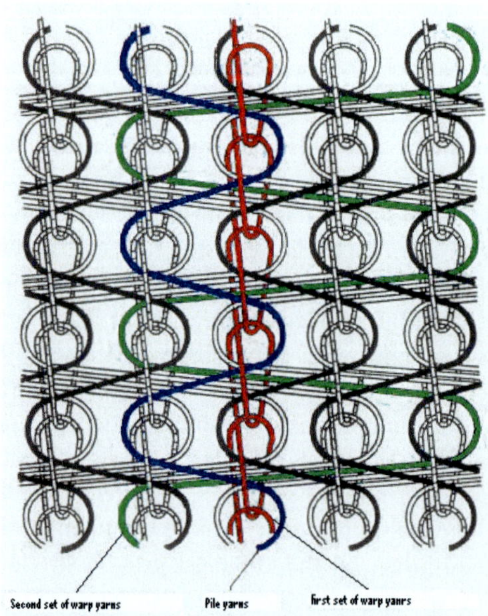

Figure 1. Illustration of warp knitted terry towelling fabric structure [8].

In practice, the ground yarns of warp-knitted terrycloth fabrics can differ in composition to that of the pile [22]. Synthetic fibers, such as like nylon and polyester, are mostly used in the ground structure – adding stability and strength [34]. Even though the percentage of synthetic fibers in the structure are is often less that 5%, but even such a small percentage can have a significant effect on the properties of the knitted fabric, as it can enhance the tensile strength without affecting the absorbency negatively. Cotton is generally used as pile yarn because of its excellent absorbency. The second set of warp yarns is often also cotton, to keep absorbency as high as possible, but also because cotton is reasonably priced and actually increases in strength when wet [14][22]. Such fabrics can be described as cotton-polyester or cotton-nylon mixtures.

According to American and European law, it is not compulsory for manufacturers of textile goods to indicate the synthetic component on the label of a product if it is less than 5% [14]. The same practice is applied in South Africa. A label may thus indicate 100% cotton, even though a synthetic component is present. As inferred from the structure of the warp knitted terrycloth (figure 2), the synthetic component plays a vital role in the durability of the towels, especially where breaking strength in the warp direction is concerned.

Care of textile fabrics plays an important role in their expected durability [6][21]. It is thus of utmost importance that the care procedures followed by industrial launderers are compatible with the fiber content and fabric construction of the laundered items.

Industrial launderers generally make use of continuous washing machines, referred to as continuous batch tunnel washers (CBTDs) [5][10]. These machines are designed to remove soiling, contamination and micro-organisms from large quantities of dirty linen. It usually entails a pre-wash, main wash and rinse, followed by tumble drying. Pre-weighed loads of laundry pass through a number of compartments (typically ten to twelve) in which they are pre-washed, washed and rinsed. A CBTD machine usually works with an internal counter-flow current system in which the wash load moves forward in the machine while the water flows in the opposite direction.

Chemical additives are added automatically to the appropriate compartments and steam can be injected into the main wash compartments to achieve the required high temperature levels. Stains and soil are removed through a combination of mechanical action, time and temperature control, detergent action and de-staining agents, which can include bleaches. For thermal disinfection, the achievement and maintenance of the required temperature areis critical.

According to the Fabricare Association [5], the most important factors that influence effective cleaning, and also the durability of the textiles being laundered, are temperature, time, level of wash liquid (dip), mechanical action during washing, additives, the tumble drying process, etc. Of these, temperature, additives and tumble drying are of particular relevance to this study.

Control of *temperature* during each phase of the laundering process is necessary, because on the one hand, temperature has a direct influence on effective soil and stain removal, and on the other, certain fabrics are sensitive to prolonged exposure to high temperatures. Control over temperature is also important when using hypochloritde bleaching agents, as bleaching at too high a temperature can damage certain fibers. In the hospitality and medical industries however, temperature plays an important role in the disinfection process. Here it is important that the critical temperature is reached to ensure effective disinfection [5]. According to Steyn [27], a washing temperature of higher than 54 °C is necessary to ensure effective elimination of micro-organisms in household laundering. In industrial laundries, a temperature of at least 71°C for three minutes or 65°C for ten minutes is recommended as safe time/temperature combinations for thermal disinfection. Additional time should be allowed for adequate penetration in large laundry loads. It is usually preferable to employ a higher temperature range to ensure thorough disinfection, and temperatures up to 80°C are often reached within certain areas in the machine [3][5].

Cotton is not thermoplastic and will not melt in the presence of heat [29]. Normal exposure to elevated temperatures, as found in routine care procedures, will not degrade cotton. White fabrics can be washed in hot water (95°C), but for colored fabrics temperatures higher than 60°C are not recommended [12]. Temperatures in the tumble dryer should not exceed 93°C to prevent the acceleration of fiber degradation [30]. The melting point of nylon varies between 215 °C and 260 °C, depending on the type of nylon [14]. Owing to its thermoplastic nature, and also because nylon does have a tendency to weaken slightly when wet, wash temperatures not exceeding 50°C and moderate drying temperatures are recommended [6][14][29].

Additives during laundering can include detergents, alkalis and bleaching agents.

Synthetic detergents are usually alkaline and are used to remove or suspend soil, reduce the effect of hard water or lower the surface tension of water with the purpose of enhancing soil removal. A pH-value of higher than ten in a detergent solution enhances the effectiveness of the wash process [5].

An alkaline medium will not damage cellulose fibers – even at high temperatures, so all detergents on the market can be used to wash cotton fabrics. Both nylon and polyester, as used in the ground structure of the terrycloth towelling fabrics, have excellent resistance to alkaline wash mediums. Polyester can be damaged by strong alkalis, but has good resistance to weak alkalis such as that found in detergents [6][14][29].

During the rinse phase, the remaining detergent and suspended soil are removed through dilution. A weak acid solution can be added to neutralise the alkalinity of the wash liquid. CSellulose fibers, like cotton, are damaged by strong acids, but even weak acids can degrade the fibers if not removed properly during rinsing [33]. Polyester has excellent resistance to acids, but nylon is sensitive to hydrochloric acid and isare damaged if exposed to sulphuric acid [14][29][20]. The effect of extended exposure to weak acids, especially in the wet state, could be questioned [20].

Bleaches are used to remove stains. The choice of a suitable bleaching agent is dependent on fiber type. The temperature of the washing liquid and the concentration of the bleaching agent should be carefully controlled, as degradation takes place more easily at higher temperatures and concentrations. Cotton is highly sensitive to, and can be degraded by, extensive use of chlorine bleach, whereas polyester is unaffected. Chlorine bleach is also known to cause damage and discoloration in nylon fabrics [14][29].

Tumble drying is the standard method of drying laundry during industrial laundering procedures. The size of the load in the dryer, the mechanical action, time and temperature have an effect on the amount of abrasion, and therefore degradation, that textile fabrics are exposed to [14]. Deans [4] wrote that tumble drying is not a homogenous process. Both the distribution of textile items in the drum and the variation in the temperatures of the air, sides of the drum and textiles being dried are dependent on textile fabric type, moisture content and packing density of the load. He explains that the nature of the textiles in the load influences the amount of moisture that migrates to the surface as well as the cohesion of items in the bundle.

The durability of textile goods is determined to a large extent by the wear and toughness of the textile yarns [31]. Durability is defined as the preservation of the physical integrity, appearance and functionality of a product under normal circumstances of usage [14][21]. This has led to the question of whether there is a difference in the effect of industrial laundering procedures on durability of the cotton/nylon and cotton/polyester warp knitted terrycloth towels.

AIM OF THE STUDY

The main purpose of the study was to compare the durability of cotton warp knitted terrycloth with a nylon ground structure with that of cotton warp knitted terrycloth with a polyester ground structure after exposure to industrial laundering procedures. In order to do so, the cotton/nylon and cotton/polyester terrycloth samples were first analysed to determine their comparability. Secondly, the durability of the fabrics was compared after exposure to industrial washing as well as to washing and tumble drying.

RESEARCH PROCEDURE

Cuprammonium fluidity tests were done on the cotton from both the unwashed cotton/nylon and cotton/polyester terrycloth samples to determine possible chemical damage to the cotton fibers as a result of the manufacturing process. The fluidity of both textile samples was <1 and the degree of polymerisation (GP-values) >3000, which showed that no chemical damage had taken place at that stage (CSIR Manufacturing and Materials Technology Centre for Fibers, Textiles and Clothing, 2004).

In order to describe the test samples and determine the similarity (and therefore comparability) of the cotton/nylon and cotton/polyester terrycloth, the percentage fiber composition, the structural properties, mass per unit area, as well as tensile strength of the unwashed cotton/nylon and cotton/polyester terrycloth were determined.

Samples were exposed to laundering procedures representative of that currently used in industrial laundries, which included washing in a continuous washing-machine and tumble drying. Test samples were exposed to 50 wash and 50 wash/tumble dry cycles and were washed and tumble dried together with other laundry. Wash bundles consisted of towels only and weighed 35kg per bundle. The laundering procedures were conducted under supervision of the researcher to ensure strict control.

Each bundle of washing was weighed in advance (35 kg) to ensure that the correct loading level and the corresponding concentration of wash additives during the washing process. Wash bundles were loaded in compartments on a conveyor belt that fed into the washing-machine. The continuous washing-machine used in this study consists of twelve different compartments into which the washing additives are added in different stages. In the first two compartments (pre-wash area) of the machine, the temperature does not exceed 30°C. In the next two compartments, still in the pre-wash area, a temperature of 78°C is reached. The main wash area, consisting of compartments five to eight, maintains the 78°C temperature. In the rinse area, the temperature gradually decreases in the ninth and the tenth compartments to 55°C and 45°C, respectively. In the last two compartments, there are again no specified temperatures to be reached. The procedure is computer-controlled to ensure adherence to specified processes and temperatures.

After washing, the excess water is pressed from the laundry and it is transported via a conveyor belt to a tumbler (not a tumble dryer) to separate the washing items from each other. The samples that were only washed were loaded directly back into the washing-machine, while the samples that were also tumble dried were transported to the tumble dryer.

The industrial tumble dryers were set to reach a maximum temperature of 120°C. During the study, temperature strips were used to measure the actual temperature inside the tumble dryer. Temperatures between 104°C and 110°C were measured. A load of towels took between 30 and 45 minutes to dry, depending on the size of the load.

In order to compare the durability of cotton/nylon and cotton/polyester warp knitted terrycloth, the tensile strength of the test samples in the warp direction was determined after exposure to 50 wash- and 50 wash/tumble dry cycles according to the SABS-standard test method. Ten specimens per terrycloth sample exposed to a specific treatment combination, were tested to ensure a representative average. All samples were conditioned according to standard procedures for 24 hours under controlled atmospheric conditions prior to testing. The tensile strength tests were only done in the warp direction because the nylon and polyester ground structure (pillar stitches) runs only in this direction. Adjacent rows of pillar stitches are not directly intertwined, but bound to each other by means of a second set of cotton warp yarns (figure 2). Since the purpose of this study was the comparison of the nylon and polyester components of the warp knitted terrycloth, it was not relevant to compare the tensile strength of the two sets of samples in the weft direction. The dimensional stability of the test samples was determined to establish whether adjustments to the tensile strength results would be necessary.

Fluidity tests were done on the cotton from the laundered cotton/nylon and cotton/polyester terrycloth samples to determine if there was any chemical damage to the cotton fiber.

Textile Testing Methods

Before any tests were done, all the test samples were conditioned according to standard procedures for 24 hours under controlled atmospheric conditions (20 ± 2°C and 65 ± 2% relative humidity). The fiber composition of the cotton/nylon and the cotton/polyester textile samples was determined according to the ISO 1833-standard test method. The structural characteristics were compared by analysing the knitted structures under a microscope. The average stitch density of the textile samples was determined according to the SABS-method 1120. The average weight of the unwashed cotton/nylon and cotton/polyester terrycloth samples was determined according to the SABS-method 79.

The average tensile strengths of the cotton/nylon and cotton/polyester warp knitted terrycloth samples were compared through conducting tensile strength tests according to the SABS-method ISO 13934-1 on an Instron Universal Testing Machine (Model 4444) after the industrial laundering procedures were completed (50 washing and 50 washing/tumble dry cycles). The dimensional stability of the warp knitted terrycloth was measured and calculated according to the SABS-method 960.

Data Analysis

To compare the results of the tensile strength tests, a two-directional cross-classification variance analysis (ANOVA) was used with fiber content and laundering procedure as factors [23]. The main effects were directly interpreted if the interaction effects were not significant; otherwise various one-directional ANOVAs were done at each of the levels of industrial laundering procedures to determine if the tensile strengths differ significantly. A Bonferroni multiple comparison procedure was used to determine where the differences between levels of the main effects were. Throughout this study a 5% significance level was used. The fiber content included cotton/nylon and cotton/polyester for the purposes of the analyses. The care procedures included unwashed, washed, and washed/tumble dried.

RESULTS AND DISCUSSION

The main purpose of this study was to compare the durability of cotton/nylon and cotton/polyester warp knitted terrycloth towel fabrics that have been exposed to industrial laundering procedures. The tensile strength results of samples that were washed 50 times or washed and tumble dried 50 times were compared to the tensile strength of the unwashed samples, as any change in tensile strength serves as a direct indication of a change in textile structure, implicating a change in durability [21].

Description of Test Samples

Fiber Composition

The average nylon content of the cotton/nylon terrycloth samples was 4,40%. The cotton/polyester terrycloth samples had an average of 4,43% polyester. The difference in synthetic content of the two groups of samples was small enough for the synthetic components to be considered as similar in proportion.

Structural Characteristics

The knitted structure of both groups of terrycloth samples was identified as warp knitted terrycloth in which three sets of warp yarns were used, similar to those in figure 4.1. From the microscope analysis it was clear that the pillar stitches that formed the ground structure of the fabric were synthetic. The other two sets of yarns, namely the second set of warp yarns and the pile threads, were both cotton. In terms of the stitch density, the cotton/nylon terrycloth samples had an average of 48,2 wales and 65,8 courses per 10cm, while the average number of wales and courses per 10cm in the cotton/polyester terrycloth samples were 46,3 and 62,7 respectively. The cotton/nylon terrycloth samples therefore had a slightly higher knit density than the cotton/polyester terrycloth samples. The average mass of the untreated cotton/nylon and cotton/polyester terrycloth samples were 560,1g/m2 en 522,2g/m2, respectively. The average tensile strength of cotton/nylon terrycloth samples before treatment was 296,24N and that of cotton/polyester was 297,84N. A variance analysis (ANOVA) was done on the above-mentioned results to statistically determine the significance of the differences, after which the cotton/nylon and cotton/polyester warp knitted terrycloth samples were considered comparable in terms of structural properties for the purpose of the study.

Durability of Cotton/Nylon and Cotton/Polyester Warp Knitted Terrycloth Samples after Industrial Washing and Wash/Tumble Drying

Of the samples that were washed only (not tumble dried), the average tensile strength of the cotton/polyester samples was 260,82 N, while that of the cotton/polyester terrycloth samples was 273,82 N. The average tensile strength of cotton/nylon terrycloth samples exposed to 50 wash/tumble dry cycles, was 222,04 N and that of cotton/polyester samples 275,43 N. These results are illustrated in fFigure 4.2. A two-directional ANOVA was done on tensile strength against laundering procedures. The interaction was significant ($p < 0.01$); and thus one-directional ANOVA's are given at each level of laundering in tables 4.2 and 4.3.

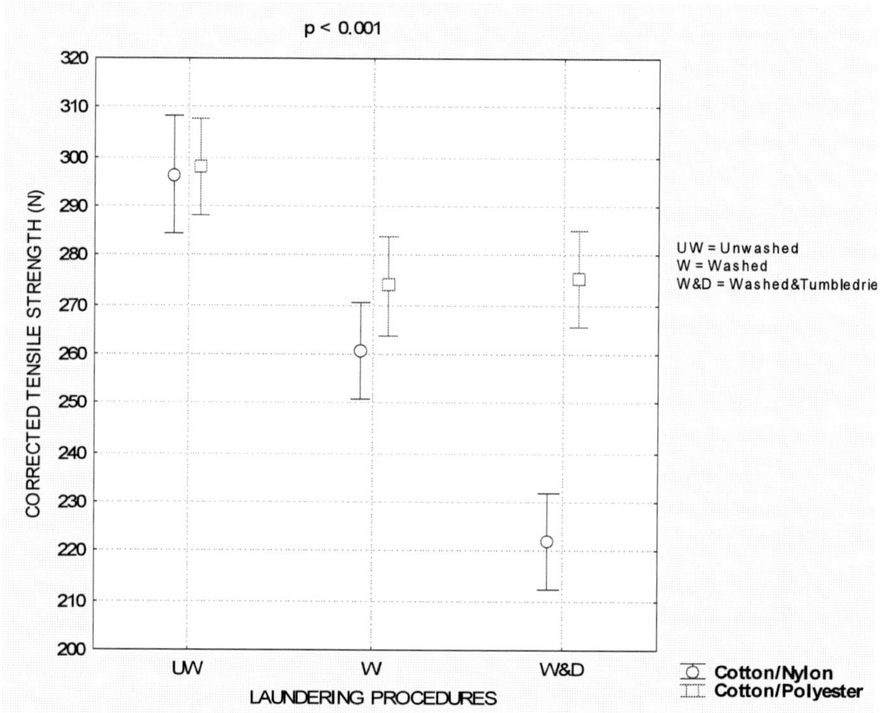

Figure 2. Comparison of the tensile strengths of unwashed (UW), washed (W) as well as washed and tumble dried (W and D) cotton/ nylon and cotton/ polyester warp knitted terrycloth samples.

Table 1. Variance Analysis of Average Tensile Strength of Terrycloth Samples Washed 50 Times

	TENSILE STRENGTH				
SOURCE OF VARIANCE	SK	VG	GK	F	P
Fiber content	2538	1	2538	5.907	0.018199
ErrorFault	24922	58	430		

Table 2. Variance Analysis of Average Tensile Strength of Terrycloth Samples Washed and Tumble Dried 50 Times

TENSILE STRENGTH					
SOURCE OF VARIATION	SK	VG	GK	F	p
Fiber content	42756	1	42756	62.537	p<0.001
ErrorFault	39655	58	684		

The tensile strengths of both cotton/nylon and cotton/polyester warp knitted terrycloth samples that were washed 50 times, are lower than those of the untreated samples, but only the decrease in average tensile strength of the cotton/nylon terrycloth samples is significant ($p < 0.05$). Statistical analyses confirmed that there is no significant difference ($p > 0.05$) between the average tensile strength of cotton/nylon and cotton/polyester terrycloth samples that were exposed to 50 washing cycles. In the case of cotton/nylon towels, this greater decrease in tensile strength can probably be attributed to the fact that nylon fibers expand/swell and weaken slightly in water as the hydrogen bonds in the amorphous areas of the fiber break when water molecules are absorbed. The thermoplastic nature of nylon also causes it to deteriorate during strenuous laundering procedures at high temperatures [6]. A temperature of 78°C is reached during the washing process that is much higher than the recommended temperature, [6][14][29]. The temperature reached during tumble drying, was also as high as 110°C, and in this case the exposure was lengthier.

The reason for the decrease in the tensile strength of the cotton/polyester terrycloth towels can probably not be attributed to the polyester component, as there is no significant difference between the wet and dry strength of polyester fibers. The abrasion resistance and strength of polyester are not influenced by the presence of water and it should remain stable after repeated washes [12][14][18][33]. It is highly improbable that the alkalinity of the detergents caused the decrease in tensile strength, as literature shows that polyester fibers are only degraded by strong alkaline solutions.

The tensile strengths of the cotton/polyester terrycloth samples that were washed 50 times and those that were washed and tumble dried, did not differ significantly ($p > 0.05$). The tensile strength of the cotton/nylon terrycloth samples did, however, decrease significantly. This decrease in tensile strength was highly significant ($p < 0.001$). There was also a highly significant difference ($p < 0.001$) between the tensile strengths of the cotton/nylon and the cotton/polyester terrycloth samples after the 50 wash/tumble dry cycles.

In terms of the results of the cuprammonium fluidity tests, the fluidity value of the cotton in the cotton/nylon terrycloth samples changed from <1 to 5,5 and the degree of polymerisation (GP-value) thereof, from >3000 to 1848 after being washed 50 times and tumble dried. According to the laboratory report this value is comparable to normal damage during washing and bleaching during manufacturing. In contrast to this, the fluidity value of the cotton in the cotton/polyester terrycloth samples changed from <1 to 19,6 and the GP-value from >3000 to 854 after being washed 50 times and tumble dried. According to the laboratory report, these values are comparable to strenuously bleached cotton that points to definite chemical damage [3]. The difference in the extent of damage of the cotton in the cotton/nylon and cotton/polyester terrycloth samples is difficult to explain, as they were exposed to the same laundering procedures. The difference in values can be attributed to the fact that the cotton used in the two samples could have differed in terms of a variety of factors, such as maturity and fineness of the fibers, which could have an effect on the amount of chemical damage during the laundering process.

As a result of the fluidity test results it could be expected that the cotton/polyester should be weaker than the cotton/nylon (which was not the case). It can therefore be expected that the most important cause of the difference in tensile strength in the washed and tumble dried samples can be attributed to degradation of the synthetic yarns in the ground structure and not to the cotton pile yarns or the second set of cotton warp yarns.

The significant decrease in the strength of the cotton/nylon terrycloth after the 50 wash/tumble dry cycles can probably be ascribed to the swelling and weakening of the nylon fibers when exposed to the severe friction action with concurrent high temperature during the tumble drying process. The fact that the tensile strength of the cotton/polyester terrycloth samples after the 50 wash/tumble dry cycles remained nearly unchanged is a good indication of the good abrasion resistance and resiliency of polyester, enabling it to withstand repeated and lengthy tumble drying cycles.

Conclusion

The purpose of the study was to compare the durability of cotton warp-knitted terrycloth with a nylon ground structure with that of cotton warp-knitted terrycloth with a polyester ground structure after exposure to industrial laundering procedures. Samples exposed to typical laundering processes were compared. There was a similar decrease in the tensile strengths of both the cotton/nylon and the cotton/polyester terrycloth samples after the 50 wash cycles. The tensile strength of the cotton/nylon warp-knitted terrycloth towelling samples decreased significantly more than that of the cotton/polyester samples after the cycles that included washing as well as tumble-drying. It can therefore be concluded that industrial washing procedures have little or no effect on the strength of the nylon and polyester ground structures of warp-knitted towels, but that the tumble-drying process has a significantly greater degrading effect on the nylon componentcontent of the cotton/nylon terrycloth fabric.

According to the literature, the tumble-drying temperatures for synthetic fabrics should be kept low to average. Although the temperature setting of the industrial tumble-dryer was 120 °C, temperatures between 104 °C and 110 °C were measured. Even such temperatures were clearly too high for the fabrics containing nylon and, as a result, adversely affected the nylon component of the cotton/nylon warp-knitted terrycloth towelling samples.

In the light of the results of the fluidity tests, on the cotton from the cotton/polyester fabrics, the important role that the polyester yarns in the ground structure play in ensuring durability of warp knitted terrycloth is underscored. The fact that the cotton content of these samples showed chemical deterioration (which should translate into a deterioration in tensile strength) would lead to the expectation that the tensile strength of the samples would be lower than that of their cotton/nylon counterparts. In fact, the opposite was found. When the results of this study, namely that the tensile strength in the warp direction of the cotton/nylon terrycloth samples was lower than that of the cotton/polyester terrycloth samples after 50 wash cycles and significantly lower after 50 wash/tumble-dry cycles, and the results of the fluidity tests are taken into account, it is clear that the polyester component plays a significant role in maintaining the tensile strength of the samples during the laundering procedures.

It is important for towel manufacturers to take note of the fact that industrial laundering procedures, and specifically tumble-drying, appear to have a much greater detrimental effect on cotton/nylon than on cotton/polyester warp-knitted terrycloth towels. The fiber content of textile products plays a critical role in their durability and required care procedures. The importance of indicating the correct fiber content on labels of textile goods, even if it is less than 5%, is clearly indicated in this study. It is strongly recommended that manufacturers of textile goods in South Africa should accurately indicate the fiber content of textile goods on labels. In this way industrial laundries and/or consumers can take note of the fiber content and adapt the care procedures accordingly.

REFERENCES

[1] Anand, SC; Smith, HM. Comparative performance of woven and warp-knitted towelling fabrics. 1994, *Kettenwirk-Praxis* 28(3):62-68.

[2] Barrie, D. How hospital linen and laundry services are provided. *Journal of Hospital Infection,* 1994, 27(3):219-235.

[3] CSIR Manufacturing and Materials Technology Centre for s, Textiles and Clothing. 2004. Laboratory Test Report. Port Elizabeth, South Africa.

[4] Deans, J. The modelling of a domestic tumbler dryer. *Applied Thermal Engineering,* 2001, 21(9):977-990.

[5] Fabric Care Research Association, 1993, *Washing technology.* SDML Consultancy and Training (RSA).

[6] Hatch, KL. Textile science, 1993, New York. West Publishing Company.

[7] Hegde, RR; Dahiya, A; Kamath, MG. Nylon fibers. Retrieved 30 August 2004. *http://www.engr.utk.edu/mse/pages/Textiles/Nylon%20fibers.htm.*

[8] Innovative Technology Makes Its Mark, 2004, Instruction sheet for Raschel warp-knitting machine: *Wirkbau-Superpol* 14123.

[9] Isaacs, M. DuPont nylon undergoes renewal at 60. *Textile World,* 1998, 148(11):41-46.

[10] Jakobi, G; Löhr, A. *Detergent and textile washing*: Principles and practice, 1987, New York. Cambridge.

[11] Johnsondiversey. 2004. Wadeville.

[12] Joseph, ML. *Essentials of textiles*. 4th ed. USA, 1988, Saunders College Publishing.

[13] Kadolph, SJ. *Quality assurance for textiles and apparel,* 1998, New York. Fairchild Publications.

[14] Kadolph, SJ; Langford, AL. *Textiles*. 9th Ed, 2002, New Jersey. Pearson Education.

[15] Kefford, C. What the textile exporter to the US and EU should know. *Textiles Unlimited,* 2001, 2(2):2.

[16] Lloyd, J; Adams, C. Domestic laundering of textiles. *Textiles*, 1989, 18(3):72-79.

[17] Lord. The serviceability of bed sheets in institutional use. *Journal of the Textiles Institute,* 1971, 62:304-327.

[18] Mason, RW. Decades later, polyester forges new image. *Textile World.* 1999, 149(1):57-60.

[19] Mccurry, JW. Towel mills modernize to compete. *Textile World,* 1999, 149(5):26-40.

[20] McIntyre, J.E. *Synthetic s: nylon, polyester, acrylic, polyolefin*. Cambridge. Woodhead Publishing Ltd. 2000
[21] Merkel, RS. *Textile product serviceability*, 1991, New York. Macmillan Publishing Company.
[22] Miller, E. *Textiles: Properties and behaviour in clothing use*, 1992, 4th ed. London. Batsford Academic and Educational.
[23] Milton, JS; Arnold, JC. *Introduction to Probability and Statistics, Principles and Applications for Engineering and the Computing Sciences*, 1990, McGraw Hill International. New York.
[24] Patel, P. Finishing of terry towels. *Journal of the Textile Association,* 1998, 58(5):195-197.
[25] Spencer, DJ. *Knitting technology*, 1983, Oxford. Pergamon Press.
[26] Steadman, RG. Cotton testing. *Textile Progress,* 1997, 27(1):1-36.
[27] Steyn, HJH.. Die invloed van wasmiddels en wastemperatuur op die groei en afsterwing van *Escherichia Coli.* 1994, PhD Thesis. University of the Free State.
[28] Taylor, MA. *Technology of textile properties: An introduction*, 1981, 2nd ed. England. Forbes. Publications Ltd.
[29] Tortora, PG. *Understanding textiles*. 4th ed. 1992, New York. Macmillan Publishing Co, Ltd.
[30] Ulrich, MM; Mohamed, SS. Effect of laundry conditions on abrasion of mercerized DP natural blend cotton/PET. *American Dyestuff Reporter,* 1982, 71(7):38-41.
[31] Trotman, ER. *Dyeing and chemical technology of textile s,* 1984, 6th ed. London. Charles Griffin and Company.
[32] Verryne, T. Cotton SA. Pretoria, 2003
[33] Wingate, IB; Mohler, JF. Textile fabrics and their selection. 1984, 8th Ed. New Jersey. Prentice-Hall.
[34] Wiska, Towel, 2003, Retrieved 19 July 2003. *http://www.wiska.co.id/e_towel.htm.*
[35] Wooten, HL. What's new in terry towel preparation? *Textile Chemist and Colorist,* 1979, 11(6):136-138.

In: Polyester: Properties, Preparation and Applications
Editors: Hina Yamashita, Yui Nakano
ISBN: 978-1-60456-753-3

Chapter 6

DEVELOPMENT OF POLYESTER TYPE SHAPE MEMORY POLYMER AND ITS APPLICATION TO COMPOSITE MATERIAL

Yong-Chan Chung[a]***, Byoung Chul Chun***[b]***, Mi-Hwa Chung***[b]***, Yong-Sik Shim***[a] ***and Jae Whan Cho***[c]
[a]Department of Chemistry and [b]Department of Polymer Engineering, The University of Suwon, Suwon, Korea
[c]Department of Textile Engineering, Konkuk University, Seoul, Korea

ABSTRACT

Polyester type shape memory polymers were synthesized to improve their mechanical and shape memory properties and used for the preparation of sandwich type composite materials. Especially, poly(ethylene terephthalate) (PET) and poly(ethylene glycol) (PEG) copolymers with shape memory ability were prepared. After selecting the best composition of PET-PEG copolymer in mechanical properties, cross-linking agent such as glycerine, sorbitol, or maleic anhydride (MAH) was included for cross-linked copolymer, followed by analysis of its effect on mechanical, shape memory, and damping properties. The highest shape recovery was observed for copolymer with 2.5 mol% of glycerine, and the best damping effect indicating vibration control ability was from copolymer with 2.5 mol% of sorbitol. With the optimum copolymers in hand, sandwich-structured epoxy beam composites fabricated from epoxy beam laminate and cross-linked PET-PEG copolymer showed that impact strength increased from 1.9 to 3.7 times depending on the type of copolymer, and damping effect also increased as much as 23 times for the best case as compared to epoxy laminate beam alone. PET-PEG copolymers cross-linked with glycerol and sulfoisophthalate (SP) were also prepared to investigate the feasibility of vibration-control of composite laminate by additional ionic interaction. Composition of glycerol and SP was varied in order to get a copolymer with the best mechanical and shape memory properties. The highest shape recovery was observed for the copolymer with 2.5 mol% of glycerol and 2.5 mol% of SP. The sandwich-type copolymer composite showed improved impact strength (3.5 times) and damping effect (2.6 times) as compared to epoxy laminate beam alone. The resultant sandwich-structured epoxy beam composite can be utilized as structural composite material with vibration control ability and its glass transition temperature can be controlled by adjustment of hard segment content and cross linking agent composition.

1. INTRODUCTION

Since its introduction as a polynorborane type by Nippon Zeon company in 1984, various forms of shape memory polymers including poly(isoprene-butadiene-styrene), polyurethane, and polystyrene series were followed [1-6]. Specifically, shape memory polyurethane, developed by Nagoya Research and Development Center of Mitsubishi Company in 1988 [7], was composed of hard and soft segment and could recover the original shape in the temperature range of –30-65 °C. Hard segment and soft segment were alternatively polymerized, and various interactions among hard segments leaded to domain formation; hard segment worked as pivoting point for shape recovery and soft segment could mainly absorb external stress applied to the polymer. In order to utilize the shape memory effect, it is very important to control the structure and physical properties of shape memory polymers. Generally, deformation of polymer is much easier at the temperature above its T_g since it is in a rubbery state and has low modulus. Moreover, since there is a more than 100 times difference in modulus with respect to T_g, after the polymer deforms at above T_g and cooled back to below T_g with the load removed, it tends to retain its deformed shape. Then the deformation becomes frozen, and it is difficult to recover its original shape even though temperature is raised. Thus, normal polymers are difficult to recover its original shape. However, shape memory polymer has fixed points that can prevent plastic flow of chains and two phase structures that can reversibly change from soft to hard state depending on the temperature, thus recovering most of its original shape.

Meanwhile, active research is undergoing on smart materials such as shape memory alloy, semiconductor, polymer, and medical supply. Shape memory material, one of smart materials, has a few characteristics such as shape memory, shape retention, and impact absorption effect. In addition, it can detect thermal, mechanical, electrical, and magnetic stimulus, and then respond via property changes in shape, location, modulus, damping, and abrasion, which can be applied to various fields [8-10]. Shape memory polymer, being superior to other materials in the points of lightness (1.0-1.3 g/cm^3), high shape recovery (maximum shape recovery ratio more than 400 %), easy processing, and high damping near its T_g, has drawn our attention and is applied to manufacture of composite laminate with vibration control ability [11-14]. In our investigation, poly(ethyleneterephthalate) (PET)/polyethyleneglycol (PEG) copolymer is used as matrix material for the ultimate purpose of developing laminate composite with vibration control ability. Cross-linking agent such as glycerine, sorbitol, or maleic anhydride (MAH) was tried and its effect on the mechanical, shape memory, and damping property was compared. Comprehensive mechanical properties of various sandwich-structured laminate composite were examined, and reasons for such high vibration control were briefly discussed. Additionally, poly(ethyleneterephthalate) (PET)/polyethyleneglycol (PEG) copolymer is used as matrix material, and gylcerol and sulfoisophthalate group were adopted as covalent and ionic cross-linking agents, respectively. The effect of two different kinds of cross-linker on the mechanical, shape memory, and damping property of the copolymers was compared. Comprehensive mechanical properties of various sandwich-type composite laminate from the above copolymers were examined, and reasons for such high vibration control were briefly discussed.

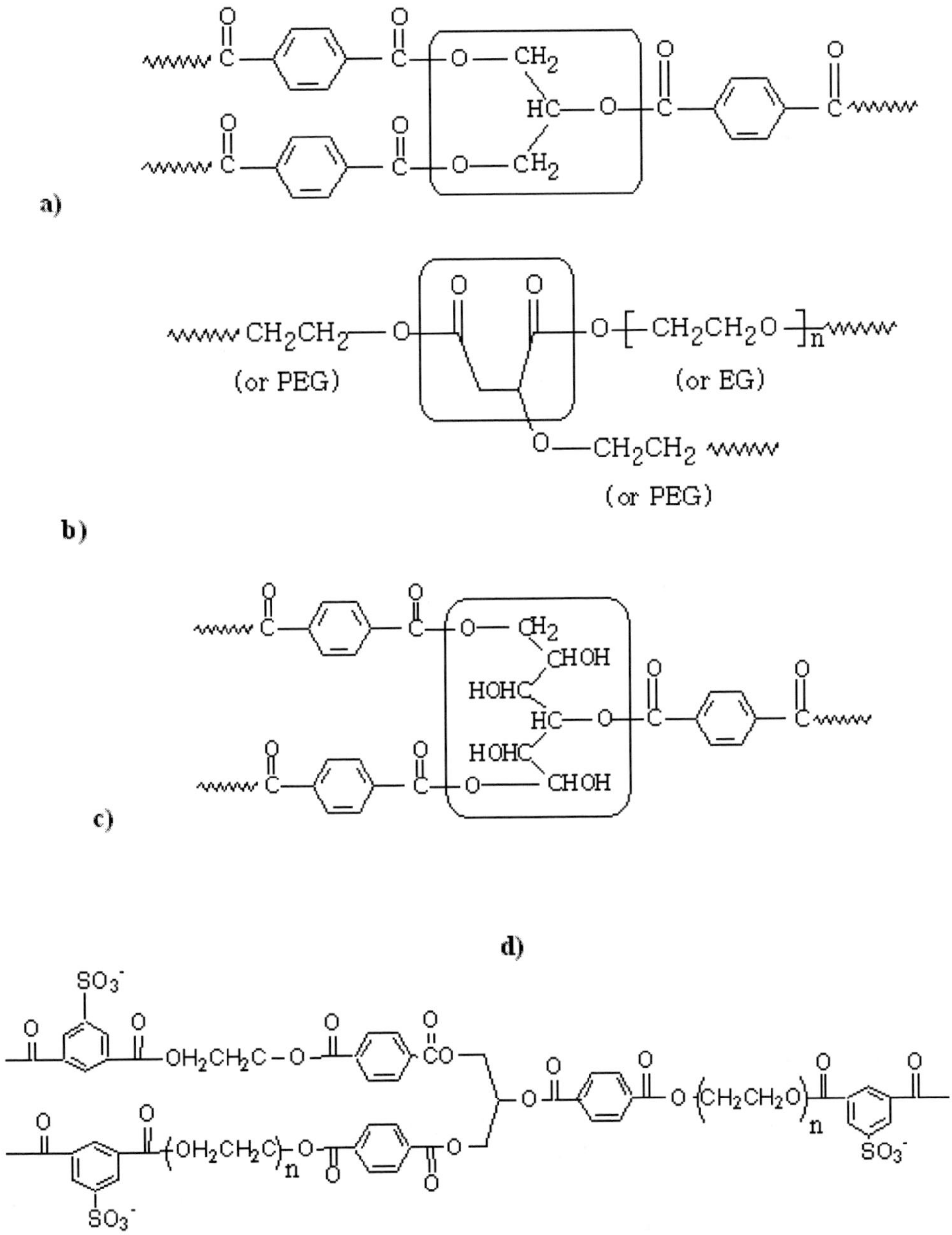

Figure 1. Schematic of PET-PEG copolymer cross-linked by (a) glycerine, (b) sorbitol, (c) maleic anhydride, and (d) glycerine and sulfoisophthalate

2. EXPERIMENTAL

2.1. Materials

DMT, EG, and PEG-200 were commercially obtained. Glycerol, sorbitol, maleic anhydride, and dimethyl 5-sulfoisophthalate sodium salt were used as obtained. Epoxy laminate beam with 1.0 mm thickness was supplied by Korea Fiber Co. and used to prepare unidirectional glass- fiber-reinforced composite with 2.7 mm total thickness.

2.2. Preparation of Cross-Linked Copolymer

PET-PEG copolymers were synthesized by melt-condensation method with a custom made reactor [13-15]. Polymerization was carried out in two steps; oligomer was prepared in the first step with DMT, EG, and PEG-200, and the oligomer made in first step was condensed and cross-linked with the agent in the second step at high temperature and high vacuum to shift reaction equilibrium further to product. Detailed synthetic procedure for PET-PEG copolymer can be found in our previous paper [13-15]. Synthetic scheme and characterization of copolymers are shown in the results and discussion section.

2.3. Intrinsic Viscosity

Intrinsic viscosity [η] of copolymer dissolved in 1,1,2,2-tetrachloroethane/phenol (4/6, w/w) mixture was measured with Ubbelohde viscometer at 35°C and 0.5 g/dl of concentration.

2.4. Thermal Analysis

T_g and T_m were measured by differential scanning calorimeter (DSC, Perkin Elmer Diamond 6). Specimen was heated to 200°C at 10°C/min of heating rate, and kept at that temperature for 3 min, and then cooled to –30°C at –10°C/min. T_g and T_m were determined from the second heating scan at 10°C/min heating rate.

2.5. Spectroscopic Analysis

NMR of the copolymer dissolved in $CDCl_3/CD_3OD$ mixture or trifluoroacetic acid (TFA) for cross-linked one was scanned by 600 MHz Bruker FT-NMR (Avance 600) at the National Instrument Center for Environmental Management, Seoul National University. FT-IR spectra were taken by Bomen MB series 104.

2.6. Preparation of Laminate Beam

Copolymer was preheated at 60°C for 24 hour to prevent any hydrolysis from moisture, and compression-molded to a sheet with 1 mm thickness and 5 mm width at 30 °C above its T_m. Sandwich-type laminate composite was prepared by compressing two 1.0 mm epoxy laminate beams and the above PET-PEG copolymer sheet in middle layer at 210 °C in a heating press.

2.7. Mechanical Properties and Shape Memory Analysis

All specimens for the mechanical property measurement were preheated at 60°C for 24 hours in order to prevent reverse reaction (hydrolysis) from the moisture during the processing, and compression molded to a sheet having 1mm thickness and 5 mm width at the temperature of 30°C above its T_m. Tensile tests were performed using universal testing machine (UTM, Lloyd LR 50K), and dumbbell type specimens according to ASTM D-638 were used with a crosshead speed of 100 mm/min. For dynamic mechanical property measurements, dynamic mechanical thermal analyzer (DMTA, Rheometric Scientific, Mark IV) was employed and specimen with 1mm thickness and 5 mm width sheet was used. All tests were conducted at a 3°C/min heating rate and 1.1 Hz.

Shape memory effect was tested using the identical specimen for a tensile test, and UTM equipped with temperature controlled thermal cabinet was used. For the shape retention rate measurement, following procedures were used. First, specimen with a length L_0 was strained to 100 % at the temperature above its T_g but below T_m, and stayed at that condition for 1min. Secondly, with the strain fixed, specimen was cooled back to the temperature below its T_g, and the load was removed and stayed at this condition for 30min, then the deformed length L_1 was measured. For the shape recovery measurement, after measuring L_1, specimen was heated again to the temperature above its T_g but below T_m, and stayed at this temperature for 10 min. Then the specimen was cooled back to the temperature below its T_g and stayed at this temperature for 30 min. After this, final specimen length L_2 was measured. Thus, shape retention rate and shape recovery rate can be defined as follows.

Shape retention rate(%) = $(L_1 - L_0)/L_0 \times 100$
Shape recovery rate(%) = $(2L_0 - L_2)/L_0 \times 100$

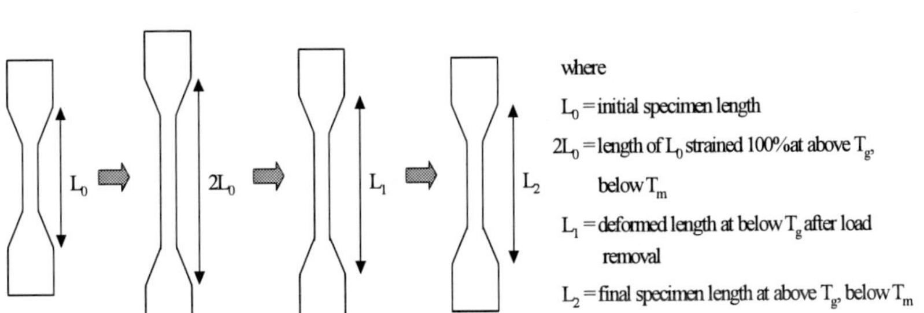

Figure 2. Specimen for testing shape retention and shape recovery rate

This whole cycle represent 1 shape memory test, and the copolymers were treated 3 shape memory tests.

3. RESULT AND DISCUSSION

3.1. Preparation of Copolymers

PET-PEG copolymers were prepared with DMT by melt-condensation method using the polymerization reactor custom made by Go Do Engineering Co. The copolymers were prepared in two different ways by either changing MW or composition of PEG. In addition, cross-linking agent was added in various ratios into the copolymer composed of 4 mol% of EG and 1 mol% of PEG. The polymerization reaction was carried out in two steps; in the first step, oligomers were prepared from DMT, EG, and PEG, and in the second step, the oligomers made in first one was condensed to polymers by esterification reaction which required high temperature, constant stirring, and high vacuum to shift the reaction equilibrium further to the products. The mole ratio of reactants was 50:80:20 (DMT:EG:PEG).

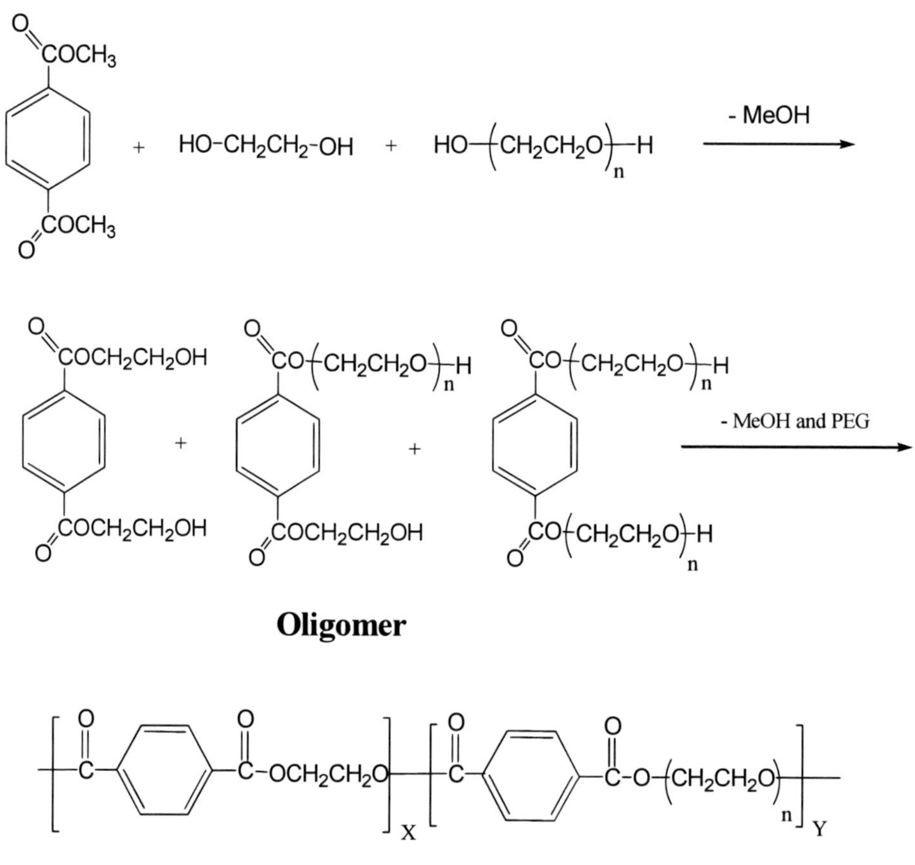

Figure 3. Synthetic scheme of PET-PEG copolymer

Into the mixture containing EG, PEG (200 g/mol), and DMT that was heated to 145°C was added calcium acetate as catalyst, and the mixture was further heated to 200 °C. Methanol was removed from the reaction mixture by distillation and the esterification was continued until the methanol was not produced any more, followed by the addition of phosphorous acid as stabilizer with an additional 10 minute stirring. After transferring the oligomer mixture into the polymerization reactor that was heated to 245°C and addition of antimony oxide as catalyst, the mixture was heated to 270°C and kept under vacuum (45 mmHg) for 30 minute to remove unreacted EG and PEG. Following the addition of cross-linking agent, the reaction was carried out for polymer condensation under high vacuum (1 mmHg) for 3 hrs. The synthetic scheme of copolymers synthesized is shown in Figure 3.

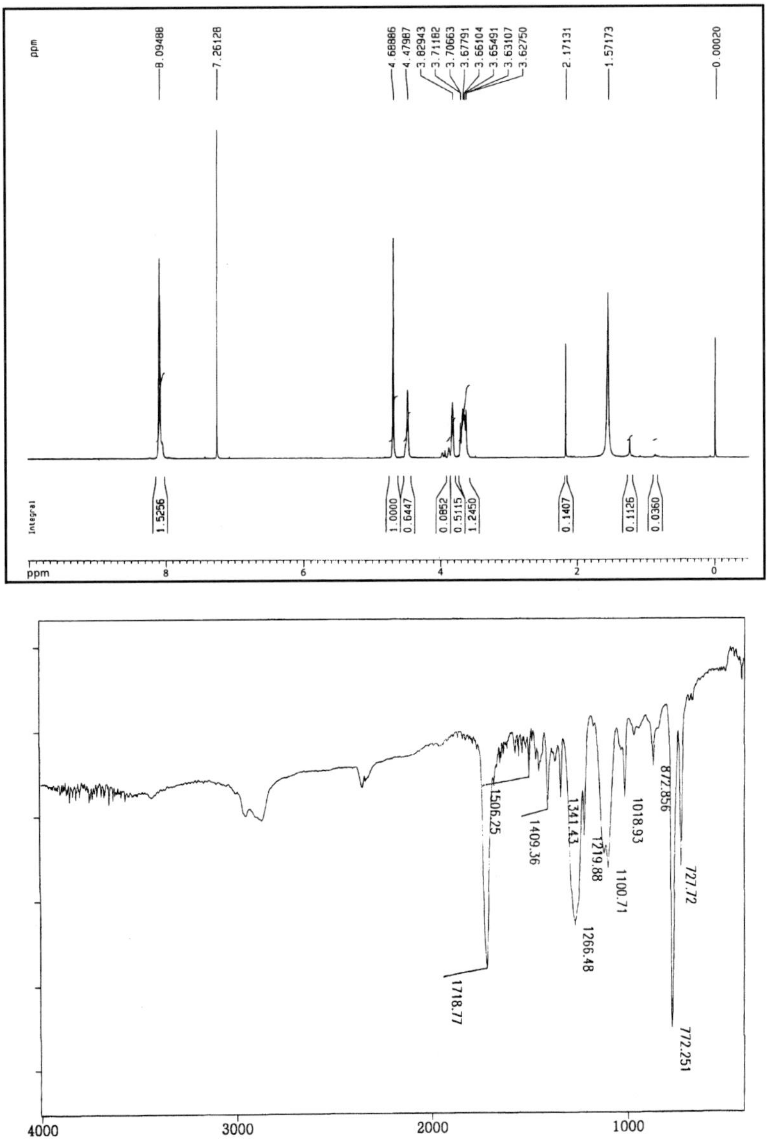

Figure 4. Spectra (NMR and FT-IR) of PET-PEG copolymers.

3.2. Spectroscopic Analysis of the Copolymers

In Figure 4, the proton NMR spectrum of a cross-linked polymer showed that chemical shift of EG moved from 3.4 ppm of free EG to 4.4~4.7 ppm of the polymer, originating from the electron-withdrawing ester formation during polymerization, and the composition of EG and PEG calculated by dividing integration ratio between EG and PEG with their number of protons was found to be 4.05:1.00 which was in excellent agreement with the theoretical ratio of 4:1. Proton NMR peaks of PEG and DMT were observed at 3.5-3.8, and 8.0-8.2 ppm, respectively, and inclusion of cross-linking agent generally shifted the above peaks downfield by 0.2-0.3 ppm. The peaks appearing at 1.57 ppm and 2.17 ppm were coming from the trace amount of methanol in CD_3OD solvent and cross-linking agent, respectively. In the FT-IR spectrum of Figure 4 was shown the complete disappearance of the –OH stretching at 3300 cm^{-1} which was found as broad band for the free EG and PEG. Judging from the NMR and FT-IR data, the copolymerization proceeded as expected.

3.3. Tensile Property

Figure 5 shows the maximum stress of PET-PEG copolymers prepared in this study. As can be seen, the maximum stress increased with the addition of PEG, but started to decrease afterwards. Especially, when the PEG with molecular weight 200 g/mol was used, highest maximum stress was obtained at 10 mol% of PEG. However, at more than 10 mol% PEG, the maximum stress started to decrease. Also, as the molecular weight of PEG increased, maximum stress generally decreased. Strain at break increased very slightly in the case of PET-PEG copolymers with the PEG molecular weight of 400, 600 and 1000 g/mol. However, in the case of PEG having 200 g/mol of molecular weight, the percent strain at break increased very slightly up to 10 mol% PEG. However, at more than 10 mol% PEG, percent strain at beak increased abruptly, and at 20 mol% of PEG, it reached more than 1000 %. Considering that the minimum percent strain at break that can test the shape memory effect is 100 %, shape memory effect can only be tested for the PET-PEG copolymers where the molecular weight of PEG is 200 g/mol and the composition is 15 to 20 mol%.

Maximum stress and percent strain at break of PET-PEG copolymers cross-linked by maleic anhydride (MAH) were shown in figure 6. Maximum stress showed an abrupt increase from about 3 MPa to 8.8 MPa at 1.5 mol% of MAH addition, and decreased to about 6 MPa at 2.5 mol% of MAH. But regardless of MA content, all PET-PEG-MAH copolymers showed higher maximum stress compared to PET-PEG copolymer. This is due to the increased resistance to deformation resulting from the formation of cross-links within the polymer chains. In the case of percent strain at break, it decreased from about 1100 % to about 750 % as the MAH is introduced. Even though there was a slight increase in percent strain at break between 1.5 and 2.5 mol% of MAH, the overall trend is obviously a decrease. This is an expected result considering that the added MAH increases the number of cross-links in the polymer chains, thus reducing the extensibility of chains.

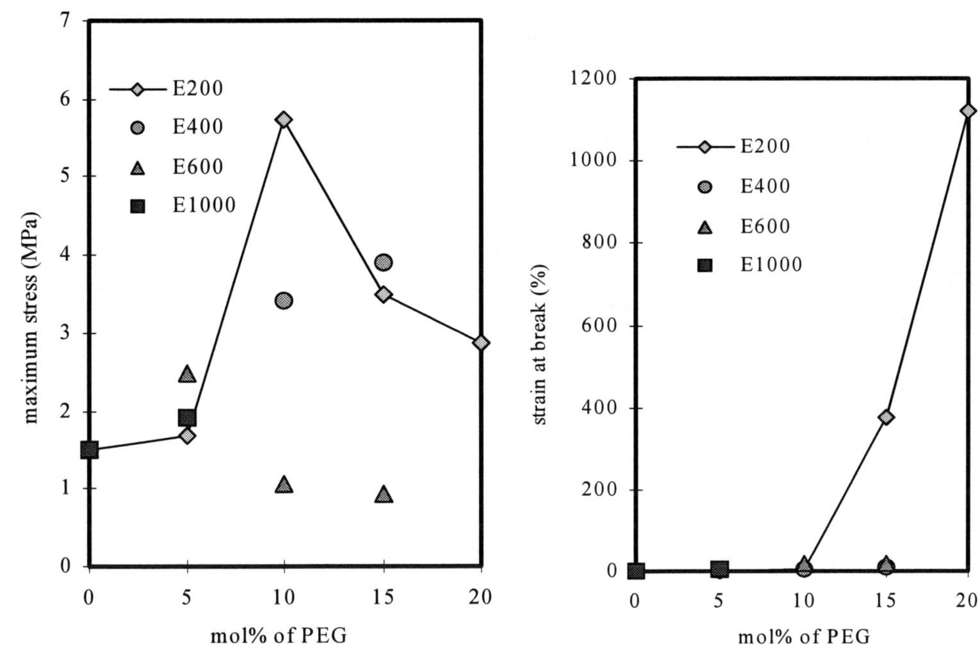

Figure 5. Maxmum stress and strain at break profiles of PET-PEG copolymers

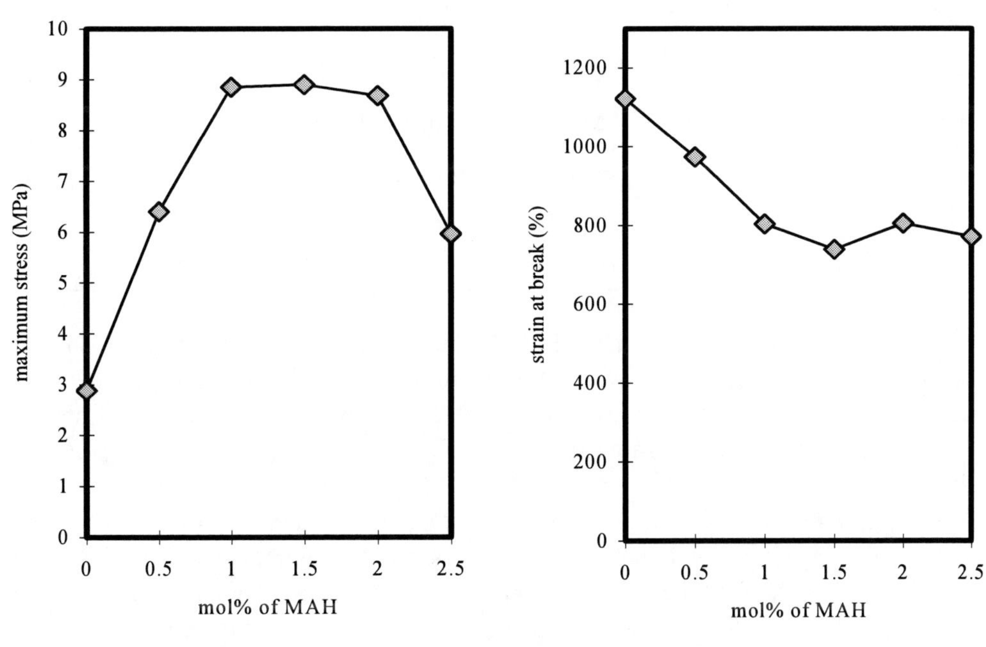

Figure 6. Maximum stress and strain at break profiles of PET-PEG-MAH copolymers.

Because PET-PEG copolymer with 20 mol% of PEG 200 (E200-20) showed the best tensile test result, the composition was used for preparation of copolymer with cross-linking agent (glycerine or sorbitol) and related shape memory test (Table 1).

Table 1. Properties of PET-PEG Copolymers

Sample Code	PEG(mol%)	PEG MW(g/mol)	T_g (°C)	T_m (°C)	[η](dl/g)
PET	0	-	79.1	252.0	0.68
E200-10	10	200	33.5	213.5	0.62
E200-20	20	200	8.1	167.3	0.60
E400-10	10	400	-12.9	200.0	0.70
E600-10	10	600	-32.4	194.6	0.67
E1000-5	5	1000	-42.5	221.2	0.55

Tensile test results of cross-linked copolymers indicated that tensile strength generally increased, but strain at break decreased as more cross-linking agent was included (Table 2).

Table 2. Comparison of Cross-linked PET-PEG Copolymers

Sample Code	PEG (mol%)	PEG MW (g/mol)	Cross-link Agent (mol%)	Maximum Stress (N/mm^2)	Strain at Break (%)	T_g (°C)
E200-20	20	200	-	5.9	1122	11.2
G-15	20	200	Glycerine (1.5)	16.5	317	14.3
G-25	20	200	Glycerine (2.5)	19.6	48	23.7
M-15	20	200	Maleic anhydride (1.5)	9.1	832	13.2
M-25	20	200	Maleic anhydride (2.5)	9.1	770	14.1
S-15	20	200	Sorbitol (1.5)	14.4	805	9.2
S-25	20	200	Sorbitol (2.5)	14.3	632	11.7

Such result undoubtedly originated from higher resistance to deformation by introduction of cross-linking in copolymer chain, and the concomitant cross-linked solid structure resulted in high tensile strength and low strain at break. Among cross-linked PET-PEG copolymers, one cross-linked by glycerine showed the highest tensile strength and the lowest strain at break.

Maximum stress of glycerol (1.0 and 1.5 mol%) cross-linked copolymers showed initial decrease with SP content and sharp increase at SP content higher than 1.5 %; copolymers with 2.5 mol% glycerol content had gradual increase with SP content. Strain at break of copolymers (1.0 and 1.5 mol% of glycerol) increased with SP contents up to 1.5 % and decreased at higher content, but 2.5 mol% glycerol copolymer showed slow decrease with inclusion of SP. Both maximum stress and strain at break results are similar to Tg data in the following section, and explanation for the combined results will appear there.

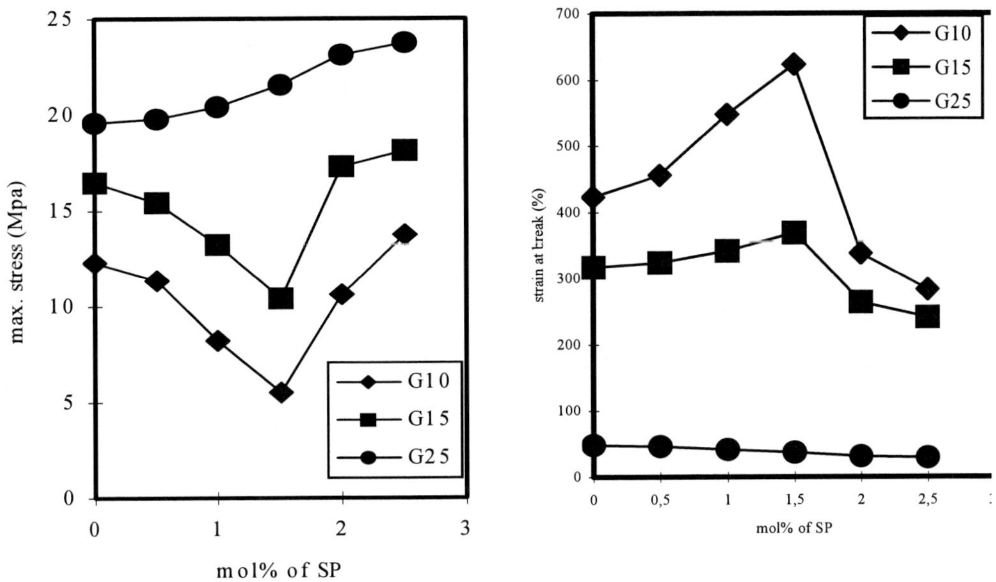

Figure 7. Maximum stress and strain at break profiles of PET-PEG-Glycerol-SP copolymers.

3.4. Thermal Property

As shown in Figure 8, T_g decreases rapidly as the PEG is introduced, but this tendency is weakened as the PEG content increases. At the same PEG content, PEG with higher molecular weight showed a larger decrease in T_g values. T_m also decreases linearly with the PEG addition. Compared to T_g behavior, T_m change with respect to molecular weight of PEG is small, but as the molecular weight of PEG decreases, T_m decreases less. Thus, T_g and T_m decrease behavior is due to the increase of chain flexibility and decrease of the chain regularity with the addition of PEG.

Table 3 shows the similar data of PET-PEG copolymers cross-linked by MAH. T_g was affected by the MAH addition, but T_m was not. In general, T_g increased from 8.1°C to 17.7°C when 2.5 mol% of MAH was introduced. The observed T_g increase with the MAH addition is due to the restricted molecular motion resulting from the increased cross-linking [16].

Table 3. Characterization of PET-PEG copolymers cross-linked by MAH

Sample Code	PEG MW (g/mole)	MAH (mole%)	T_g (°C)	T_m (°C)	[η] (dl/g)
E200-20	200	0	8.1	167.3	0.60
MAH05	200	0.5	8.3	165.1	0.62
MAH10	200	1.0	9.8	164.2	0.71
MAH15	200	1.5	11.6	166.3	0.65
MAH20	200	2.0	13.6	164.3	0.79
MAH25	200	2.5	17.7	164.2	0.67

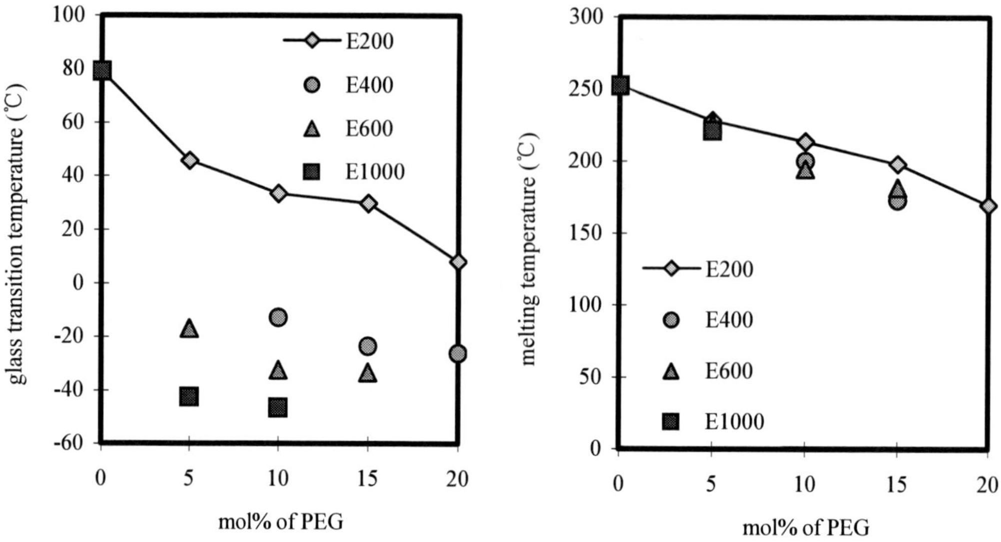

Figure 8. Phase transition temperatures of PET-PEG copolymers.

T_g of the copolymers with different combination of glycerol and SP was compared in Figure 9.

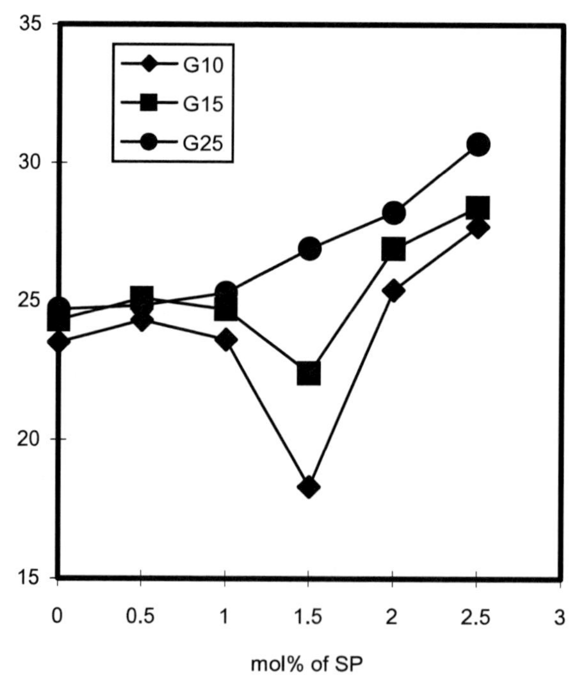

Figure 9. Glass transition temperatures of PET-PEG-Glycerol-SP copolymers.

For the copolymers with 1.0 and 1.5 mol% of glycerol, addition of SP decreased the transition temperature up to 1.5 mol%, and steep increase of T_g was observed from 1.5 to 2.5 mol% region. Unlike the above series, the copolymers with 2.5 mol% of glycerol showed constant increase of T_g with the SP content. Such result can be understood from the fact that the copolymer chain is not tightly cross-linked at lower mol% of glycerol (1.0 and 1.5 mol%) and the presence of SP along the chain hinders ordered chain packing, thus resulting in reduced T_g. But, as the SP content increases, inter-chain ionic interaction among sulfonyl sodium salt groups additionally improves inter-chain attraction and, therefore, T_g value increases at SP content higher than 1.5 mol%. Highest inclusion of glycerol (2.5 mol%) dominates the copolymer cross-linking, and the decrease of T_g at lower mol% of SP observed in the above case do not happen because covalent cross-linking of glycerol can subdue the disrupting effect of SP. As more SP was included, T_g of 2.5 mol% glycerol copolymer also increased due to the additional ionic inter-chain interaction originating from sulfonyl groups. The above explanation can be applied to the tensile test results. Such trend can be consistently observed in the following mechanical and shape recovery data, and combination of covalent cross-linking by glycerol and ionic interaction by SP can manipulate the desired transition temperature of copolymers.

3.5. Shape Memory Effect

Shape memory experiments were performed on the PET-PEG-MAH copolymers. As shown Figure 10, PET-PEG-MAH copolymers generally exhibited a slightly lower shape retention value compared to PET-PEG copolymer, but all specimens showed an excellent shape retention of more than 90 %. The overall trend is a decrease in shape retention with the MA addition, but shape retention increased again at more than 2 mol% of MAH. Maximum shape recovery rate is obtained at 1.5 to 2.0 mol% of MAH, which is due to the effect of cross-links formed by the MAH on the recovery of the soft segment: PET acts as physical cross-link point and MAH does as chemical cross-link point. And it is dispersed evenly within the matrix as physical cross-link point, and results in the improvement of shape memory effect in PET-PEG-MAH copolymers. This behavior can be observed in the Figure 10, which shows the shape recovery rate of PET-PEG-MAH specimens after 3 cycles of shape memory tests. Best shape recovery effect is observed for the PET-PEG-MAH copolymer with 2mol% of MAH, which is due to the high resistance to plastic deformation of chain segment to the stress by chemical cross-linking of MAH. In contrast, PET-PEG copolymer fractured after only 2 cycles of shape memory tests due to the severe plastic deformation of chains to the stress.

As for glycerol or sorbitol cross-linked copolymers, shape retention rate slightly decreased, but shape recovery significantly increased after cross-linking. Especially, PET-PEG copolymer cross-linked by glycerol showed the highest shape recovery rate (Figure 11). For example, shape recovery rate of E200-20 decreased from 50.5 % to 21.5 % after two cycles of shape memory test, but G-15 and G-25 maintained 51.0 % and 79.9 %, respectively, under the same condition. PET-PEG copolymer cross-linked by sorbitol also exhibited high shape recovery rate. The specific shape memory test results are summarized in table 4. As for all cross-linked copolymers shape retention rate was more than 95 %, which told us that deformed shape of copolymer could be almost perfectly kept below transition temperature.

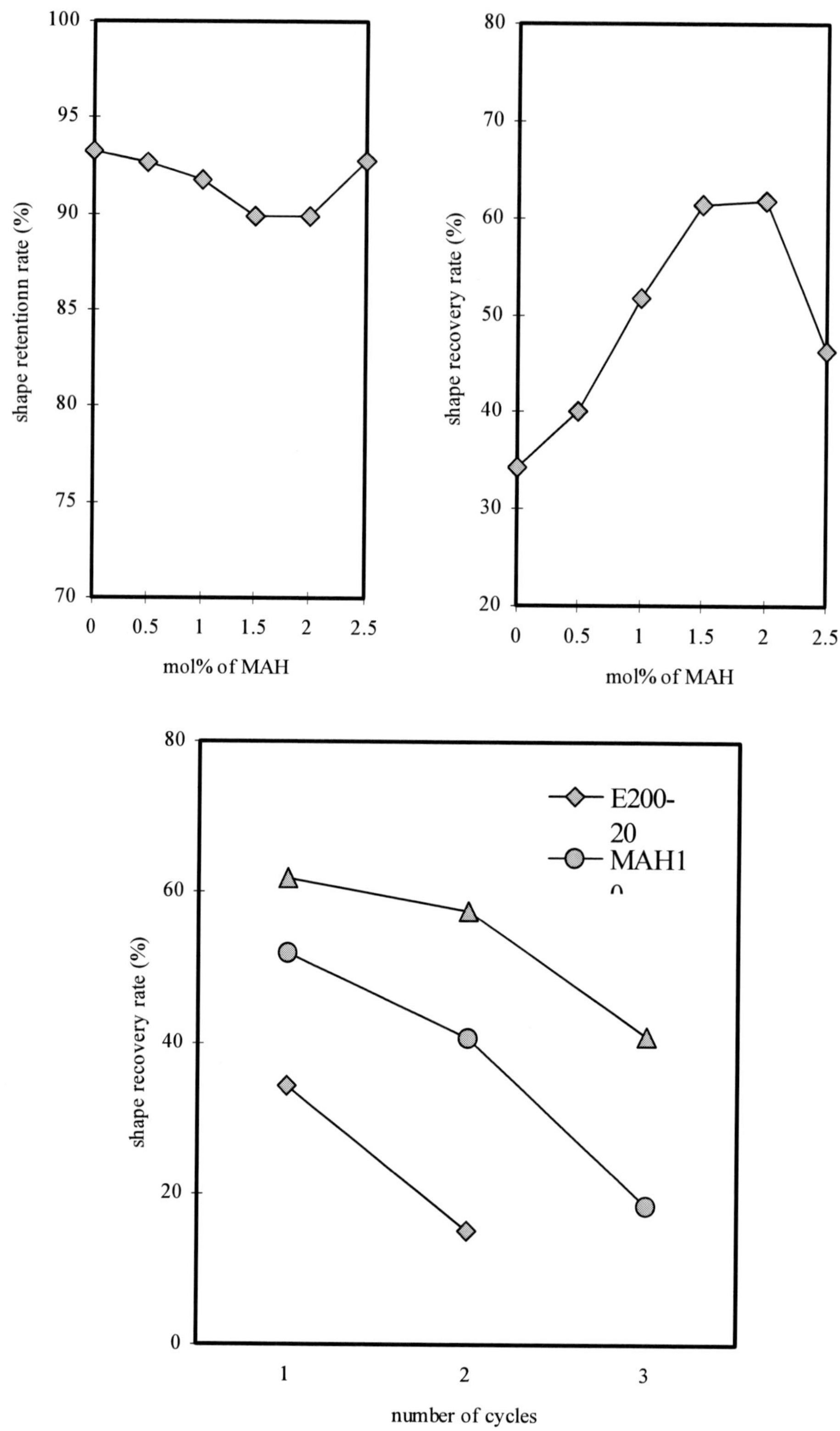

Figure 10. Shape memory profiles of PET-PEG-MAH copolymers.

Table 4. Shape Memory Test Results of Copolymers

Sample Code	First Cycle		Second Cycle		Third Cycle	
	Recovery (%)	Max Stress (Mpa)	Recovery (%)	Max Stress (Mpa)	Recovery[1] (%)	Max Stress (Mpa)
E200-20	51.8	0.85	23.5	0.56	-	0.37
G-15	69.4	0.90	53.7	0.75	-	0.60
G-25	85.8	1.00	82.2	0.85	-	0.52
M-15	70.3	0.97	41.7	0.49	-	0.46
M-25	76.1	1.01	66.0	0.77	-	0.33
S-15	52.3	0.76	36.1	0.68	-	0.42
S-25	68.7	0.92	34.7	0.78	-	0.53

[1] Recovery for the third cycle was not proceeded because of the very low value of E200-20

Shape retention rates of the glycerol and SP cross-linked copolymers generally maintain 97-99 % of the original shape, but shape recovery rates are significantly dependent on cross-linking agents. In Figure 11, cyclic shape memory test results of 1.0 mol% glycerol copolymers with various SP contents were compared; shape recovery was getting worse with inclusion of SP up to 1.5 % and was substantially improved at 2.5 mol% of SP to a level that was better than the starting one without SP. Cyclic tests were done with 1.5 mol% glycerol copolymers with various SP contents; shape recovery again decreased up to 1.5 mol% of SP and went up at 2.5 mol% of SP, a result showing similar trend as 1.0 mol% case. Finally, 2.5 mol% glycerol copolymers were not much affected by the content of SP: there are some drops in shape recovery for the cases of 0.5 and 1.0 mol% of SP, but 1.5, 2.0, and 2.5 mol% of SP all showed higher shape recovery. From above results, shape recovery rates of the copolymers at low mol% of glycerol can be controlled by the content of SP, and high glycerol content dominates the shape recovery effect with very minor dependence on SP content. Again, shape memory results remind us of the differences in inter-chain interactions coming from glycerol and SP as explained above.

In addition to chemical cross-linking, following characteristics can be responsible for high shape recovery of cross-linked copolymer: (1) molecular interactions among hard segments (PET part), such as hydrogen bonding, dipole-dipole interaction of carbonyl groups of PET, and induced dipole-dipole interaction of aromatic rings, work as pivoting point for shape memory after several deformation-recovery processes; (2) soft segment (PEG part) effectively and entirely absorb the applied stress so that relative position of hard segment or original shape of copolymer can be safely preserved under severe experimental conditions; (3) glycerol, shorter cross-linking agent than sorbitol, is more effective for shape memory, because, although both are connecting hard segment, short and rigid structure of glycerol is more favored for shape retention of hard segment: (4) MAH which cross-links soft segment is inferior to glycerol or sorbitol, because soft segment's role in shape memory copolymer is to absorb external stress and cross-linking by MAH hinders flexible movement of soft segment. The mechanism for the higher shape memory properties of glycerol-cross-linked copolymers can be schematically explained as in Figure 12: hard segments connected through cross-linking and inter-chain interactions maintains the original shape, while soft segments reversibly absorb external stress.

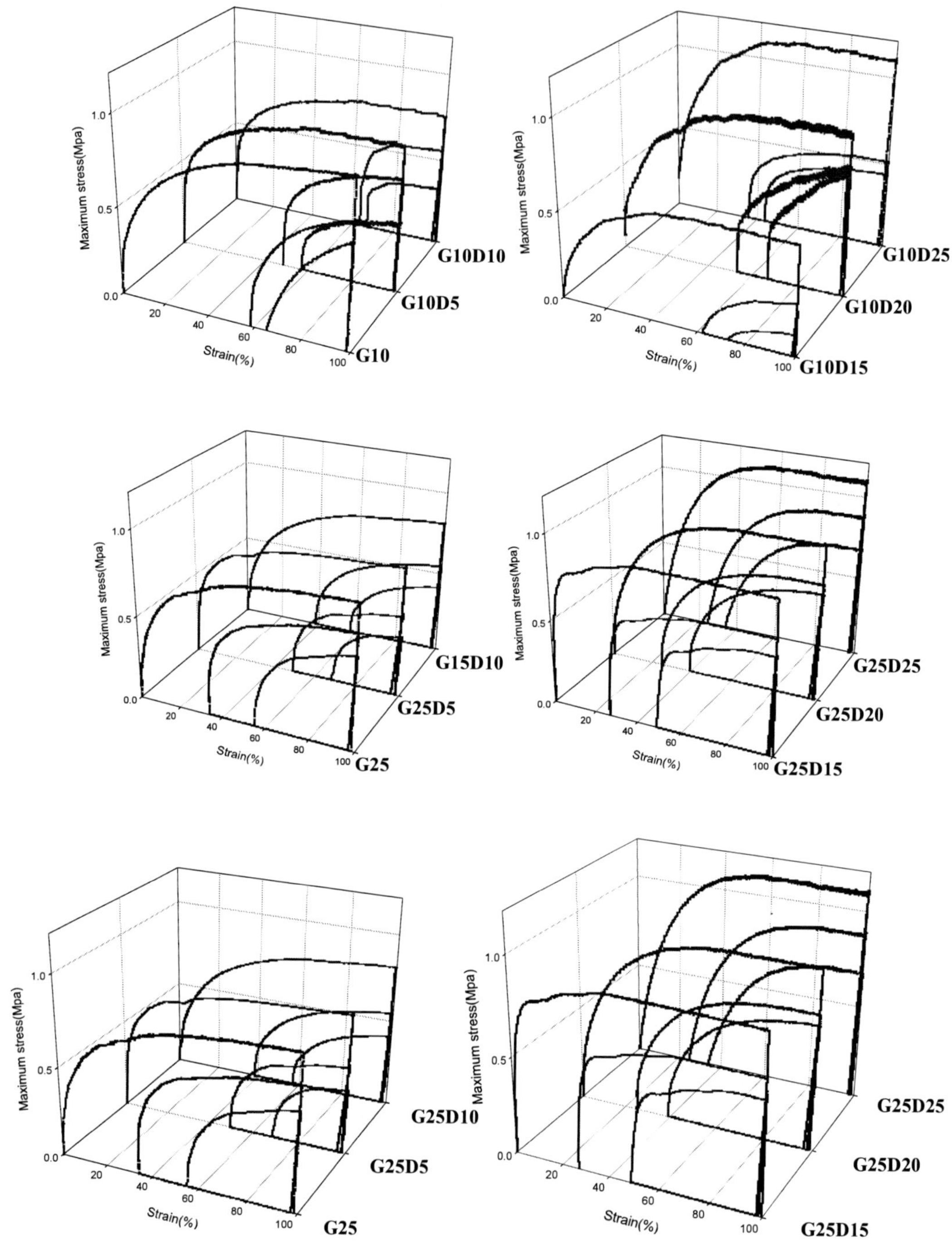

Figure 11. Shape memory profiles of PET-PEG-Glycerol-SP copolymers: G and D designate the mole % of glycerol and sulfoisophthalate, respectively.

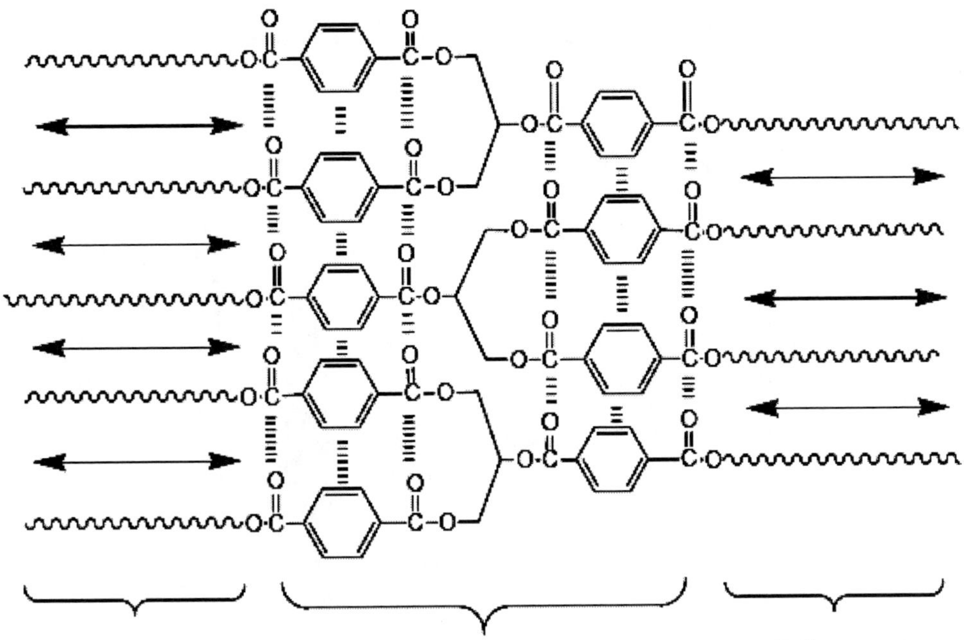

Figure 12. Schematic of shape memory effect of glycerol cross-linked copolymers.

3.6. Charpy Impact Strength Test

Unnotched Charpy impact strength of composite laminate was compared in Figure 13. Impact strength normal to the laminate layer of sandwich structure direction was measured with unnotched specimens [17]. As compared to epoxy beam laminate by itself, composite laminate showed higher impact strength; laminate with uncross-linked copolymer (E200-20) or cross-linked one by sorbitol (S-25) or maleic anhydride (M-25) had more than 4 times higher impact strength than epoxy laminate; laminate with glycerol (G-25) was the lowest in impact strength, but still 2 times higher than epoxy laminate. Above results are closely related to the stiffness of copolymer which differentiates overall impact strength of laminate; from the transition temperature of copolymer (E200-20: 8.1 °C, G-25: 23.7 °C, S-25: 11.7 °C, and M-25: 14.1 °C), relative softness at room temperature where impact strength is measured should be reversely proportional to transition temperature. Therefore, G-25 itself and the laminate made of G-25 get the most solid structure among compared copolymers, but such rigid structure is very weak in absorbing external impact and shows low impact strength. Compromise of shape memory and impact strength of copolymer is necessary depending on what is the required property under various surrounding conditions.

Laminates made from the copolymers with glycerol and SP showed 1.3-2.6 times higher impact strength than epoxy laminate. Inclusion of glycerol lowered impact strength of the laminate and presence of SP additionally decreased the value; the higher rigidity of cross-linked copolymer is mainly responsible for the decrease of impact strength and the desired impact strength can be adjusted by control of the composition of glycerol and SP.

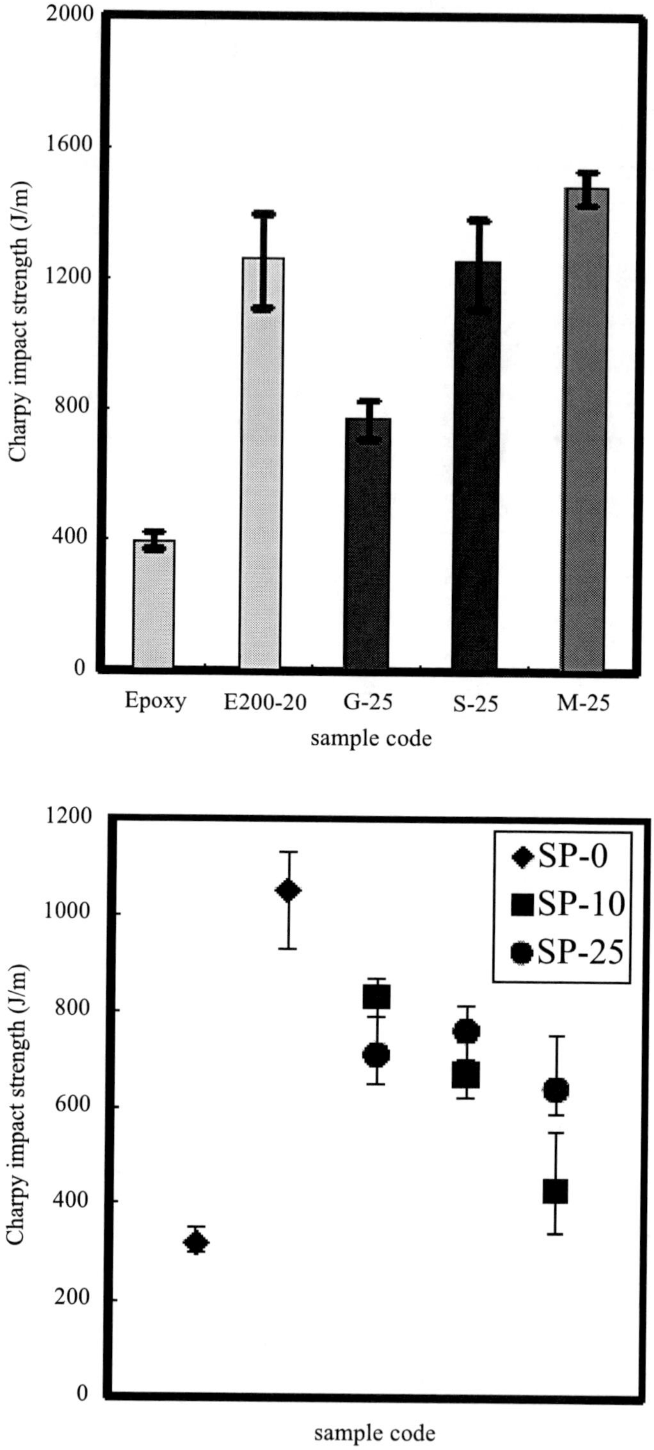

Figure 13. Charpy impact strength of cross-linked copolymers.

The impact strength results are closely related to the transition temperature of copolymers (E200-20: 8.1 °C , G10D10: 23.6 °C, G10D25: 27.7 °C, G15D10: 24.7 °C, G15D25: 28.4 °C, G25D10: 25.3 °C, G25D25: 30.7 °C); relative softness at room temperature where impact strength is measured should be reversely proportional to transition temperature. Therefore, laminates made of G-25 series get the most solid structure among compared ones, but such rigid structure is very weak in absorbing external impact and shows low impact strength. Compromise of shape memory and impact strength of copolymer is necessary depending on what is the required property under various surrounding conditions.

3.7. Dynamic Mechanical Property of Sandwich Laminate Beam Structure Composite

Storage modulus and tan δ of sandwich laminate beam composites were compared in Figure 14. Storage modulus of epoxy laminate beam was in high contrast with all other copolymer laminates; all of copolymer laminates except G-25 showed similar shape over the temperature range, significant drop at transition temperature and immediate recovery at higher temperature

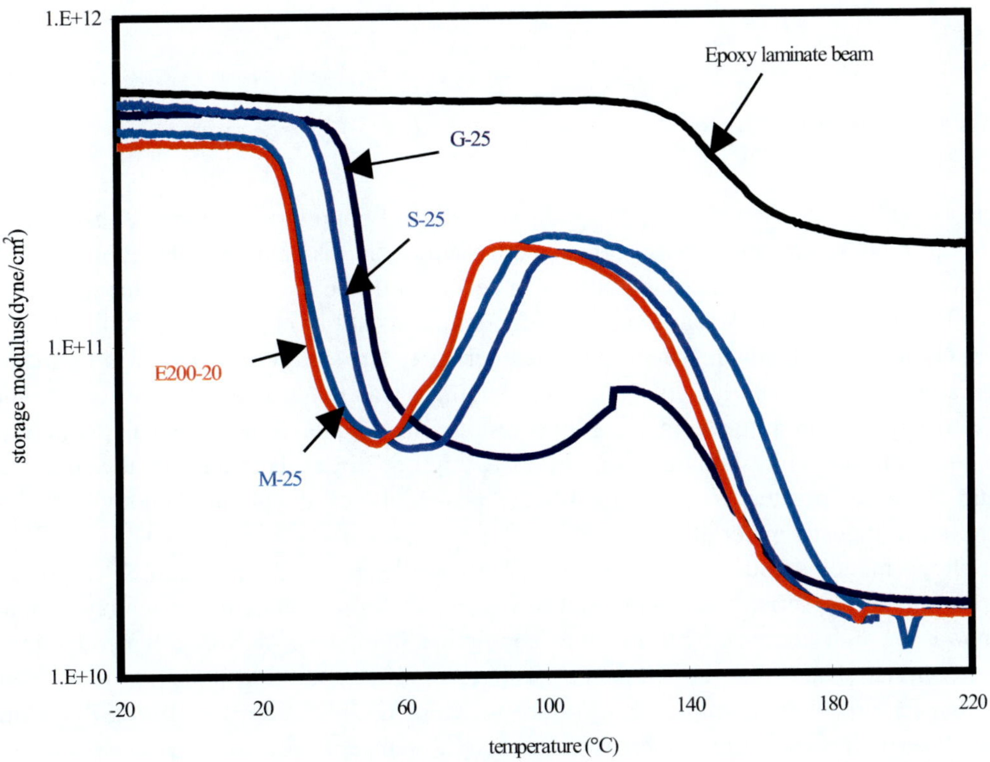

Figure 14. Storage modulus and tan δ of composite laminates with cross-linked copolymers.

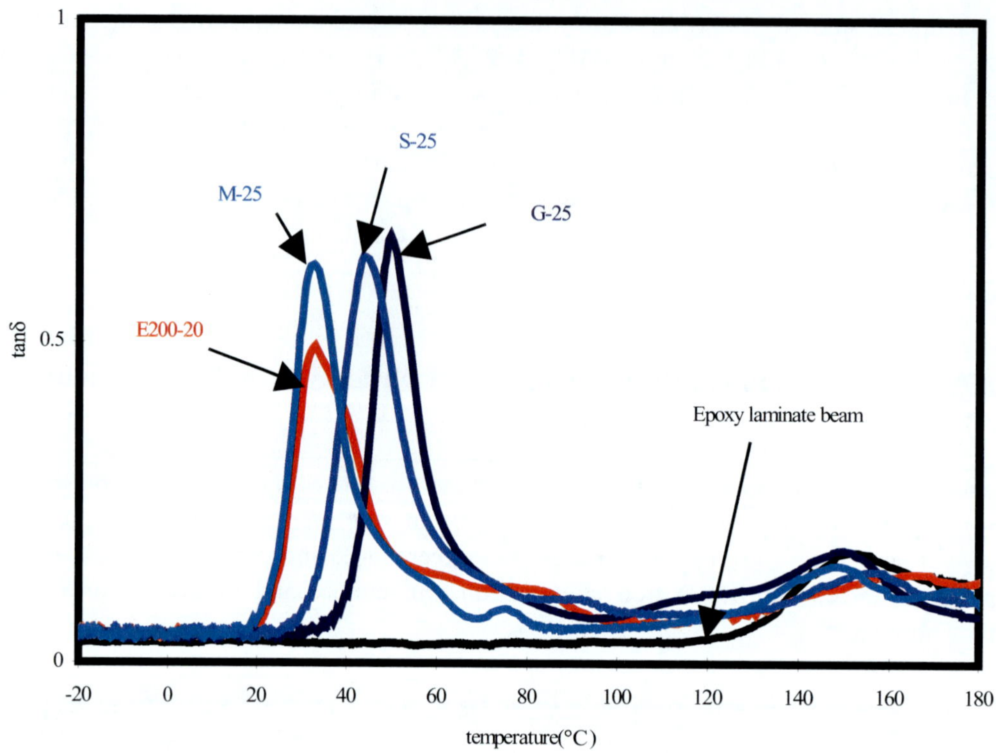

Figure 14. (Continued).

Laminate with G-25 with its rigid and well cross-linked copolymer is hard to rearrange its chains at high temperature once organized structure is dismantled. Deviation of G-25 laminate from other copolymer laminates is in accord with the above Charpy impact strength result. Tan δ sharply changed at transition temperature for all of copolymer laminate with no exception of G-25 laminate; their peak temperatures are in good order with transition temperatures of each copolymers. But epoxy laminate was not responsive over temperature ranges as expected from the storage modulus results. Because tan δ indicates damping ability, high tan δ of copolymer laminates (23 times as high as epoxy laminate for the best case) around room temperature range suggests the possibility of developing highly vibration-controlling composite material.

Storage modulus and tan δ of sandwich laminate beam of glycerol and SP copolymer composites of G25 series were compared in Figure 15. Storage modulus of epoxy laminate beam was in high contrast with all other copolymer laminates; G25 copolymer laminates showed similar shape over the temperature range, significant drop at transition temperature and immediate recovery at higher temperature. Additional ionic interaction from SP, together with glycerol cross-linking, increased transition temperature and the increase was proportional to SP content. Tan δ sharply changed at transition temperature for all of copolymer laminate. But epoxy laminate was not responsive over temperature ranges as seen in the above copolymers. There are still more rooms in fine-tuning of copolymer structure through modification of cross-linking method, PEG chain length, and hard segment structure for the control of transition temperature, vibration absorption, and mechanical properties.

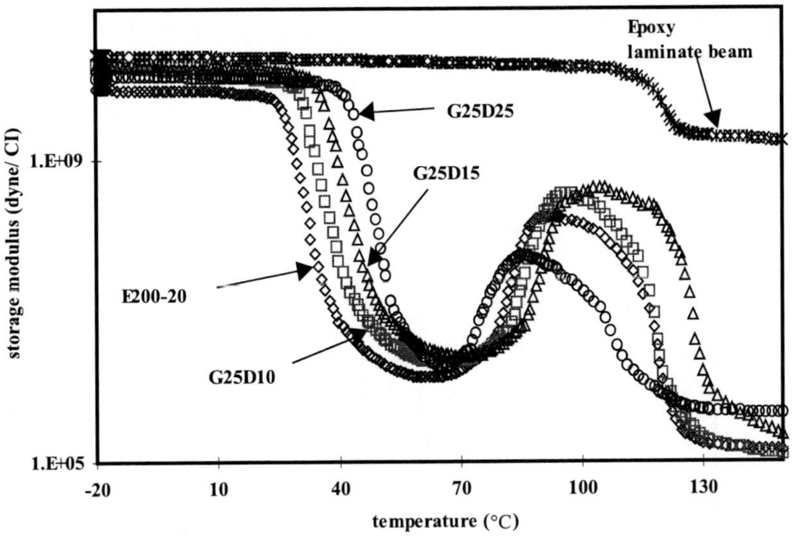

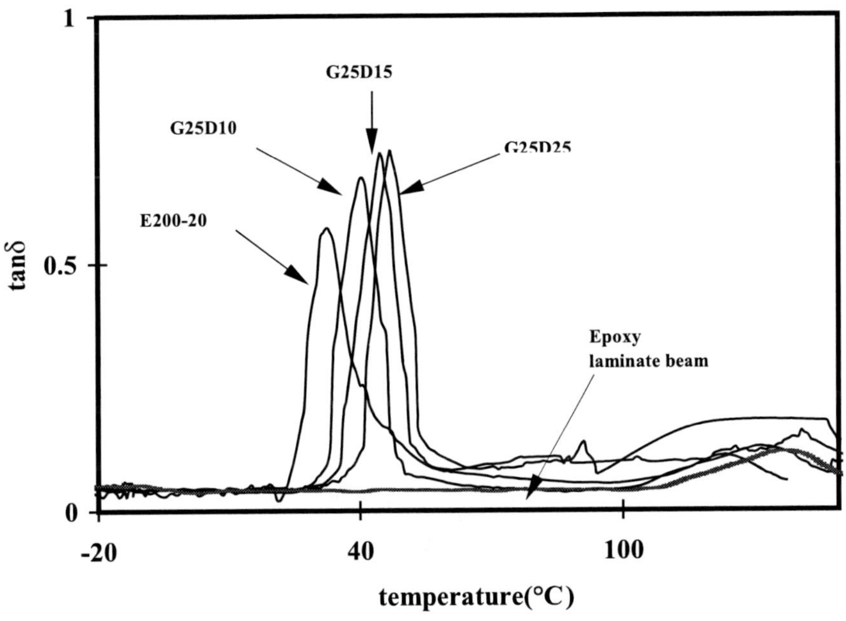

Figure 15. Storage modulus and tan δ of composite laminates with SP cross-linked copolymers.

3.8. Dynamic Mechanical Thermal Property

Figure 16 shows loss tan δ and storage modulus E′ of PET-PEG (PEG MW= 200 g/mol) and PET-PEG-MAH copolymers (1.0 and 2.0 mol% of MAH). All specimens exhibit a order of 2 or 3 decrease in storage modulus around their T_g. Storage modulus below T_g becomes higher as MAH is introduced into PET-PEG copolymer, originating from the reduced chain mobility by MAH cross-linking. As a shape memory copolymer, PET-PEG copolymer will retain the deformed shape when it is deformed in the rubbery plateau region and cooled below

T_g. And it will recover its original shape, after heating to the rubbery plateau region. At the temperature below T_g, copolymer with higher E′ shows better shape retention, and the copolymer with smaller E′ becomes easily deformed at the temperature above T_g. In general, modulus drop at T_g is about 10~1000 [18].

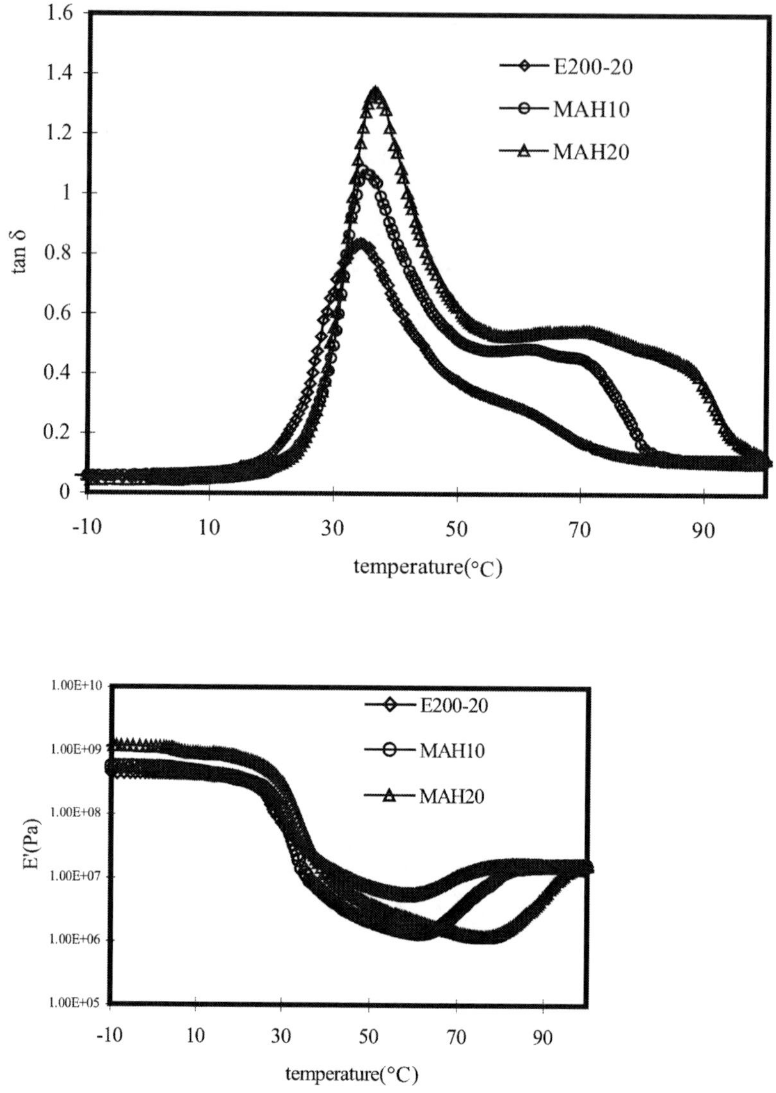

Figure 16. Storage modulus and tan δ of PET-PEG copolymers.

The storage modulus drop width (transition width) is narrower for the PET-PEG-MAH copolymers compared to PET-PEG copolymer, indicating that MAH cross-linking conferred temperature sensitivity to the copolymer by inducing more organized structures [19]. Higher tan δ was observed as MAH is introduced, and also the peak was found at the higher temperature. It is known that better shape recovery is achieved as tan δ becomes higher since it can deform like an elastomer. Thus, it can be predicted that PET-PEG-MAH copolymer will get better shape recovery since it has higher tan δ value. Another point to note is that as

MAH is introduced, the width of rubbery plateau region above T_g increases. Because it is well known that the width increases with molecular weight, the above result confirms that MAH cross-linking increases molecular weight of the PET-PEG copolymer. Similar behavior is also observed in the intrinsic viscosity of PET-PEG-MAH copolymer (Table 3).

Storage modulus and tan δ of three types of cross-linked PET-PEG copolymers (glycerol, sorbitol, and MAH) were compared. Storage modulus abruptly decreased around T_g, and increased later at about 60 °C above their T_gs for all three kinds of cross-linked copolymers, which could be coming from disorganzation at T_g and recrystallization at higher temperature of PET-PEG copolymer chains during heating process in dynamic mechanical property measurement, and such result was not uncommon as previously observed for poly(trimethyleneterephthalate) [20-23]. As more cross-linking agent was included in the copolymer chain, storage modulus profile generally shifted toward higher temperature, and thus recrystallization of copolymer chains occurred at higher temperature, which originated from combined interactions from additional chemical cross-linking, dipole-dipole interaction of carboxyl groups of PET, and induced dipole-dipole interaction between phenyl rings [24,25]. Based on the result, it can be suggested that variation of dynamic mechanical property over temperature can be overcome by drawing process of specimen before heating.

Similar thermomechanical trend was observed in the tan δ profile at around T_g and small relaxation shoulder immediately after T_g was probably coming from temporary disruption of chains during recrystallization. Addition of cross-linking agent resulted in broader tan δ profile, because restricted chain movement by cross-linking agent made chain rearrangement slower and more energy-consuming process than uncross-linked copolymer; therefore, both disorganization and recrystallization were observed at a broader temperature range. In addition to broadening of transition area in tan δ vs. temperature profile, and temperature at peak transition shifted toward higher region, which could be interpreted by the same mechanism. As high tan δ means better deformation and shape recovery, cross-linked PET-PEG copolymers are promising candidates for shape recovery material based on tan δ results. Rubbery plateau region observed at higher temperature narrowed with introduction of cross-linking agent such as glycerine or sorbitol, but MAH-cross-linked copolymer had similar width of rubbery plateau region as uncross-linked one; such difference which is presumed to be coming from the cross-linking point, hard-segment linking (glycerine and sorbitiol) or soft-segment linking (MAH), tell us once again the importance of selection of cross-linking agent in developing highly reliable shape memory copolymer.

It is generally known that shape memory polymer retains its deformed shape, if stressed at rubbery plateau region that is above T_g and then cooled to a temperature below T_g, and it recovers original shape if heated to rubbery plateau region again. Therefore, shape memory polymer has better shape retention property at temperature below T_g with the resulting high storage modulus, and its deformation becomes easier at temperature above T_g because of low storage modulus. From a different perspective, cross-linked PET-PEG copolymer, especially one by glycerine or sorbitol, showed significant modulus change around T_g which was above the necessary modulus change (order of three) as a temperature sensitive material, resulting in high hope for application in that area. Among the copolymers, comparatively sharper transition in glycerine-cross-linked one suggests that cross-linking at hard segment (PET), and short and rigid cross-linking agent are also advantageous in developing highly thermosensitive material [26].

Dynamic mechanical test results of copolymers with three different cross-linking agents are summarized as follows: (1) cross-linking by glycerine provides higher T_g, storage modulus, tan δ, and apparent change of storage modulus plus sharp transition of tan δ around T_g; (2) cross-linking by sorbitol also shows similar characteristics as glycerine, but transitions for both storage modulus and tan δ are blunt, which comes from the fact that flexible cross-linking of hard segment by sorbitol may mitigate strong interaction and shape retention among hard segments as observed in shorter glycerine; (3) cross-linking by MAH is least successful in obtaining improved thermomechanical properties, and interconnection of soft segments like MAH cross-linking is not desirable for our purpose, because soft segment should absorb external stress instead of blocking free movement of PEG chains.

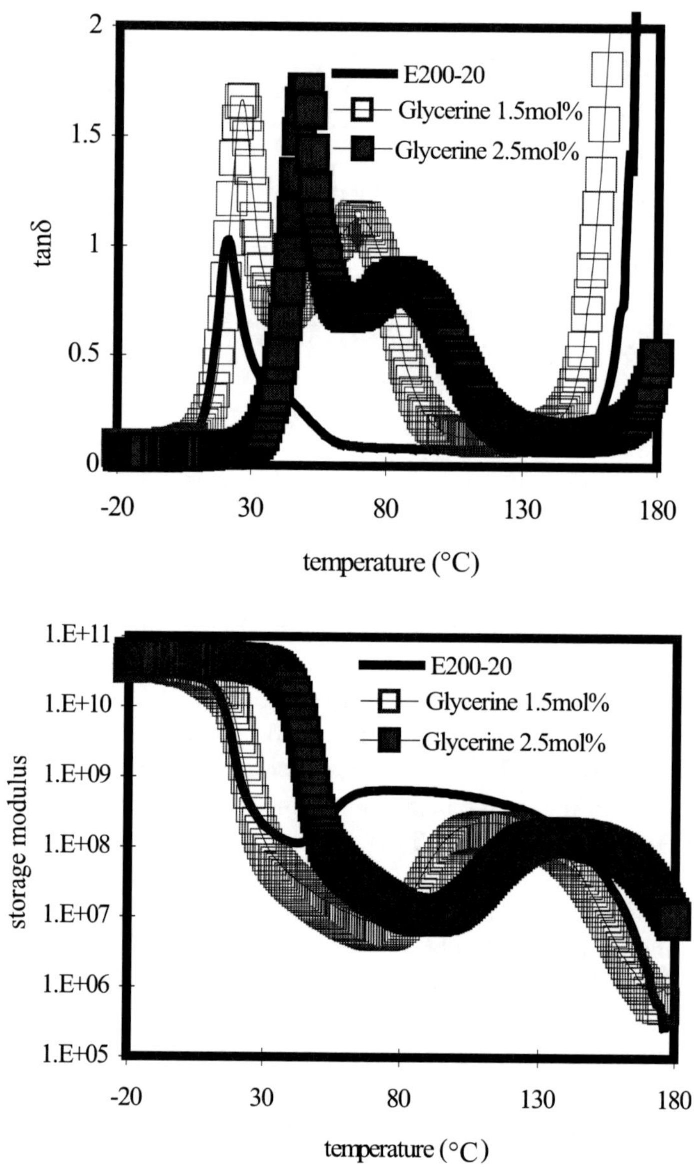

Figure 17. Storage modulus and tan δ of glycerine cross-linked copolymers.

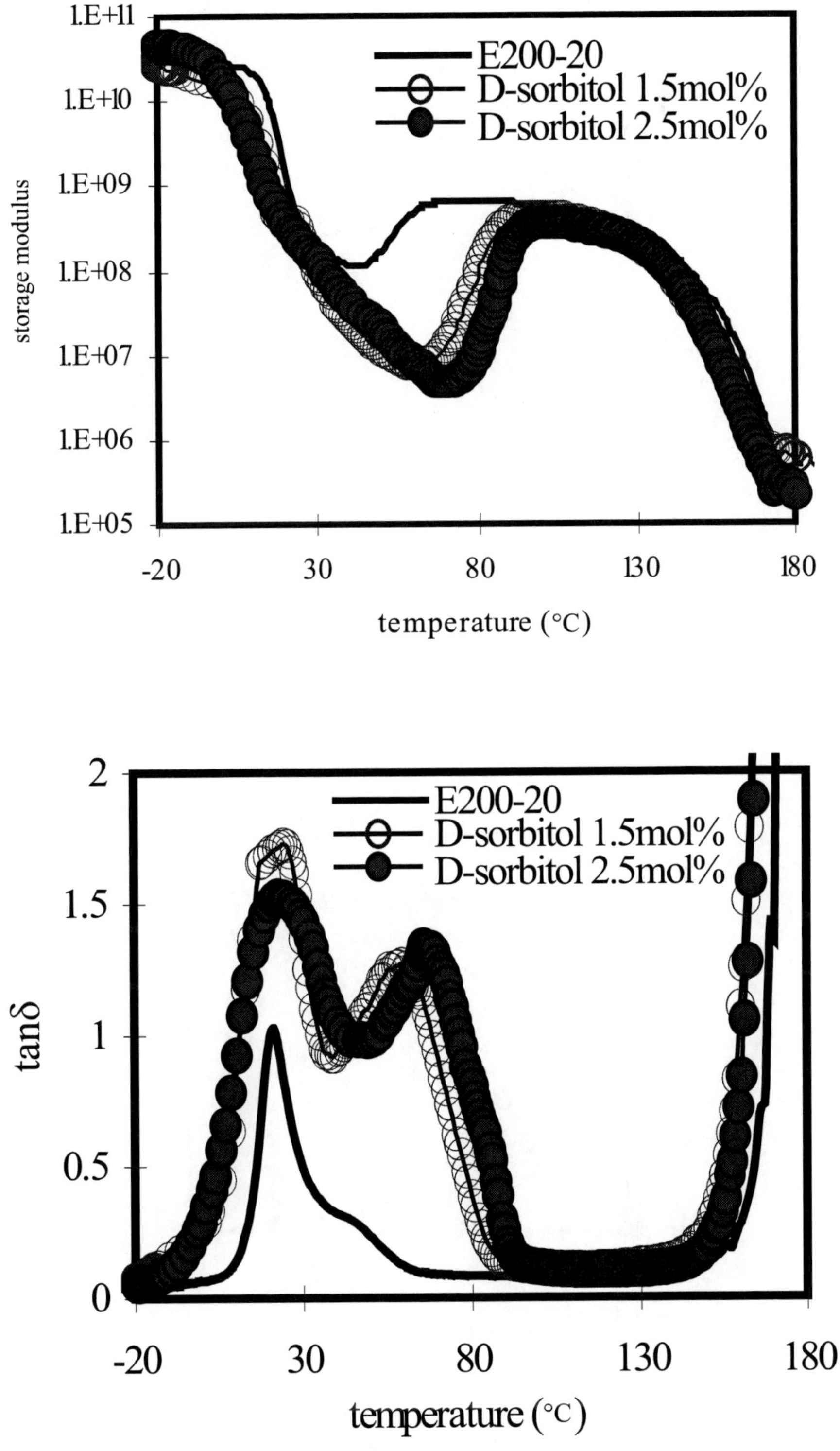

Figure 18. Storage modulus and tan δ of sorbitol cross-linked copolymers.

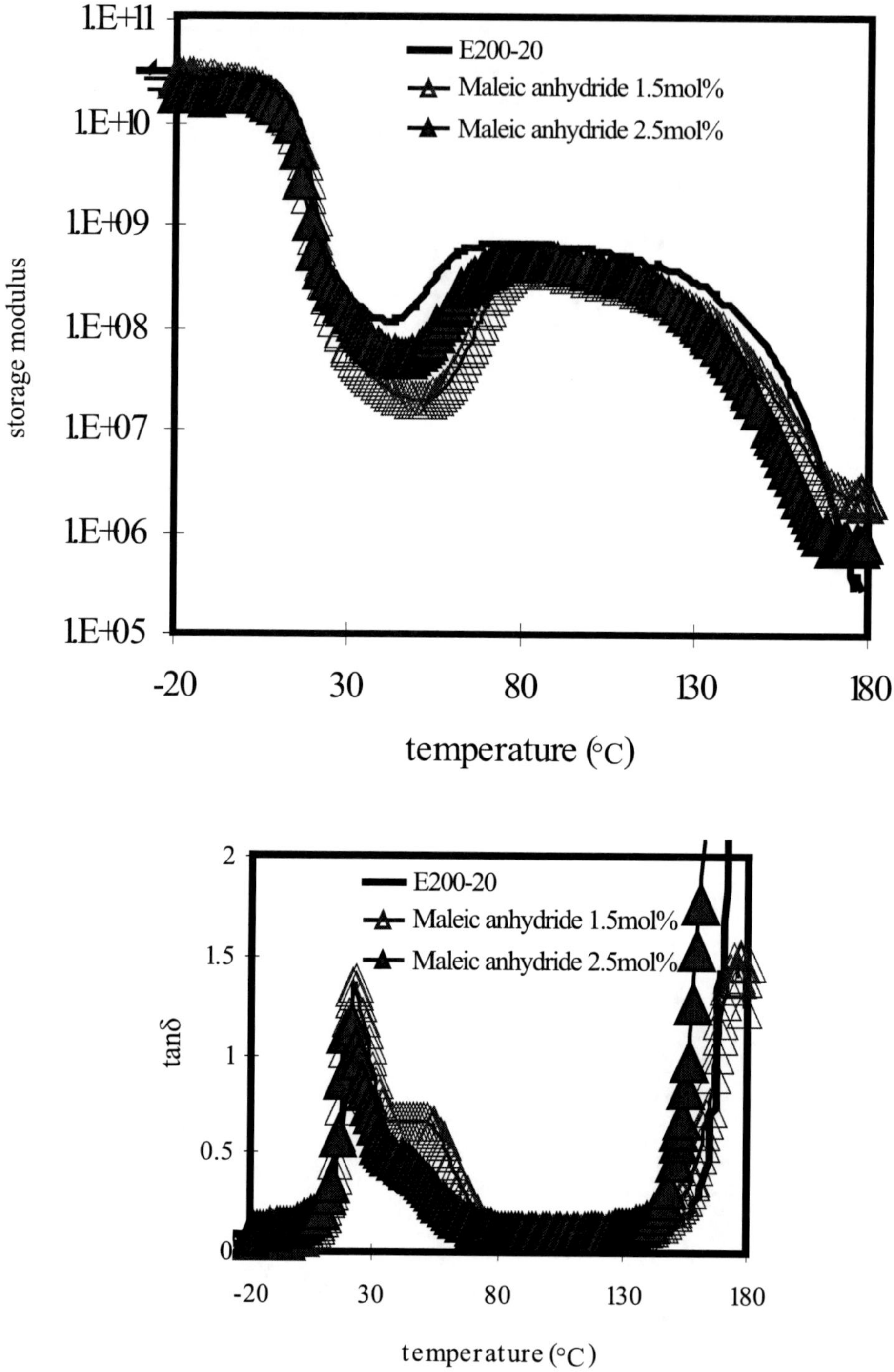

Figure 19. Storage modulus and tan δ of MAH cross-linked copolymers.

4. Conclusion

As a preliminary search for developing composite laminate material with vibration control ability, various cross-linked shape memory copolymers were investigated for damping effect. PET-PEG copolymers consisting of PEG soft segment and PET hard segment was synthesized along with PET-PEG copolymers cross-linked by maleic anhydride (MAH). Experimental results showed that highest tensile properties were obtained in the case of PET-PEG copolymer with thc mol ratio of 80 mol% EG and 20 mol% PEG having 200 g/mol of molecular weight. Addition of MAH resulted in decrease of strain at break and increase of maximum stress, together with increase of shape memory effect and damping (1.5 to 2.0 mol% of MAH). Cross-linking agents such as glycerine and sorbitol were also tried, and each cross-linked copolymer showed its own merits in different tests. Addition of glycerine, or sorbitol as a cross-linking agent also improved both shape memory and damping effect. Highest shape recovery rate was obtained with 2.5 mol% of glycerine, and highest tan was from 2.5 mol% of sorbitol. When sandwich-type composite laminate was prepared with glass-fiber-reinforced epoxy beam and cross-linked copolymer, Charpy impact strength increased 1.9 ~ 3.7 times and damping effect also increased 23 times for the best case at room temperature as compared to pure epoxy beam. Thus, cross-linked PET-PEG copolymer sandwiched to laminate composite, enables developing composite materials with controlled T_g and high damping effect. Copolymers having covalent cross-linking by glycerol and ionic interaction from SP group were compared with the above cross-linked copolymers. Shape retention rate of the copolymers showed very high value (97-99 %); shape recovery at low mol% of glycerol showed initial decrease and sharp increase with SP content, but 2.5 mol% glycerol copolymer series had gradual increase with SP content. Such results originated from the balance between strong covalent cross-linking by glycerol and weak ionic-interaction from SP, and the additional fine control of shape recovery by ionic interaction helps in developing custom-made smart material. Similar trend was observed for the glass transition temperatures of copolymers. Composite laminates prepared with the copolymer (glycerol and SP) showed high impact strength and loss tangent δ, and their value can be adjusted with SP content. Finally, physical properties of the sandwich-type laminate with shape memory copolymer can be controlled by proper selection of cross-linking agents, which is very helpful in developing composite materials with controlled T_g and high damping effect.

References

[1] Gandhi, M. V.; Thompson, B. S. Smart Materials and Structures, Chapman and Hall, New York, 1992.

[2] Takagil, T. *J. Intelligent Mater. Sys*. 1990, 1, 149.

[3] Varadan, V. K. *Smart Structures and Materials*, 1994, 2189.

[4] Riley, P. J.; Wallace, G. G. *J. Intell. Mater. Syst. And Struct*. 1991, 2, 228.

[5] Wallace, G. G.; *Materials Forum*, 1992, 16, 111.

[6] Granqvist, C. G.; Azwns, A.; Isidorsson, J.; Ronnow, D.; Mattsson, M. S.; Veszelei, M. *J. Non-crystalline Solids*, 1997, 218, 273.

[7] Shirai, Y.; Hayashi, S. *Development of Polymeric Shape Memory Material*, MTB184, Mistubishi Heavy Industry, Inc., Japan, December, 1988.

[8] Gandhi, M. V.; Thompson, B. S. Smart Materials and Structures, New York: Chapman and Hall, 1992.

[9] Takahashi, T.; Hayashi, N.; Hayashi, S. *J. Applied Polym. Sci.* 1996, 60, 1061.

[10] Wei, Z. G.; Sandstrom, R. *J. Material Science*. 1998, 33, 3743.

[11] Liang, C.; Rogers, C. A.; Malafeew, E, *J. Intellingent Material Systems and Structures,* 8, 380.

[12] Takagil, T. *J. Intelligent Mater. Sys*. 1990, 1,149.

[13] Chun, B. C.; Cha, S. H.; Chung, Y. C.; Cho, J. W. *J. Applied Polym. Sci*. 2002, 83, 27.

[14] Oh, P. R.; Kim, K. J.; Kim, Y. H. *J. Korean Fiber Soc.* 1999, 36, 132.

[15] Chun, B. C.; Cha, S. H.; Park, C.; Park, M. J.; Chung, Y. C., Cho, J. W. *J. of Applied Polym. Sci.* 2002, 90, 3141.

[16] Spathis, G.; Niaounakis, M.; Kontou, E.; Apekis, L.; Pissis, P.; Christodoulides, C. *J. Appl. Polym. Sci.* 1994, 54, 831.

[17] Dutra, D.; Soares, B.; Campos, E.; Silva, J. *Polymer* 2000, 41, 3841.

[18] Lee, S. Y.; Kim, B. K.; Xu, M., *Polymer*, 1996, 37, 5781.

[19] Choi, Y. J.; Kim, B. K.; Jeong, H. M. *Polymer (Korea)*, 1988, 22, 1.

[20] Oh, P. R.; Kim, K. J.; Kim, Y. H. *J. Korean Fiber Soc.* 1999, 36, 132.

[21] Hoffman, D. M. *J. Applied Polym. Sci*. 2002, 83, 1009.

[22] Wu, C.; Yanagishi, T.; Nakamoto, Y.;Ishida,S.;Nitta,K.;Kubota, S. *J. Polym. Sci*. 2000, 38, 2285.

[23] Cascone, M. G. *Polym. Intnl*. 1997, 43, 55.

[24] Cho, D.; Choi, Y.; Drzal, L. T. *Polymer* 2001, 42, 4611.

[25] Krumova, M.; Lopez, D.; Benavente, R.; Mijangos, C. *Polymer* 2000, 41, 9265.

[26] Lee, S. Y.; Kim, B. K., Xu, M. Polymer 1996, 37, 5781.

In: Polyester: Properties, Preparation and Applications
Editors: Hina Yamashita, Yui Nakano
ISBN: 978-1-60456-753-3

Chapter 7

DEGRADATION OF AROMATIC CO-POLYESTERS DERIVED FROM N-OXYBENZOIC, TERE- AND ISOPHTHALIC ACIDS AND DIOXYDIPHENYL

E. V. Kalugina, K. Z. Gumargalieva* and V. G. Zaikov**
Polyplastic Co., 14A, General Dorokhov St., Moscow 119530, Russia
*N.N.Semenov Institute of Chemical Physics, 4, Kosygin St., Moscow 119991, Russia
**N.M.Emanuel Institute of Biochemical Physics, 4, Kosygin St., Moscow 119991, Russia

Special attention to these polymers is defined by their specific feature, which is orientation in the melt, mostly associated with the intense development in computer technologies. Owing to this property such polymers are devoted to the "family" of liquid-crystal polymers. The liquid-crystal properties are also observed for PAI with an uneven number of CH_2-groups [1]. It should be noted that polyalkanimide (PA-12), discussed in [2 - 14], also displays liquid-crystal properties under definite processing modes.

Liquid-crystal aromatic copolyesters (LCP) were studied. They were derived from dioxydiphenyl diacetate, acetoxybenzoic, iso- and terephthalic acids (IPA and TPA, respectively): 100/0, 75/25, 50/50, 25/75, 0/100.

$n = 50$, $m = 25$, $f = 25$

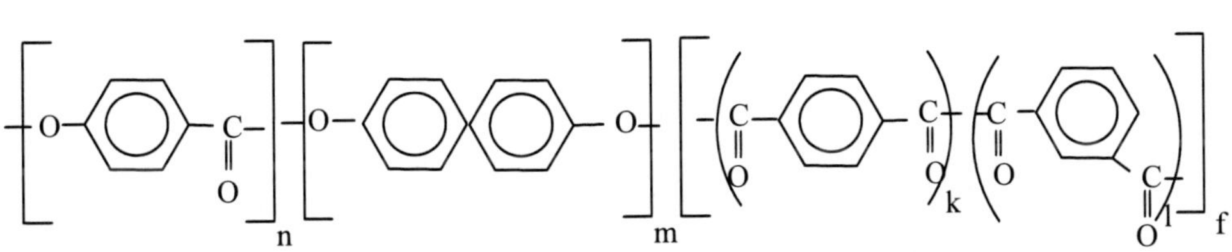

International analogue – *Xydar* (Amoco).

Thermal stability of LCP with different TPA/IPA ratio was studied by dynamic TGA/DTA techniques. Table 1 shows DTA/TGA data obtained in argon flow. Data obtained in air are shown in Figure 1.

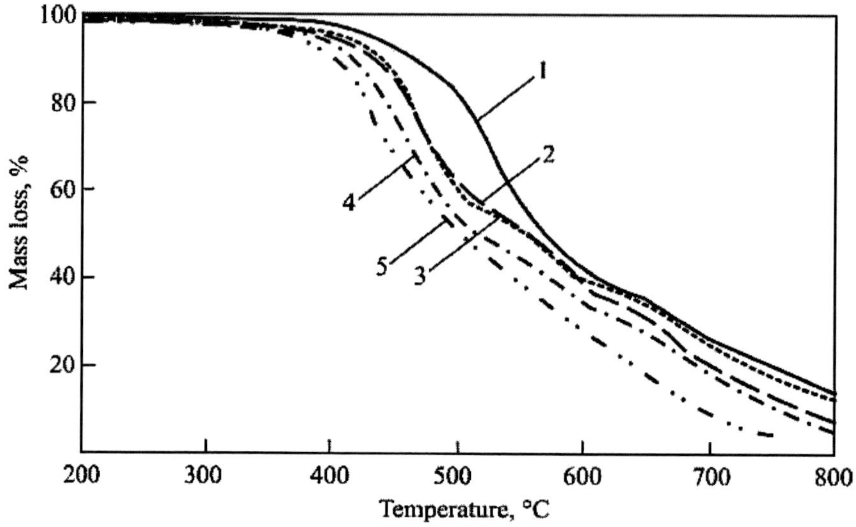

Figure 1. TGA data obtained at the heating rate of 5°/min in air for liquid-crystal polymers (LCP): KI-0 (1), KI-25 (2), KI-50 (3), KI-75 (4), and KI-100 (5).

Table 1. LCP sample characteristics

Sample	TPA/IPA ratio, mol%	TGA/DTA data in argon flow (with heating rate 10°/min)	
		Melting range, °C	Degradation initiation temperature, °C
KI-0	100/0	410 - 415	490
KI-25	75/25	360 - 365	480
KI-50	50/50	340 - 350	472
KI-75	25/75	300 - 315	455
KI-100	0/100	315 - 325	450

Without air oxygen LCP degrade in one stage, forming significant amount of coke residue (up to 40wt.% at 700°C). Two endothermic heat effects were observed on DTA curves: a low one in the melting range and quite intense one in the polymer degradation zone. Calculation of the heat effect gave ΔH = 1 – 2.5 kJ/mol. As decided from the studies of 4-hydroxybenzoic and 2,6-hydroxynaphthoic acid copolymer by DSC method [15], the heat effect of about 1 kJ/mol relates not to real melting, but to changes in the order strength at transition from crystal to mesophase. Apparently, due to superimposition of heat effects which accompany degradation, and transition to the isotropic melt, nobody succeeded in detecting temperature transition associated with LCP changing the isotropic degree.

Thermal stability of LCP in air is significantly (by 25 - 30°C) lower than in argon. According to dynamic TGA data in air (Figure 1), mass losses of studied LCP are observed in the temperature range of 350 - 800°C. The degradation proceeds in two stages: the first stage at 350 - 550°C is accompanied by mass losses up to 40 wt.%; the second stage is slower and proceeds in the temperature range of 550 - 800°C up to full degradation of the polymer. The coke content at 750 - 800°C equals 1 – 3 wt.%. As shown by DTA data, LCP degradation stages are accompanied by exothermal heat effects. As tested in the air, a low endothermic heat effect is absent in the melting range, apparently, due to overlapping by exothermal effects of degradation reactions. LCP rating in the sequence with thermal stability decrease is the following: KI-0 > KI-25 ≥ KI-50 > KI-75 > KI-100.

The increase in IPA content shifts the melting range towards lower temperatures and reduces LCP thermal stability. The study of LCP phase transitions by X-ray analysis in the temperature range of 20 - 400°C indicates similar changes in all LCP. Annealing at 300°C causes an insignificant increase of the main crystalline reflex. As an example, Figure 2 shows the difractogram for powder-like and mold samples of KI-75 LCP. No phase transitions (reflex occurrence and elimination) were detected in the studied temperature range in LCP. This may prove the DTA results and suggestions about closeness of degradation temperatures and transitions to mesophase [15]. The ability of studied materials to transit to the so-called "liquid-crystal state" characterizes their behavior at processing temperatures. At softening temperature (as regards to the structure, this range falls within 300 - 400°C) a jump-like viscosity decrease is observed in all polymers. Hence, extremely strong fibers are formed from the melt. This effect is explained [16] by cooperative orientation of large macromolecule axes along the flow direction (viscosity anisotropy), which is realized only in LCP.

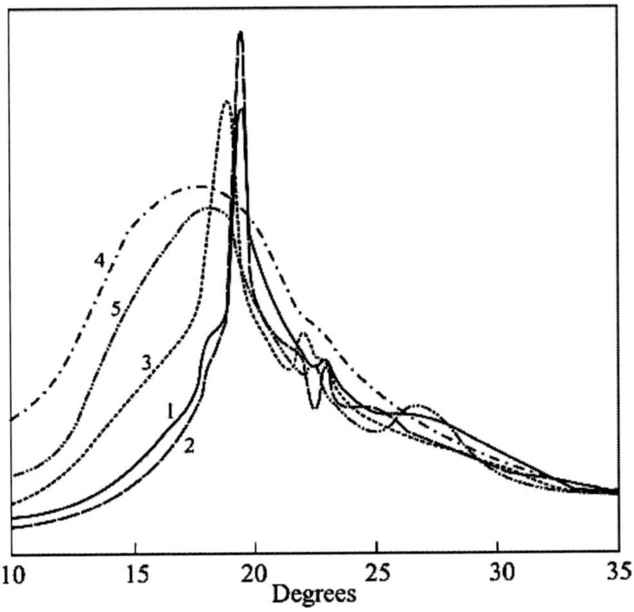

Figure 2. Diffraction patterns for LCP-2 powder at heating 1 – powder-like sample (20°C) 2 - powder-like sample (300°C) 3 - powder-like sample (350°C) 4 powder-like sample (400°C) 5 – mold sample.

Thermal stability of polymers depends upon several factors: structure, molecular-mass parameters, content of macrochain defects, labile end groups (currently, hydroxyl ones), low-molecular organic (non-reacted, residual monomers) or inorganic (increments of metal ions from the raw material and equipment) additives in the macromolecules.

The composition of inorganic additives in monomers and LCP were studied by plasma-emission spectroscopy technique. The example of KI-75 LCP, studied by TGA, shows the effect of some metal increments on thermal stability. The additive content was increased by injection of inorganic salts of appropriate metal into the polymer. Table 2 shows comparative data on Cu, Fe, Ni, Ca, and Al content and temperatures of degradation initiation by TGA.

Table 2. The effect of metal increments on the thermal oxidative stability of KI-75 LCP

Metal	Content, wt.%	Degradation initiation temperature (T_d) by TGA in air, 10°/min heating rate
Fe	1.3×10^{-3}	320
	1.3×10^{-2}	300
Al	1.4×10^{-3}	320
	1.4×10^{-2}	320
Ca	4×10^{-3}	320
	4×10^{-2}	320
Ni	1.0×10^{-3}	320
	2.0×10^{-2}	325
Cu	1.3×10^{-5}	320
	2.0×10^{-3}	330

The results obtained show different effect of metal increments on LCP thermal oxidative stability. In the studied range of concentrations, aluminum and metals of the alkaline sequence (Ca, Na, K, etc.) do not practically affect the thermal stability. Iron causes the negative effect, whereas Cu and Ni, vice versa, increase thermal stability of LCP. It should be noted that injected concentrations are quite corresponded to usual content of metal increments in industrial samples of engineering, bulk polymers, such as polycarbonate, aliphatic polyamides, polystyrene, etc.

The composition and content of organic additives to LCP were studied by the mass-spectrophotometry technique. Phenol and sioxydiphenyl (94 and 186 m/e, respectively) were identified. They represent the hydrolysis products of the initial monomer of dioxydiphenyl diacetate and heavy fragments of the following structure:

m/e=214

m/e=306

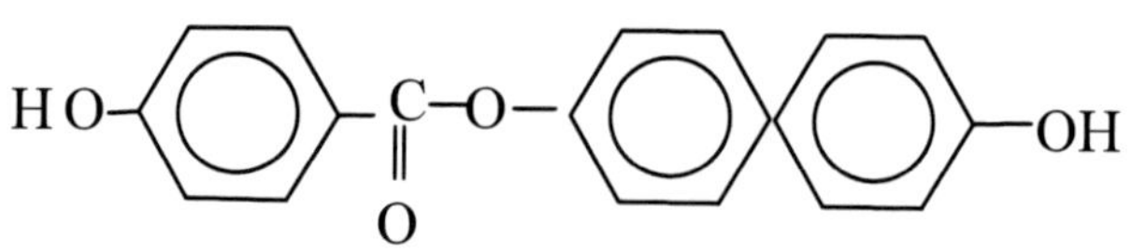

Total amount of organic additives in different samples has not exceeded $(1.0 - 2.0)\times10^{-2}$ wt.%, which does not practically affect thermal stability of the polymers.

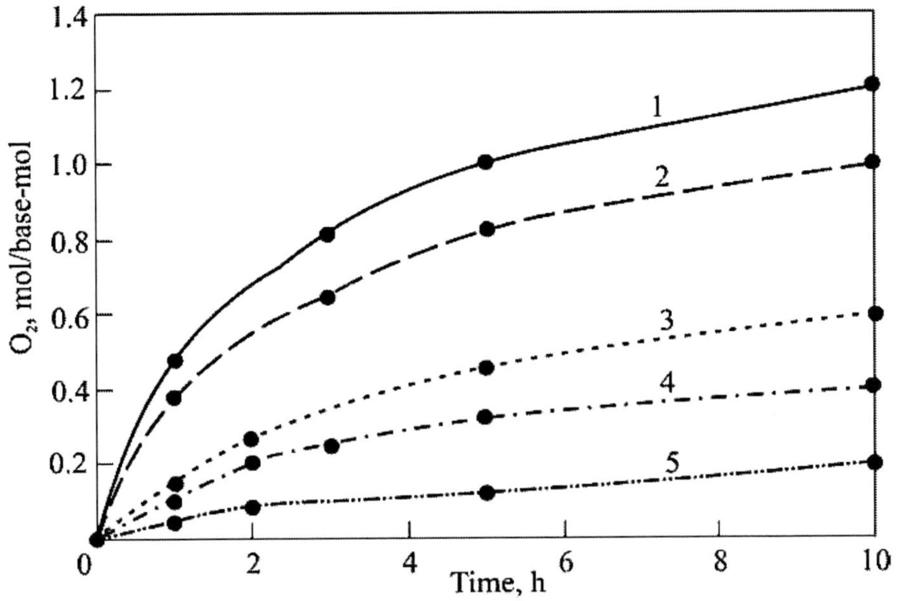

Figure 3. Kinetics of O_2 absorption by LCP KI-100 (1), KI-75(2), KI-50 (3), KI-25 (4), and KI-0 (5).

As shown by kinetics of oxygen absorption at 350°C (the processing temperature), thermal stability of LCP is reduced with increase of IPA content (Figure 75). This result confirms the TGA data. The kinetics of O_2 absorption is the two-stage process with absorption of 1 mole of O_2 per monomeric unit at the first, quick stage (2 – 3 h), and 0.2 mol/base-mol during following 7 – 8 h of thermal oxidation. Analogous dependencies are displayed by CO_2 release (Figure 4) - the main gas product of the studied LCP degradation product. Shown below are heavy, highly boiling LCP degradation products, identified by NMR and MS techniques [17 – 20].

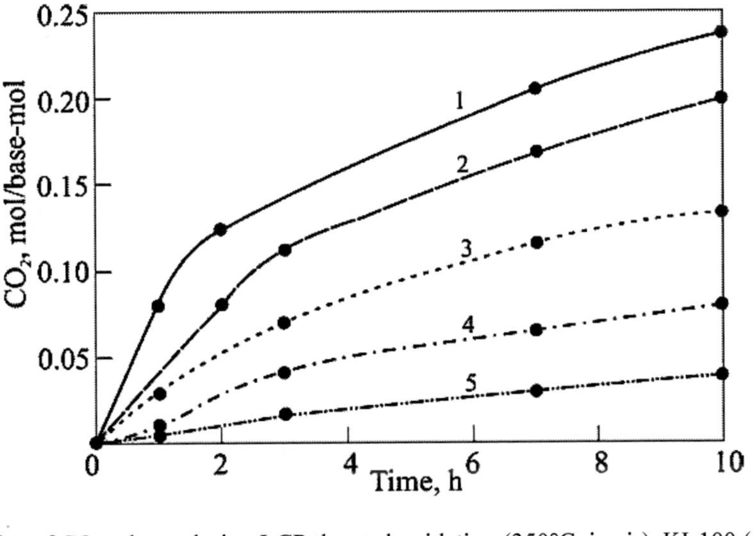

Figure 4. Kinetics of CO_2 release during LCP thermal oxidation (350°C, in air): KI-100 (1), KI-75 (2), KI-50 (3), KI-25 (4), and KI-0 (5).

Heavy Products of LCP Degradation

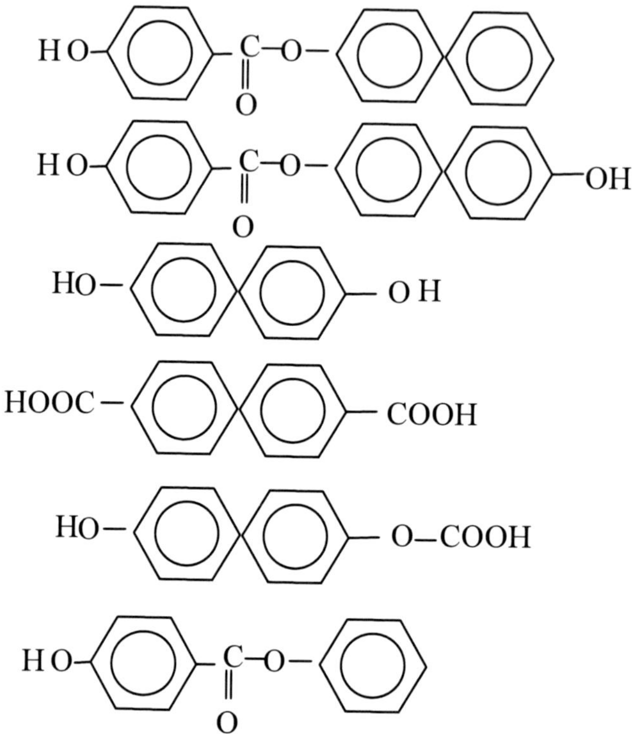

As observed from the composition, heavy products of LCP degradation, precipitated on cold zones of the ampoule, represent a mixture of the initial monomer (dioxyphenyl diacetate), products of its hydrolysis (dioxydiphenyl – DODP - for example), and products of DODP and *p*-OBA interaction. It should be noted that the composition of the above-mentioned products is identical for all studied LCP and slightly differs just by the ratio of components.

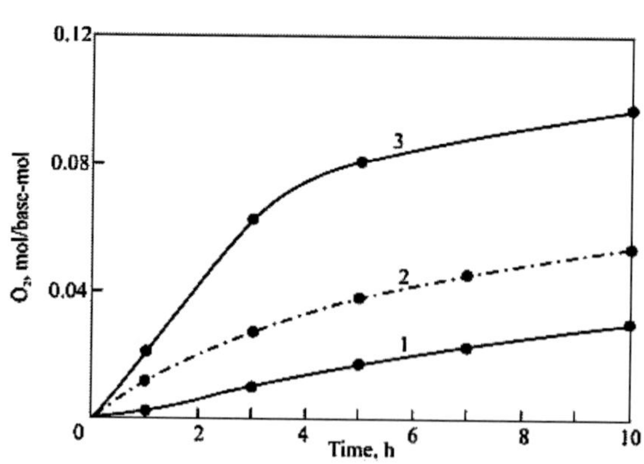

Figure 5. Oxygen absorption kinetics for KI-75 at 300 (1), 320 (2) and 350°C (3) in air.

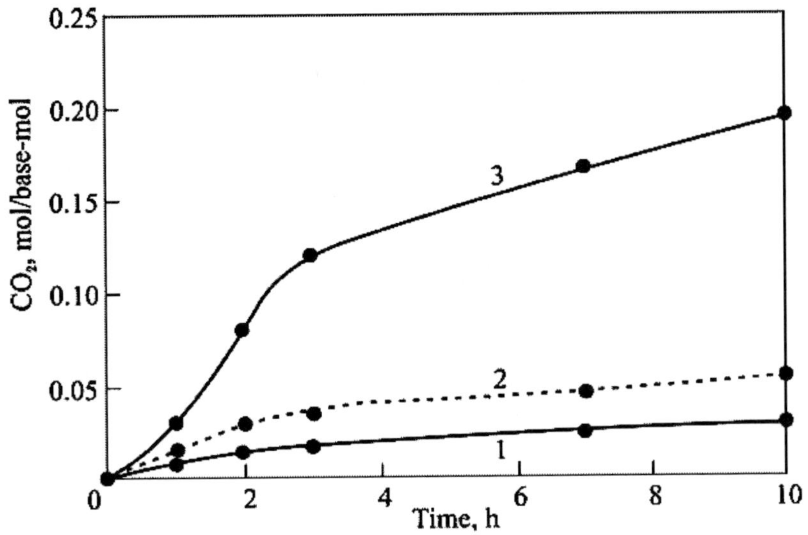

Figure 6. Carbon dioxide release kinetics at KI-75 thermal oxidation at 300 (1), 320 (2) and 350°C (3) in air.

Degradation transformations were studied on the example of KI-75 LCP in the processing range. As shown (Figures 5 - 6), at the melting point or during melting (300 and 320°C) the oxidation rate is much lower than at 350°C, when according to the X-ray diffraction analysis the whole polymer transits to isotropic melt. Besides the main gas product (CO_2), hydrogen (at early oxidation stages at 0.5 – 1 h exposure) and water (at 4 h exposure) were also detected. As observed from dynamics of the elemental composition change, hydrogen content decreases and carbon content increases in KI-75 during its thermal oxidation, e.g. a graphite-like structure is formed. This process proceeds intensively at 350°C. IR-spectroscopy data [21 - 23] show that initial changes happen in the absorption range of ester aromatic fragments: absorption band intensity at $\nu_{C=O}$ = 1740 cm^{-1}, ν_{C-O} = 1270 and 1160 cm^{-1}, $\nu_{C=C}$ = 1600 and 1500 cm^{-1}, $\delta_{C=C}$ = 720 cm^{-1} is reduced. At maximal exposure (thermal oxidation at 350°C during 10 h) only ether absorption bands at ν_{C-O-C} = 1080 cm^{-1} and aromatic structure bands are preserved. The spectral background significantly decreases, which is caused by formation of intermolecular crosslinks. At 350°C a great amount of oligomers is formed. They precipitate in the ampoule near the reaction zone at temperature ≈ 150°C. The structure and ratio of oligomers with appropriate end groups, identified by ^{13}C NMR technique in the oligomer degradation products, are shown on the next page.

The amount of oligomers was estimated by the ratio of reflex squares with appropriate chemical shifts: 118.5 and 115.82 ppm (*product a*); 115.43 ppm (*product b*), and 120.72 ppm (*product c*). According to these data *product a* gave 15 mol%, *product b* – 31 mol%, and *product c* – 10 – 12 mol%.

The observation of *p*-oxybenzoic acid in thermal oxidation products of neighboring units allows a suggestion about simultaneous proceeding of copolycondensation and homopolycondensation of *p*-acetoxybenzoic acid. Free *p*-oxybenzoic acid output is 3 – 4 times higher than in the bound state in the form of end groups of oligomers. Apparently, free *p*-oxybenzoic acid is formed in thermal reactions at degradation of labile bonds in structural *p*-oxybenzoic blocks.

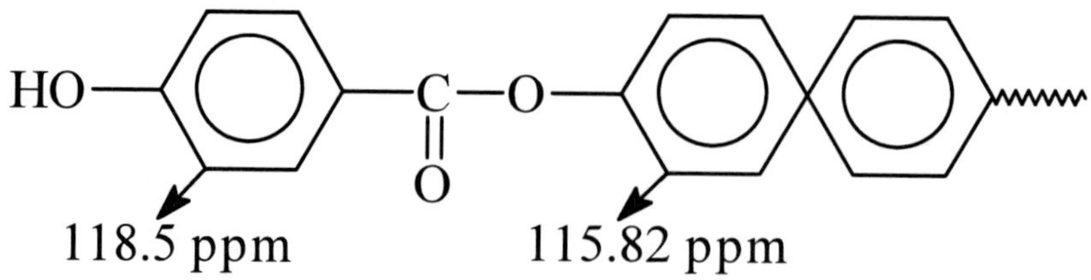

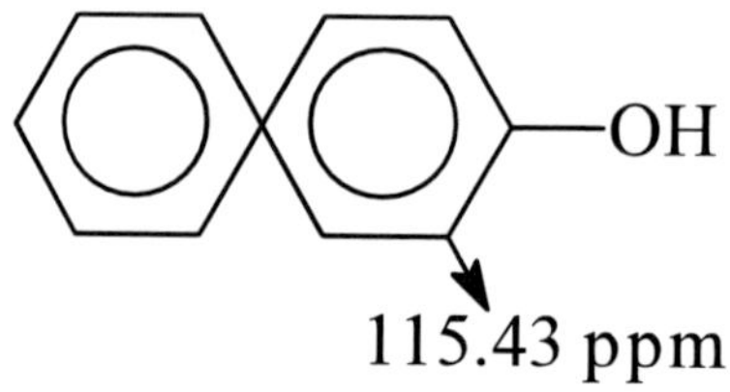

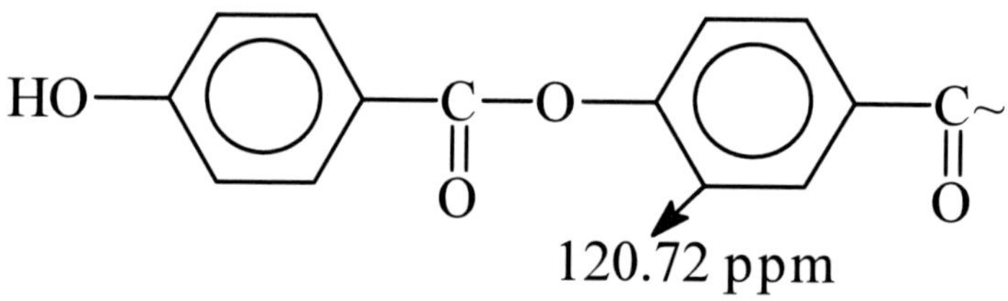

Analysis of kinetics and LCP degradation products in the processing temperature range allowed detection of some general features, observed in degradation behavior of heat resistant polyheteroarylenes [24]: structure graphitization, H_2 release, thermal oxidation stability increase at transition metal injection, etc. The idea of their stabilization is based on the following suggestions about degradation mechanisms:

- classical radical-chain thermal oxidation mechanism;
- formation of a molecular complex with oxygen;
- molecule transition to electronically excited state.

Injection of additives is the common method for investigating the mechanism of chemical reactions. It was found that of high effectiveness is the mixture stabilization by the triple system of copper compound, phenol antioxidant and phosphite in polyalkanimide, polyphthalamide and other heat-resistant polymers. The idea of such mixture is based on the action mechanism of such additives:

- phenol antioxidant inhibits thermal oxidation by interacting with peroxy radicals $ROO^{\bullet}$;
- phosphite, secondary antioxidant, destroys hydroperoxides;
- Cu^{2+}-containing compound acts differently, for example, forms complexes during inhibition of peroxy radical or macrochain defects.

Table 3 shows comparative data on thermal oxidative stability for non-stabilized and stabilized KI-75 LCP.

Table 3. Thermal oxidation of KI-75 LCP in air at 350°C during 30 min

Stabilization compounding	Oxygen absorption, mol/monomeric unit	CO_2 release, mol/monomeric unit
Nonstabilized	0.22	0.033
0.005% $CuSO_4$ + 0.3% *Irgafos 126* + 0.1% *Irganox 1010*	0.08	0.019

The results obtained show that stabilizer injection causes a significant (over two times) deceleration of thermal oxidation in LCP. Of interest is the effect of additives on polymer morphology, determined during studying stabilized and non-stabilized samples (before and after thermal oxidation in air) by the X-ray diffraction analysis. It is found that crystalline reflex is preserved in stabilized polymers, whereas it disappears in non-stabilized samples. The stabilization effect on the physical structure of polymers was not studied well with respect to chemistry of degradation processes. Only complex consideration of the problem (chemistry + change of physical permolecular structure) may cause the increase of thermal stability of prepared product and extension of the material lifetime in articles.

REFERENCES

[1] *US Patent No. 2,944,993*, Glass-filled thermoplastic composites derived from linear polypyromellitimides, Publ. Jun. 14, 1955 (cl. 260-37).

[2] Semenov N.N., *On Some Problems of Chemical Kinetics and Reactivity (Free Radicals and Chain Reactions)*, 2nd Ed., Moscow, *AN SSSR*, 1958, 686 p. (Rus)

[3] Kandratiev V.N. and Nikitin E.E., *Chemical Processes in Gases*, Moscow, *Nauka*, 1981, 262 p. (Rus)

[4] Voevodsky V.V., *Physics and Chemistry of Elementary Chemical Processes*, Moscow, Nauka, 1969, 414 p. (Rus)

[5] Neiman M.B., Aging and Stabilization of Polymers, Moscow, *Nauka,* 1964, 332 p. (Rus)

[6] Kuz'minsky A.S., Oxidation of Caoutchoucs and Rubbers, Moscow, *Goschimizdat,* 1957, 319 p. (Rus)

[7] Grassi N., Chemistry of Polymer Degradation Processes, Moscow, *Inostrannaya* Literatura, 1959, 252 p. (Rus)

[8] Madorsky S., Thermal Degradation of Organic Polymers, Moscow, *Mir,* 1967, 320 p. (Rus)

[9] Emanuel N.M., Denisov E.T., and Maizus Z.K., Chain Reactions of Hydrocarbons Oxidation in the Liquid Phase, Moscow, *Nauka*, 1965, 375 p. (Rus)

[10] Emanuel N.M. and Buchachenko A.L., Chemical Physics of Polymer Aging and Stabilization, Moscow, *Nauka*, 1982, 359 p. (Rus)

[11] Shlyapnikov Yu.A., Kiryushkin S.G., and Mar'in A.P., Antioxidant Stabilization of Polymers, Moscow, *Khimia,* 1986, 252 p. (Rus)
[12] Popov V.A., Rapoport N.Ya., and Zaikov G.E., Oxidation of Oriented and Stressed Polymers, Moscow, *Khimia*, 1987, 232 p. (Rus)
[13] Zaikov G.E. and Moiseev V.V., Chemical Resistance of Polymers in Aggressive Media, Moscow, *Khimia,* 1979, 216 p. (Rus)
[14] Minsker K.E. and Fedoseeva G.T., Degradation and Stabilization of Polyvinylchloride, 2nd Ed., Moscow, *Khimia*, 1979, 272 p. (Rus)
[15] Mark H., Atlas S.M., and Ogata N., *J. Polym. Sci.*, 1962, vol. 61, p. 849.
[16] Ed. N.A. Plate, Liquid-Crystal Polymers, Moscow, *Khimia*, 1988. (Rus)
[17] Slonim I.Ya. and Urman Ya.G., NMR-Spectroscopy of Heterochain Polymers, Moscow, *Khimia*, 1982, 232 p. (Rus)
[18] Johnson L.F., *Carbon-13 NMR Spectra*, N.Y., *Wiley,* 1972.
[19] Breitmaier E., Haas G., and Voelter W., *Atlas of Carbon-13 NMR Data, Heyden,* London, 1979.
[20] Formacek V., Desnoyer L., Kellerhals H.P., Keller T., Clerc J.T., *^{13}C Data Bank,* Bruker Physik, Karlsruhe, 1976.
[21] Nakanisi K., IR-Spectra and Structure of Organic Compounds, Moscow, *Mir,* 1965, 215 p. (Rus)
[22] Hummel/Scholl, *Atlas of Polymer and Plastics Analysis.*
[23] Dekhant I., Danu V., Cimmer V., and Schmolke R*., Infrared Spectroscopy of Polymers*, Moscow, *Khimia*, 1976. (Rus)
[24] Kalugina E.V., NovoTortseva T.N., and Andreeva M.B., 'Thermal Oxidation Features of Heat-Resistant Heterochain Polymers', *Obzor Polim. Mater.*, 2001, No. 6, pp. 29 - 37. (Rus)

In: Polyester: Properties, Preparation and Applications
Editors: Hina Yamashita, Yui Nakano
ISBN: 978-1-60456-753-3

Chapter 8

THERMAL STABILITY AND FIRE PERFORMANCE OF UNSATURATED POLYESTER RESINS

E. Kicko-Walczak
Industrial Chemistry Research Institute Rydygiera 8,
01-793 Warszawa, Poland; e-mail: Ewa.Kicko-Walczak@ichp.pl

ABSTRACT

The thermal decomposition of halogenated and non-halogenated unsaturated polyester resins (UPR`s), fire retarded by zinc hydroxystannate (ZHS) and cross-linked with styrene, has been investigated by thermogravimetry (TG) and TG coupled on-line with Fourier transform infra red spectroscopy (TG-FTIR) or mass spectroscopy (TG-MS).

In this Chapter, thermal analysis of the decomposition process has been performed – hence, the flame retardancy and thermal stabilization of halogenated and non-halogenated polyester resins by ZHS may be explained by the formation of surface-localized spherical barriers which are growing according to the nucleation growth mechanism and which attenuate the transfer of heat from the decomposition zone to the substrate. This effect was found as dominating in the flame-retardancy mode of action.

Key words: unsaturated polyester resin (UPR), flame retardation, zinc hydroxystannate (ZHS), thermal decomposition of UPR

INTRODUCTION

Flame retardants are a crucial element in an enormous variety of products, from plastics and textiles to building and electronic materials. At the same time they are increasingly regulated, particularly in the U.S. and the European Community. Every manufacturer needs to be aware of the constantly shifting regulatory landscape, and the products and processes that will meet them. Moreover, new materials and formulations are rapidly changing the economic equation, and companies that adopt the latest technology will have the edge in providing their customers with the balance of properties at the lowest possible price.

The rapid progress in the field of flame retardancy of polymers in the last 10 years expressed itself in the emergence of new additives, new application systems to an ever increasing diversity of products for which flame retardancy is a dominant requirement and in new standards and testing methods and instruments. Of particular importance became the environmental considerations which focused attention on the need for new environmentally friendly flame retardant systems. These developments were accompanied by a pronounced effort to gain a better understanding of the underlying principles and mechanisms governing flammability and flame retardancy and to develop new mechanistic approaches for the emerging new flame retardancy systems.

Flame retardancy due to physical effects usually requires relatively large amounts of additives: 50-65% in the case of aluminium trihydrate ($Al(OH)_6$) and magnesium hydroxide ($Mg(OH)_2$). The activity of these additives consists in: (a) dilution of polymer in the condensed phase; (b) decreasing the amount of available fuel; (c) increasing the amount of thermal energy needed to raise the temperature of the composition to the pyrolysis level, due to the high heat capacity of the fillers; (d) enthalpy of decomposition – emission of water vapour; (e) dilution of gaseous phase by water vapour – decrease of amount of fuel and oxygen in the flame; (f) possible endothermic interactions between the water and decomposition products in the flame; (g) decrease of feedback energy to the pyrolyzing polymer; (h) insulative effect of the oxides remaining in the char.

Considerable attention is given to the particle size – surface area, as it is known to influence greatly the melt flow of the treated polymer, and to the crystal size of the additive. Hydrophobic coatings on the particles of the additives are applied to facilitate dispersion of the highly hydrophilic metallic oxides in the polymer. The possible effect of these coating on the combustion behaviour of the polymers is not clear.

Another important property appears to be the "surface free" energy which determines the reactivity of the additive with acidic groups in polymers, which might produce crosslinks and reduce melt flow, as well as with acidic groups in air, i.e. carbon dioxide, or in acid rain, which might cause the "chalking effect" and decrease flame retardancy.

Recently evidence has been presented, pointing to the possibility of an interaction between the endothermic additives and several polymers during pyrolysis and combustion, in the presence of some metallic catalysts, such as nickel(II) oxide, manganese borate, 8-hydroxychinolino-copper and ferrocene. These interactions are believed to be the cause of a highly significant increase in char and Oxygen Index (OI). Other compounds, i.e., nickel(III) oxide and metal acetylacetonates gave negative results. Specific effects of transition metals, when coprecipitated with magnesium hydroxide as solid solutions, included: (a) facilitation and lowering the temperature of the dehydration of the additive; (b) catalysing dehydrogenation of the polymer; (c) promoting carbonization; (d) improving acid resistance. The mechanism of the catalytic and antagonistic activities of metallic compounds has not yet been clarified. There are qualitative and quantitative differences between the effects of various ions and effect of the ionic radius has been proposed. The elucidation of the mechanisms governing the activity of the metallic compounds in flame retardancy of polymers, with endothermic additives as well as with intumescent formulations, appears to be a highly intriguing challenge for polymer science today.

The unsaturated polyester resins (UPR`s) belong to the most used thermosetting materials for composite applications, structural parts of automobiles, building and coating materials and electrical parts [1, 2]. Since UPR`s are extensively used in enclosed spaces, it is essential from safety reasons to possess detailed knowledge about their thermal behaviour with particular attention paid to the kinetic aspects of decomposition. In that respect, kinetic parameters give an additional insight into the mechanism of degradation and make possible prediction of the system's behavior in an extrapolated range of degree of conversion, time and temperature.

A literature survey on kinetic of decomposition of UPR`s reveals that in spite of the large amount of work that has been done in polymer chemistry of polyesters, there has been less work on their decomposition features - their better knowledge is important from both a fundamental and a technological point of view. From the side of fundamental research, the thermal decomposition of a polymeric materials is a complex, heterogeneous process, consisting of several partial reactions. The course of decomposition is affected by different factors, connected with the sample, with the degradation process alone and with the surroundings determined by analytical equipment. A valuable tool towards a better understanding of degradation mechanisms offer methods for the evaluation of kinetics parameters, provided that the basic relationships and assumptions on which they are based, are properly taken into account.

From the technological point of view, investigation of decomposition processes has a dual meaning. The first concerns stabilization of a polymer in order to obtain novel materials with a desired set of thermal properties which will be able to fulfill demands of contemporary materials engineering. The role and action of a stabilizer may be more effective, if a better understanding of polymer decomposition schemes is provided during its development.

The second meaning of the decomposition studies concentrates on safe disposal or recycling of plastic wastes which become a social problem of growing importance nowadays [3]. Both aspects are not contradictory, but rather complementary, particularly in case of widely used polymers - such as unsaturated polyester resins.

In recent years [4-6], results of studies of the thermal behaviour of UPR have been published. Baudry et al. have studied the thermal degradation of unsaturated polyester resins with cyclopentadiene end cap [7]. It has been found that at the beginning of thermal degradation energy of activation (E) was about 100 kJ/mol, which corresponds to polystyrene depolymerisation; for degree of conversion (a) equal to 0.5, which corresponds to UP network degradation, E was 170 kJ/mol. In an another study, Agraval et al. [8] investigated kinetics of decomposition of three unsaturated polyesters, based on diethylene glycol and isophthalic/adipic/maleic or fumaric acid, by isothermal TGA.

In this paper the flame retardancy and thermal stabilization of halogenated and halogen-free polyester resins by zinc hydrostannate has been investigated by termogravimetry (TG) in both dynamic and isothermal mode, and TG coupled with Fourier Transform Infrared Spectroscopy (TG-FTIR) or Mass Spectroscopy (TG-MS) [9].

Experimental Part

Materials

- Unsaturated polyester resin Polimal 1201P, produced by "Organika-Sarzyna" Co. Poland, was of the izophthalate/maleate/propylene glycol type. UPR (styrene solution) was cured by mixing with methyl ethyl ketone peroxide and cobalt naphthenate.
- Unsaturated polyester resin Polimal 163-53, produced by "Organika-Sarzyna" Co. Poland, was of the maleate/phthalate/brominated glycol type. UPR (styrene solution) was cured by mixing with methyl ethyl ketone peroxide and cobalt naphthenate.
- Zinc hydroxystannate (ZHS) (Alcan Co., UK).

Techniques

Limiting Oxygen Index (LOI)

Limiting Oxygen Index (LOI) analyses were carried out using a Stanton Redcroft analyzer. LOI was determined on 10x150mm specimens with 3mm thickness using NFT 51071 procedure. Analyses were conducted at ambient temperature, with a 40mls^{-1} gas flow in the quartz column. The oxygen ratio variation in the gas was 0,2%. Results were obtained on three samples tested in the same conditions. LOI were determined for samples crosslinked at ambient temperature.

Thermogravimetric Analysis

Thermogravimetric analysis was performed on a Netzsch TG 209 thermal analyser, operating in a dynamic mode at a heating rate of 2.5, 5, 10 and 20 K/min. The conditions were: sample weight - ~ 5 mg, atmosphere - argon or air, open α-Al_2O_3 pan. The raw data were converted to ASCII files and kinetic analysis was carried out using an in-house program and a Netzsch Thermokinetic Program (v. 99/10) on an IBM-compatible computer with Pentium III processor.

Thermogravimetric Analysis Coupled with Fourier Transform Infra-Red Spectroscopy

Thermogravimetric analysis coupled with Fourier transform infra-red spectroscopy was carried out using a Mettler-Toledo TGA/SDTA 851 thermal analyser (heating rate = 20 Kmin^{-1}, sample weight ~ 2 mg, air flow = 50 cm^3/min) and a JASCO 610 FT-IR spectrometer. The thermogravimetric analyser and spectrometer were suitably coupled to enable the passage of evolved products from the furnace to the gas cell over a short path, to minimise secondary reaction or condensation on cell walls. Moreover, the experimental conditions have been chosen to ensure that the condensable products form a submicron aerosol mist. This size aerosol has two advantages: (i) the particles follow the gas stream lines, thus minimizing condensation, and (ii) the particles produce little scattering in the mid IR, so the condensable products can be analysed online in the FT-IR cell.

Thermogravimetric Analysis Coupled with Mass Spectroscopy

TG-MS measurements were made on a TA Instruments DTA-TGA (SDT 2960) coupled with a Balzers Thermostar quadropole mass spectrometer. The quadropole mass spectrometer was connected to the thermobalance via a capillary coupling. The ionizing voltage of the cross-beam electron impact ionozation source amounted to 70 eV and the flow rate of the purge gas (helium) was 100 cm^3/min. Mass of the sample was ca. 4 mg and the heating rate was 20 K/min.

RESULTS AND DISCUSSION

The determination of oxygen index (LOI), which characterized ignition of polyester resins, indicated that are much less flammable when the amount of ZHS content in composition is greater than 5% weight (figure 1). When UPR`s characterized by the oxygen index are greater than 22 – which means that materials do not burn in air.

Sample	$ZnSn(OH)_6$, %	LOI, %
1	-	18,2
2	2	.
3	5	22,1
4	10	23,4
5	40	26,7

Figure 1. LOI index determined on samples 1-5 crosslinked at ambient temperature.

The goal of this study was do determine, using thermogravimetric analysis, the eritical temperatures of thermal degradation of typical unsaturated polyester materials crosslinked by styrene, then to identify, by coupled TG technique, products released during the pyrolysis. Tests with different thermal degradation speeds also allowed to us calculate activation energies of thermal degradation of these materials and to classify the behaviour of the UPR [9].

Results of thermogravimetric analysis of samples 1-5 are shown in figure 2.

Thermal decomposition of unmodified polyester resin starts at ca. 280^0C and proceeds in two steps up to ca. 420^0C. The maximal decomposition rate occured at 360°C, as evidenced by DTG profile. Samples containing zinc hydroxystannate undergo a two-step degradation; the first step was found to be operational at the temperature range of 240-340^0C, while the second and specific occurs above 430^0C.

The data from TG analysis show that degradation proceeds with the char residue at 500^0C and was considerably higher for flame – retarded resins.

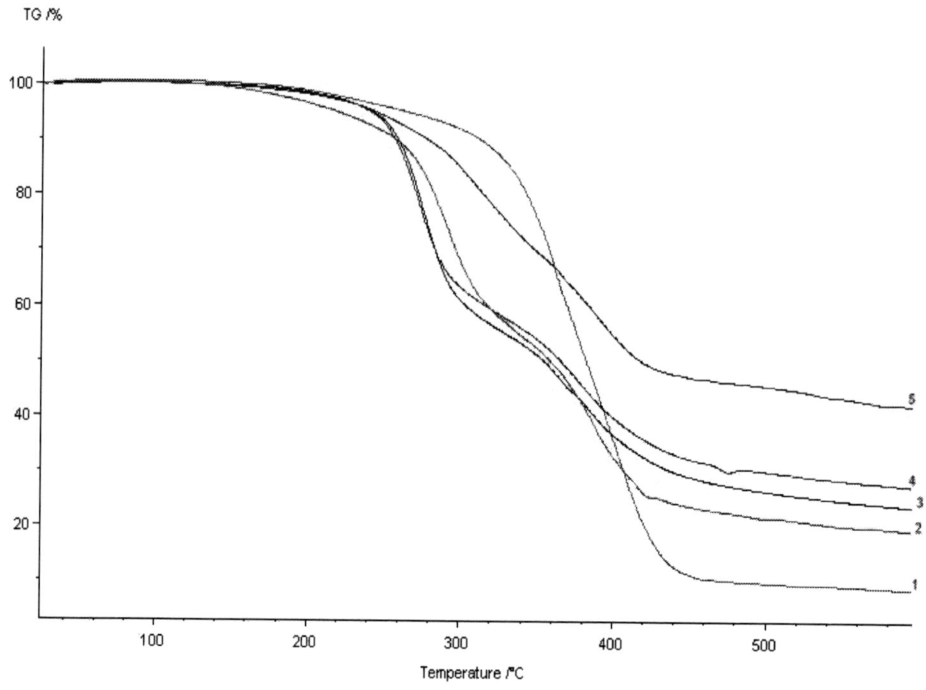

Figure 2. Thermogravimetric curves of samples 1-5 at 10 deg/min under air atmosphere. 1 – UPR; 2 – UPR + 2% $ZnSn(OH)_6$; 3 - UPR + 10% $ZnSn(OH)_6$; 4 - UPR + 5% $ZnSn(OH)_6$; 5 - UPR + 40% $ZnSn(OH)_6$.

Figure 3 and figure 4 present results of thermogravimetric analysis coupled with Fourier transform infrared spectroscopy (TG-FTIR) and TG coupled with mass spectroscopy (TG-MS).

To have an additional insight into the mechanism of decomposition, hyphenated methods were used – results of TG-FTIR of sample 1 revealed that characteristic absorption bands at ~1750, ~2460 and ~1200cm^{-1}, are observed, which correspond to the vibrational frequencies of CO_2, CO groups, respectively.

Analysis of TG-MS data yields additional information about emission of volatiles. Analysis of the volatile decomposition products by FT-MS spectroscopy revealed that during controlled heating emission of phtalic anhydride takes place at ca. 330^{0}C, than, at ca. 380^{0}C styrene and a complex mixture of other aromatic compounds are evolved.

Further studies by TG-FTIR revealed that evolution of CO_2 is considerably lowered by ZHS, whereby at a higher stage of decomposition evolution of H_2O ad HBr can be identified.

It should be noted that for the heating rate of 20 deg/min all the thermal events are slightly shifted towards higher temperatures. Such a relatively high heating rate was, however, necessary to produce volatiles at concentrations high enough to be detected on-line in the FT-IR measuring cell and to provide continuous monitoring of the i.r./mass spectra of evolving products as well as quantitative analysis of gases [3] . In contrast to pyrolysis GC, in which all the gases produced by heating to a given temperature are separated and analyzed as a batch; TG-FT/IR/MS offers the great advantage of sequentially identifying the gases for a comprehensive vapour-phase analysis.

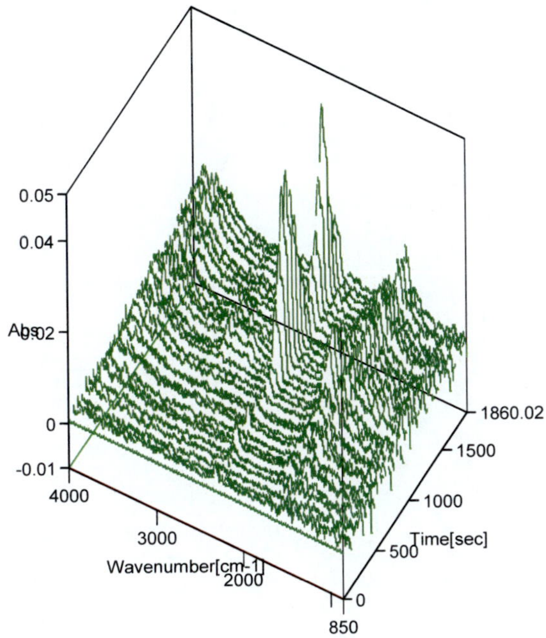

a.

Stacked plot of FT-IR spectra of the decomposition process of sample 1 at heating rate of 20 K/min ($T_{initial}$ = 30ºC)

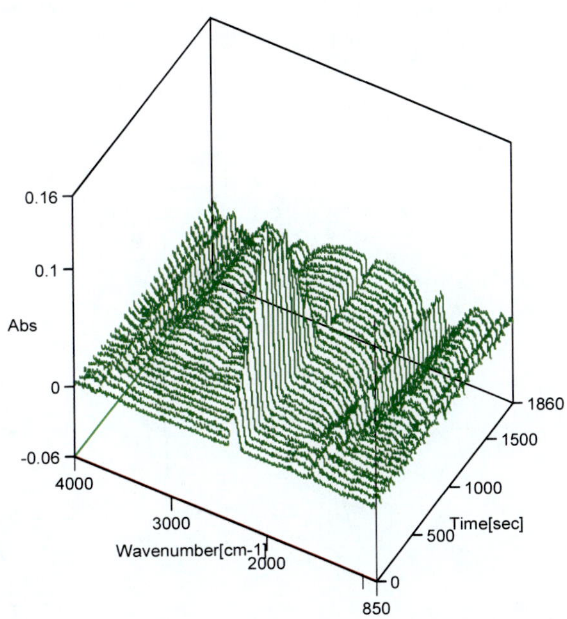

b.

FT-IR spectra taken at different temperatures during the decomposition of sample 3 at heating rate of 20 K/min ($T_{initial}$ = 30ºC)

Figure 3. Results of thermogravimetric analysis (TG-FTIR).

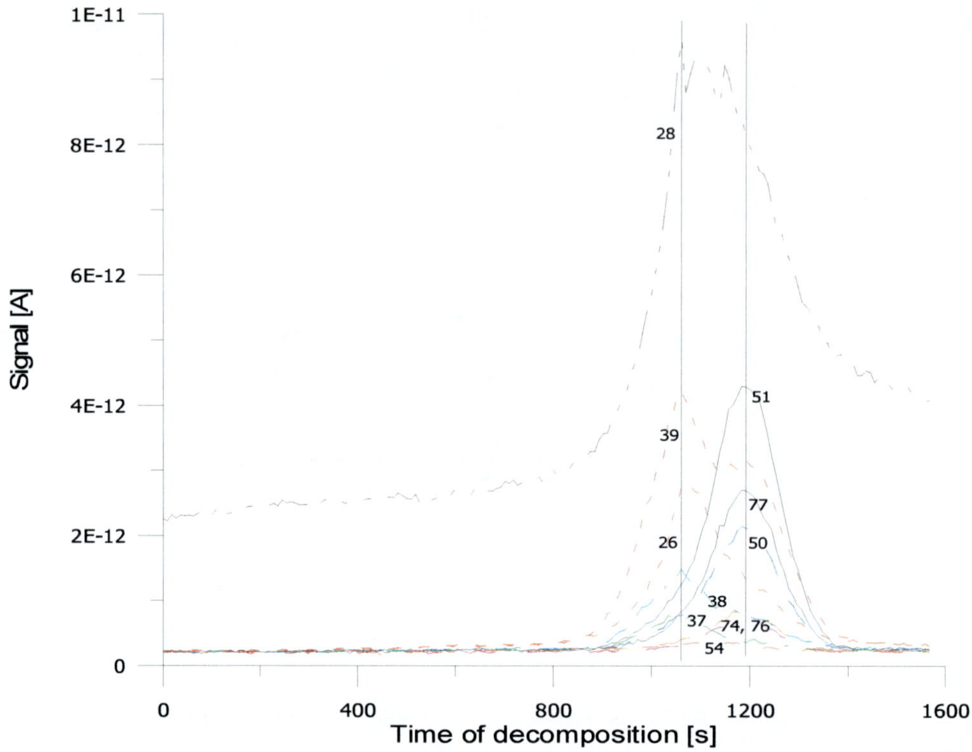

Figure 4. Intensity of m/e signals for the decomposition of sample 1 (without additives) followed by TG-MS method at heating rate of 20 K/min ($T_{initial} = 30^{0}C$).

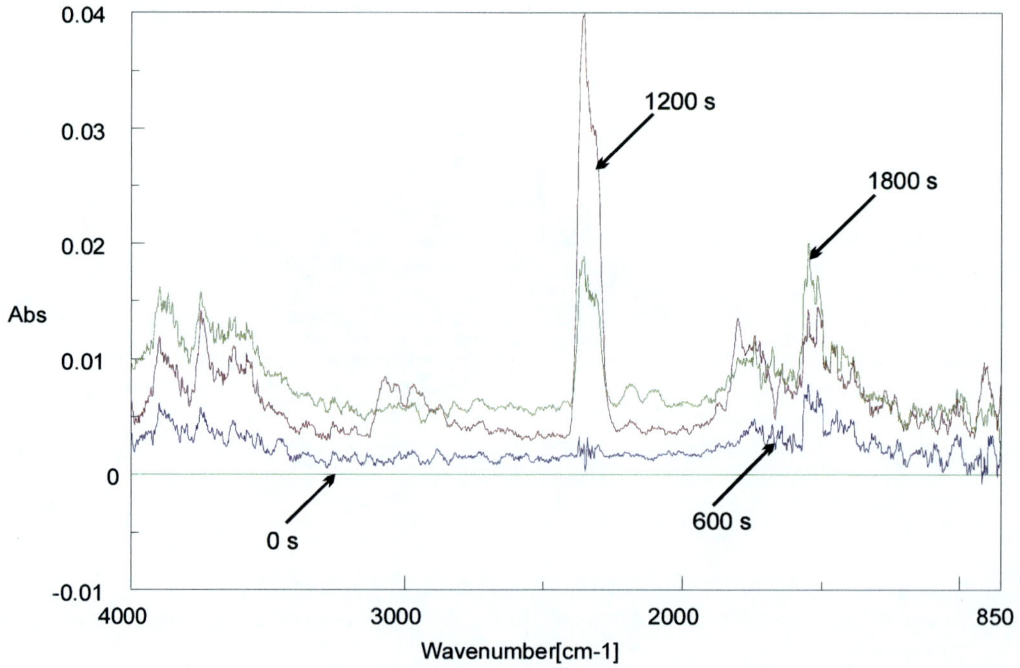

Figure 5. Stacked plot of FT-IR spectra of the decomposition process of sample 3.

CONCLUSIONS

Knowing literature data on thermal degradation mechanisms and on the basis of the obtained results we can say that in the case of UP thermal degradation one or more of at least three different general mechanisms may be responsible, as demonstrated for thermal degradation of other materials:

1. a high degree of random scission, but this required an high energy for thermal degradation;
2. intramolecular transfer during unzip; this corresponds with the degradation of polystyrene phase in the UP network, in the beginning of thermal degradation;
3. secondary reactions of primary products, corresponding with degradation induced by radical products, for example RO^{o} products coming from ester linkage rupture reaction.

Stabilizing action of ZHS antypirenes may be explained by the formation of surface-localized spherical barriers which are growing due to the nucleation growth mechanism and which attenuate the transfer of heat from the decomposition zone to the substrate – this results was qualify as predominating in the flame – retardant of action.

An additional role may be ascribed to the metal ion-polymer functional groups interaction that should bad to the formation of the long-range the system [9].

REFERENCES

[1] Kicko-Walczak, E., New ecological poliester resins with reduced flammability and smoke evolution capacity, *J. Appl. Polym. Sci.,* 74, 379 (1999).

[2] Kicko-Walczak, E., Cone calorimetric studies on the mechanism of action of new fire-retardants used for polyester resins, *Polimery,* 45, 808 (2000),.

[3] Pielichowski, J.; Pielichowski, K., Application of the thermal analysis for the investigation of polymer degradation processes, *Journal of Thermal Analysis,* 43, 505 (1995).

[4] Weil, E. D.; Hirschler, M. M.; Patel, N. G.; Said, M. M.; Shakir, S., Oxygen index: correlation to other fire tests, *Fire and Materials,* 16, 159 (1992).

[5] Cusack, P. A.; Heer, M. S.; Monk, A. W., Zinc hydroxystannate as an alternative synergist to antimony trioxide in polyester resins, *Polym. Degrad. Stab.,* 58, 229 (1997).

[6] Atkinson, P. A.; Haines, P. J.; Skinner, G. A., Inorganic tin compounds as flame retardants and smoke suppressants for polyester thermosets, *Thermochim. Acta,* 360, 29 (2000).

[7] Baudry, A.; Dufay, J.; Regnier, N.; Mortaigne, B., Thermal degradation and fire behaviour of unsaturated polyester with chain ends modified by dicyclopentadiene, *Polym. Degrad. Stab.,* 61, 441 (1998).

[8] Agraval, J. P.; Sarvade, D. B.; Makashir, P. S.; Mahajan, R. R.; Dendage, P. S., Thermal degradation studies of novel diethylene glycol based unsaturated polyesters in air, *Polym. Degrad. Stab.*, 62, 9 (1998).

[9] Kicko-Walczak, E., Kinetics of thermal decomposition of unsaturated polyester resins with reduced flammability, *J. Appl. Polym. Sci,* 88 (13), pp2851-2857 (2003).

INDEX

A

D

E

I

N

Q

R

T

U

V

W

X

Y

Z